D1130697

HANDBOOK OF ELECTRONICS MANUFACTURING ENGINEERING

HANDBOOK OF ELECTRONICS MANUFACTURING ENGINEERING

Second Edition

Bernard S. Matisoff, P.E., CMfgE

 VAN NOSTRAND REINHOLD COMPANY
_____ *New York*

Copyright © 1986 by Van Nostrand Reinhold Company Inc.

Library of Congress Catalog Card Number: 85-17777
ISBN: 0-442-26072-5

Manufactured in the United States of America

Published by Van Nostrand Reinhold Company Inc.
115 5th Ave.
New York, New York 10003

Van Nostrand Reinhold Company Limited
Molly Millars Lane
Workingham, Berkshire RG11 2PY, England

Van Nostrand Reinhold
480 Latrobe Street
Milbourne, Victoria 3000, Australia

Macmillan of Canada
Division of Gage Publishing Limited
164 Commander Boulevard
Agincourt, Ontario M1S 3C7, Canada

15 14 13 12 11 10 9 8 7 6 5 4 3 2 1

Library of Congress Cataloging-in-Publication Data

Matisoff, Bernard S.
   Handbook of electronics manufacturing engineering,

   Includes index.
   1. Electronic industries.   2. Production
engineering.   I. Title.
TK7836.M37   1986   621.381   85-17777
ISBN 0-442-26072-5

*To my wife Louise, my two sons Marty and Glen,
and two very dear friends, Nancy and Stephanie,
whose faith and support made it all worthwhile.*

PREFACE TO THE SECOND EDITION

The *Handbook of Electronics Manufacturing Engineering* has increased in popularity throughout the world since it was published in 1978. Today, it is used extensively as a reference in the manufacture of electronic products, both commercial and military, in many countries. The primary goal of this second edition is to make this handbook of even greater practical value than the fist edition. This has been accomplished by the addition of new material vital to practicing manufacturing engineers in their everyday assignments.

The new material covers a large variety of subjects essential to engineers, technicians, production foremen and supervisors involved in the production of electronic products from home entertainment to computers for high performance military aircraft and space vehicles.

The selection of materials from the vast pool of available data pertaining to the manufacture of modern electronic equipment has been based on the need and requirements of meeting the high quality standards of today's electronic systems and the goals of achieving higher productivity.

I have received many practical suggestions from friends in the field of manufacturing engineering; this cooperation has proved to be invaluable and is sincerely appreciated.

The handbook user who directs the author's attention to some defect or omission of material considered of general value renders a valuable service to his or her fellow engineers. For this reason I gratefully accept criticisms and suggestions about revisions and the inclusion of new materials.

BERNARD S. MATISOFF, P. E., CMFGE

PREFACE TO THE FIRST EDITION

Abraham Lincoln said, "Let us have faith that right makes might; and in that faith let us to the end dare to do our duty as we understand it."

Benjamin Franklin started with a silk handkerchief and two crossed sticks . . . and brought lightning down from the clouds. With nothing but a penknife for his tool, James Ferguson made a wooden clock that accurately measured the hours. The beginning of Sir David Wilkie's famous drawings was a barn door on which he drew with a burnt stick.

Confidence is essential to success. Andrew Carnegie once said, "Immense power is acquired by assuring yourself in your secret reveries that you were born to control affairs." Confidence in yourself makes other people confident in you. It makes them want to cooperate with you, makes them enthusiastic about you. It gives you poise, balance, and steadiness.

You don't have to talk about how good you are. Just bring the best you have to work every day, and carry it with confidence. You'll find you're better than you thought you were, and others will discover this too. To know your tools is to have confidence in yourself. To be a master of your job, no matter what it is, and to be a servant to your task: study, practice, have confidence.

With confidence in yourself and your tools, you can't help but build a successful career. Now is the time to start.

CONTENTS

1
MANUFACTURING ENGINEERING:
Definition and Purpose

Manufacturing engineering is that branch of professional engineering requiring such education and experience as is necessary to understand and apply engineering procedures in manufacturing processes and methods of production of industrial products. It requires the ability to plan the practices of manufacturing; to research and develop tools, processes, machines, and equipment; and to integrate the facilities and systems for producing quality products with the optimal expenditure of capital.

Manufacturing engineering embraces the activities in the planning and selection of the methods of manufacturing, development of the production equipment, and research and development to improve the efficiency of established manufacturing techniques and the development of new ones. These activities include but are not limited to:

1. Facilities planning, including processes, plant layout, and equipment layouts.
2. Tool and equipment selection, design, and development.
3. Value analysis and cost control relating to manufacturing methods and procedures.
4. Feasibility studies for the manufacture of new or different products with respect to the possible integration of new items into existing facilities.
5. Review of product plans and specifications, and the possible changes to provide for more efficient production.

1

6. Research and development of new manufacturing methods, techniques, tools, and equipment to improve product quality and reduce manufacturing costs.
7. Coordination and control of production within a plant and between separate plants.
8. Maintenance of production control to assure compliance with scheduling.
9. Recognition of current and potential problems and the implementation of corrective action to eliminate them.
10. Economic studies related to the feasibility of acquiring new machinery, tools, and equipment.

Manufacturing engineering is divided into four basic functional areas:

1. *Manufacturing planning* is the preliminary engineering work relative to the establishment of a manufacturing system for the production of a product. It includes the selection and specification of the necessary facilities, equipment, and tools, as well as the plant layout to provide the most efficient operation.

2. *Manufacturing operations* is the engineering work involved in the routine functioning of an existing plant or facility to provide efficient and economical production output to quality standards. It includes the improvement of existing layouts, procedures, tooling, and product plans and specifications.

3. *Manufacturing research* is the pursuit of new and better materials, methods, tools, techniques, and procedures to improve manufacturing processes and reduce costs. It includes the creation of concepts and innovative uses of existing items.

4. *Manufacturing control* is the management of manufacturing operations to assure compliance to required schedules. It includes coordination of all manfacturing departments and support departments, such as purchasing and materials.

This handbook has been prepared with the goal of setting standard practices within the area of modern manufacturing. In order to present the most current practices, many commercial, trade-association, and military requirements have been taken into consideration. When the practices presented herein are applied properly, they will resolve most situations that arise in everyday routines.

Since the proper application of the practices presented can only be indicated in a general way, you must always ask yourself, Who will use it and for what?, before useful work can begin. Then, once started in the right direction, these practices can be useful as tools to accomplish the desired result—that is, a completed product that is accurate, complete, and readily usable as intended.

This handbook has not been prepared as a textbook for beginners. Nevertheless, the chapters are arranged in progressive order to guide new engineering personnel into accepted practices. It also provides a convenient reference for all other personnel who work in and with the manufacturing department.

The basic function of a manufacturing engineer is to bridge the gap between the engineering design and the actual building of the hardware. Manufacturing engineers determine the "how" in the manufacturing process: how should the product be manufactured in the most efficient and economical method? how should the materials and manufacturing processes be implemented to insure the maximum in productivity? They are responsible for translating highly technical product presentation into simplified instructions so that nontechnical production workers can easily understand each task they must perform. Manufacturing engineers also train the production workers in the proper use of tools and methods.

To accomplish their tasks, manufacturing engineers must consider all aspects: what is to be manufactured? how many? what schedule must be maintained? how will it be built? where will it be built? what is required to build it?

Modern business, whether private or government, must be competitive in order to survive and grow. To operate with any amount of success, the manufacturing engineer must clearly define the objectives, establish plans and a system of procedures and methods to accomplish them, define the responsibilities of each functional group, and establish standards of performance and evaluate results. In addition, the manufacturing engineer must maintain control at all times. The effective engineer is aware of all problems, both real and potential, that might affect the output of the task. Success depends on the quality of the control maintained in the procedures guiding the smooth-flowing operation and on the method of disseminating information.

Today's business, with its many complexities, requires a constant check on controls and performance. In addition to improving the accuracy and reliability of control, management is constantly seeking better means of achieving the most efficient operation through continuous improvement of policies and procedures. This approach requires that methods be reviewed and appraised by individuals with the required training and experience to perform such tasks. Management must periodically take a close look at its overall program to know when changing conditions have necessitated a change in approach or emphasis.

Manufacturing engineering involves deciding things, doing things, and then evaluating what has been done. Systems and procedures are an integral part of manufacturing engineering. The person who directs the activities of others is responsible for their performance, the ways of accomplishing the tasks assigned, and the methodology of the processes used. Coordinating and controlling the various tasks of any enterprise is far from easy.

The three prime elements in manufacturing engineering are (1) Plan, (2) Organize, and (3) Direct. Planning is the basis for achieving the desired results with the least expenditure of time, effort, and funds. *Organization* is employed to define the lines of responsibility and authority and to coordinate all efforts toward a smooth flowing of operations for the attainment of established goals. *Direction* is the acting to enforce the methods and procedures employed—that is, getting the various work routines to move along together toward a common goal.

No one need reinvent the wheel to achieve good control. Methods and procedures have been developed at great expense and have been tried and improved over the years. All good engineers should have this experience available to them, and should be well versed and capable of tailoring existing procedures to fit any organization. To achieve their goals in a most expeditious manner, manfacturing engineers will first perform these specific functions:

1. Coordinate all operating departments to insure that the product configuration is maintained and the maximum in producibility is achieved.
2. Make or buy determinations—that is, define which tasks would be more economically advantageous purchased from outside sources, based on evaluating in-house capabilities.
3. Define and provide tooling and fixtures.
4. Prepare assembly standards to be used as a basis for maintaining and monitoring production performance.
5. Define the facilities and capital equipment required to manufacture the selected product lines.

A manufacturing engineer is, by definition of the Society of Manufacturing Engineers, an engineer "who has acquired certain standards of professional accomplishment." The Society of Manufacturing Engineers has initiated a program for the certification of manufacturing engineers. The advantage of such certification is that it indicates an engineer's ability to meet a certain set of standards related to the many aspects of modern manufacturing engineering. These standards pertain not only to the academic requirements needed; more importantly, they pertain to the experience required of the manufacturing engineer.

To be certified, an applicant must have a good record of 10 years or more in responsible charge of important manufacturing engineering projects, which have included:

Planning and selecting methods of manufacturing
Designing equipment for manufacturing
Research and development leading to the creation of new manufacturing methods or the improvement of existing processes
Administration of manufacturing operations covering long-range planning, materials, production control, product configuration, manufacturing facilities, and production management and systems.

2

FUNDAMENTALS OF SUPERVISING

SUPERVISING

To supervise is usually defined as *to oversee* or *to direct work*. However, there is a lot more to supervising than that. Indeed, there are the matters of communications, hiring and firing, representing management, training new and old employees, helping to develop better methods of production, keeping records, and so on. As a supervisor, one deals with directing people rather than dealing with things or machines, as one's employees do.

Although there are many facets to supervising that tend to complicate the picture, the basic fundamentals of supervising remain the same. These are:

1. Planning
2. Assignments
3. Follow-up
4. Evaluation.

Planning

Planning simply means thinking an activity through before starting it. Whether you are going to do a job or take a vacation, it will be more successful if you plan it, step by step, from beginning to end before you start it. Therefore, each employee's first assignment should be thought out and prepared before the start of each shift. Overtime hours spent at the end of a shift are often due to hours lost at the start.

There are three basic steps to good planning:

1. Find out *what* must be done.
2. Find out *how much* must be done.
3. Find out *when* it must be done, and always schedule things to be done in order of their importance.

The most careful planning in the world is wasted, of course, if the plans are not carried out effectively. This leads to the next subject, assignments.

Assignments

The *assignment of work* means letting employees know specifically what they are supposed to do and how much time they have in which to do it. For many supervisors, telling people what to do is the most difficult aspect of their job. This is because they lack confidence.

An easy way to acquire such confidence is through good, careful planning, for the planning and assigning of work go hand in hand. Thus, you should assign work only when you have carefully planned the tasks to be done.

The way in which an assignment is given is important. Supervisors should neither act like dictators nor demand results. However, they can accomplish their goals by firmly letting employees know what is expected of them and by offering to help them achieve the desired results.

Quite often, employees don't understand their assignments but are ashamed to admit it. Sometimes, they think they understand when they really don't; this is particularly true of those who not do comprehend English well or who are hard of hearing. Therefore, make sure that your employees understand their assignments before they begin working. Don't let them "spin their wheels."

Make assignments for short periods of time, and do not make too many at the same time. Generally, give assignments that last for 2 hours or less, if applicable. When work is estimated and related to time, and when the assignment is for a short period, it is much easier to follow up.

Follow-up

Follow-up is checking to see that an assignment is proceeding according to plan. The best planning and assignment in the world can fall through if there is no follow-up, yet it is the most neglected aspect of supervising. Many supervisors don't follow up because they believe that their employees resent feeling "spied upon." However, this is true only for people who are not accustomed to follow-up. Once people become used to follow-up, they realize that it is a necessary part of the job.

Whenever you let employees "goof off" and get away with it, you lose their respect. In fact, you may be considered "a soft touch." Moreover, your employees will rarely repay you for your oversight. Therefore, conduct effective follow-ups. You will earn the respect of your employees, who will realize that you are "on the ball."

Do not accept excuses for lack of performance. Try to figure out possible excuses in advance, and then eliminate them. In dealing with schedule misses and the reasons given for them, take care of the most serious ones first.

Do not be afraid to ask for help. When you cannot handle something yourself, see your immediate supervisor. In dealing with other sections or departments, it is especially important to ask the foreperson or supervisor for assistance.

All supervisors make honest mistakes, which show up during the follow-up. Never cover up a mistake, which might only perpetuate it. It is better to bring it out into the open.

Evaluation

Evaluation means reviewing performance reports to check results. The current productivity should be compared to past performance and to the targets set for each person and for each machine. The purpose of evaluation is to determine if planning, assignments, and follow-ups are effective. Performance reports should help toward that end by spotlighting any problem areas.

Evaluation of performance is necessary for making decisions. You can determine which direction to go in only by first determining where you stand. In this sense, the performance reports serve as your map and compass.

Find out who the low performers are, and concentrate on them. Your good producers will take care of themselves.

If goals are not being met, proper evaluations of performance should determine what corrective action is required. Before considering corrective action, check your planning, assignments, and follow-ups. Are they effective? If you are convinced that they are adequate, then look for other reasons for schedule misses and consider corrective action.

Definition of *Supervision*

Plan:

Think through ahead of time what you want to do, the steps you must take, the resources you need.

- Forecast. Look ahead to estimate the problems and opportunities of the future.

- Objective. Spell out in concrete terms the goals you hope to reach.
- Policy. Establish and explain standing decisions that will apply to repetitive questions.
- Program. Decide the steps you will follow to reach your objectives.
- Schedule. Establish time limits within which work will be completed.
- Procedure. Standardize the methods by which work is to be done.
- Budget. Allocate resources to carry out programs and reach objectives.

Organize:

Coordinate the work of your team toward a common objective.

- Develop organization structure. Arrange and group the work to form sound, balanced, organizational units.
- Delegate authority and responsibility. Entrust responsibility and authority to others and establish accountability for results.
- Establish effective working relationships. Promote the conditions necessary for effective teamwork.

Lead:

Cause other people to take effective action.

- Initiate. Start the actions of your team.
- Decide. Make decisions applying to two or more members of your team.
- Communicate. Establish understanding between yourself and other people.
- Motivate. Inspire and encourage people to take action.
- Develop employees. Improve attitudes, knowledge, and skills of your employees.

Control:

Measure and regulate results.

- Develop performance standards. Establish yardsticks for performance based on the plans you have developed to guide your employees.
- Report. Records and measures results.
- Evaluate results. Appraise accomplishments by comparing actual performance with standards.
- Corrective action. Improve results by eliminating variances from established procedures or processes.

MANUFACTURING CONTROL

The basic function of manufacturing control is *control*. The effective manager is at all times aware of the problems, both real and potential, that will affect the output of the organization. The success of an organization depends on the quality of the control maintained in the procedures guiding a smooth-flowing operation. The manager must understand the ills of the operation in order to improve it and to eliminate problems. A feedback system of reporting must be used to provide all available information back to management on a timely basis.

Following is a simplified method of scheduling, which provides a feedback system to management to flag problems in a timely manner. This method is a three-part system designed to provide information on product availability to the sales/marketing department, materials requirements to the purchasing department, and labor and machine data to the production department. It includes techniques for the tracking of manufacturing runs and the recording of production problems on a feedback basis. This is important since the reporting of problems can be isolated and statistical analysis can be performed in order to initiate corrective actions rapidly.

Product Schedule

The product schedule is a method of scheduling production using marketing/ sales data and inventory data to establish realistic need dates for product completions. In addition to being used for scheduling production, the schedule informs the marketing/sales department of available merchandise. Both the finance and materials departments will use this information for forecasting and planning. The product schedule is the basis for all production-line schedules.

Procedure

Using forecasts and data from marketing/sales, materials, accounting, and inventory, a plan is established for production to meet all requirements in conformance with manufacturing capabilities. The product schedule form provides a means of presentation of the planned performance of the production department and the availability of marketable product. It also provides actual output data for the marketing department and for the management analysis of performance against planned output. Thus, management is informed of slippage and problems in a timely manner.

See Fig. 2-1. The following information is entered:

1. The month of the schedule.
2. The revision level each time a change is made.
3. Dates of workdays each month. Block out weekend and holidays.
4. Product description and/or number. Planned quantity for each run.
5. Quantity by days. First line is for planned quantity. Second line is used when necessary to show actual quantity when deviations from plan occur.

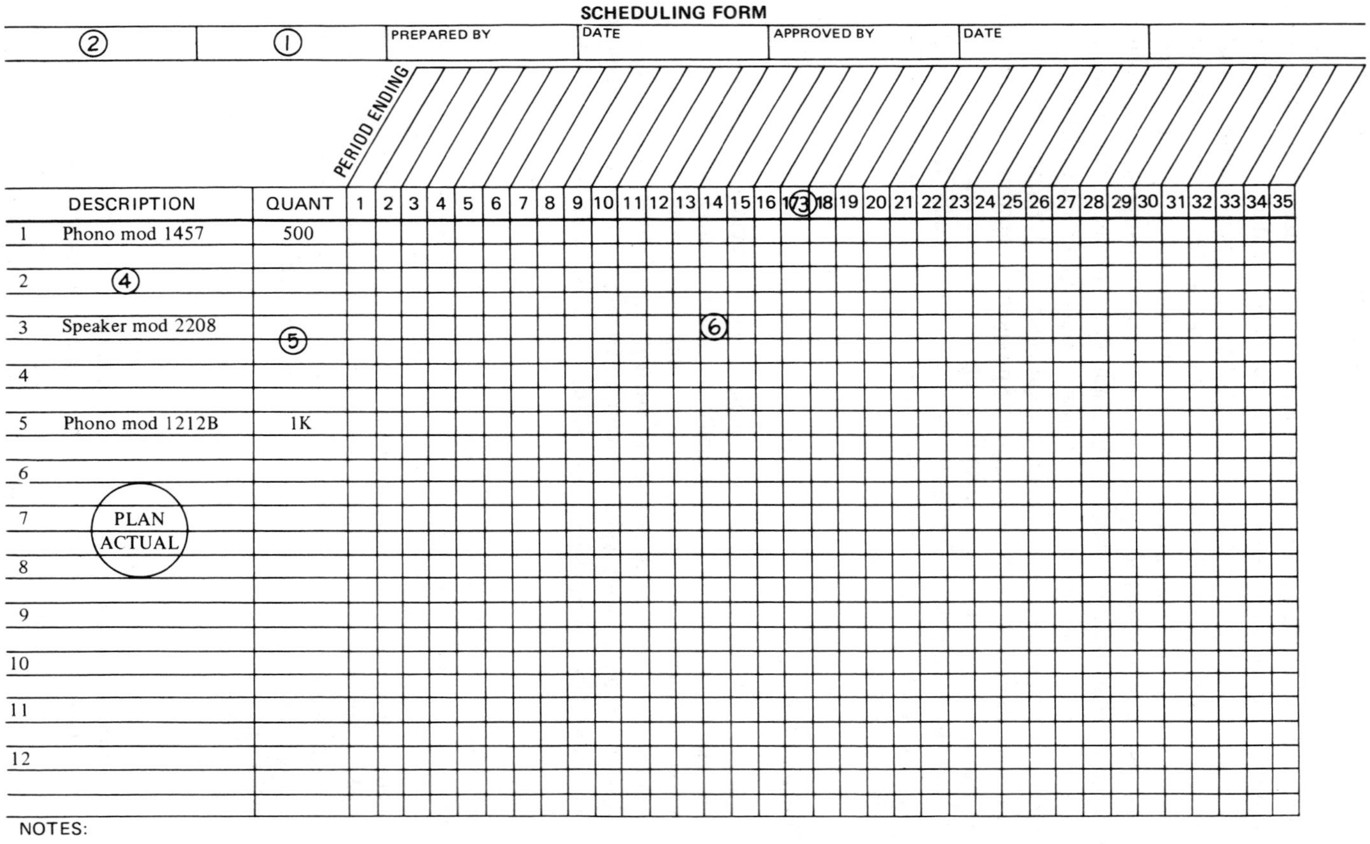

Fig. 2-1. Sample production scheduling form.

Parts and Materials Schedule

The availability of parts and materials is critical to the maintenance of production-line schedules. Thus, open communications must be available between production and purchasing/materials at all times. The availability of hardware will be used as input data in establishing production-line schedules; any changes in the parts/materials schedule must be implemented into the production schedules immediately. This information when handled efficiently can allow considerable savings as ample time is provided to make necessary changes and line balance requirements can be instituted without lost time.

Procedure

During the planning phase of each production run, all parts and materials are scheduled, and receipt dates are verified. Copies of the parts/materials schedule are routed to production for their planning. Any potential problems should be indicated on the schedule form so that production can plan an alternate run if it should be required. Constant comparisons with the production line schedules should be maintained, and the changes in parts/materials schedules must be communicated to the production department immediately.

See Fig. 2-2. The following information is entered:

1. Job number or run number.
2. Quantity required for the entire run.
3. Part name or material description (i.e., screw-4-40, flat head or aluminum alloy-5052-T4, .06 thick).
4. Part number when available.
5. Date the order is placed.
6. Purchase-order number. (This is often required for tracking parts and materials.)
7. Name of vendor or supplier. (This is often needed for follow-up.)
8. Final date that the parts/materials can be received in time for production commitments.
9. Date the supplier promised delivery. (To be used for planning for handling or stores.)
10, 11. These are for tracking parts and materials and for handling partial shipments.

Product/Labor Line Schedule

The line schedule is a method of (1) scheduling each operating department within the production department and (2) implementing a system of generating feedback

PARTS CONTROL

ITEM	QTY	DESCRIPTION	PART NO.	DATE ORDERED	PURCHASE ORDER	VENDOR	DATE ORDER	DATE PROMISE	DATE PARTIAL QUANT	NOTES

Fig. 2-2. Sample parts control tracking form.

on production status to management with a minimum of recordkeeping. Production control will generate and maintain all product/labor schedules, which the head of each operating department should maintain and enforce at the department level.

Product/labor schedules should be made for every line department in the production department. For example:

Mill
Paint shop
Case shop
Assembly line
Printed-circuit shop
Final assembly and test.

Procedure

Using the product schedule to establish the dates required for completion of various products, the labor line schedule is developed based on the manufacturing capability of the organization. This should include available personnel and special machines and equipment. This plan should give careful consideration to historically difficult runs, such as pilot runs, first production runs, and runs of products known to have difficulties. Such runs should not be planned back-to-back or sequentially, in order to prevent a severe production monthly output, but rather spaced in a manner that allows the most effective production to be met for any given month.

See Fig. 2-3. The following information is entered:

1. Department number. (This can be used for accounting and tracking labor hours by department.)
2. Date of issue of the schedule and dates of all revisions.
3. Working days of the month.
4. Date of the superseded schedule for history only.
5. Model or part number, run number (normally sequential, i.e., first run is 1, tenth run is 10; tracks the number of times a given product is scheduled for manufacture), quantity to be made this run.
6. This is used for tracking (see Fig. 2-4).
7. Planned total hour for this run. Daily hour breakdown is shown in bar chart.
8. This is used for tracking (see Fig. 2-4).

Figure 2-3 contains only the production planned output. The tracking is done on a daily basis and is defined as shown in Fig. 2-4.

LINE SCHEDULE

PRODUCT/LABOR

SUPERSEDES SCHEDULED
DATED: _____
APPROVED: _____

MONTH: _____

WORKING DAYS
DATE ISSUED: 11 - 11 - 74
REVISED: _____

DEPT. 04-Assy

	MODEL	RUN	QTY	1	2	3	4	5	6	7	8	9	10	11	12	13	14	15	16	17	18	19	20	21	22	23	24	25	26	27	NOTES
PLAN	1875K	02	150																												
ACT.																															
LABOR		192														96	96														
ACT.																															
PLAN	1420K	20																													
ACT.																															
LABOR		216																72	72		72										
ACT.																															
PLAN	1450K	18	750																												
ACT.																															
LABOR		504																			72	72	72	72	72	72		72	72		
ACT.																															

SAMPLE

Fig. 2-3. Sample production-line schedule—planning.

LINE SCHEDULE

MONTH: NOV WORKING DAYS 19

DATE ISSUED: 11-11-74 REVISED: ②

SUPERSEDES SCHEDULED ④
DATED: ⑤
APPROVED:

PRODUCT/LABOR

DEPT. 07 ①

MODEL	RUN	QTY	1	2	3	4	5	6	7	8	9	10	11	12	13	14	15	16	17	18	19	20	21	22	23	24	25	26	27	NOTES
PLAN 1845K	01 ⑤	25 ⑥								25																					
ACT.																															
LABOR	216 ⑦							72	72	72																				⑨
ACT.	192 ⑧							64	64	64																				
PLAN 1875	02	150									50	100																			
ACT.																															
LABOR	144										72	72																		
ACT.	128										64	64																		
PLAN 1420K	20	1K													300	700															148I2-I CAP
ACT.																															
LABOR	144													72	72	0														
ACT.	128													64	64	64														
PLAN 1450K	18	750																		100	100	300	100	50							17467 ¢ 17468 XSISTOR
ACT.																72	72	72	72	72	72	72	0							
LABOR 504																64	64	64	62	64	64	56	64							
ACT. 438																														
PLAN 1155	12	150																										100	50	
ACT.																															
LABOR 360																							72	72	72	72	72		
ACT.																															

SAMPLE

Fig. 2-4. Sample production-line schedule—plan versus actual.

See Fig. 2-4. The following information is entered:

6. At the end of each day, the actual quantities are recorded.
8. Record the actual number of hours charged each day.
9. Make note of problems encountered for management use in making corrective actions (i.e., machine down, no material, staffing short). All problems entered in this column must be routed to management immediately.

Production control should be solely responsible for the maintenance of all schedules. Changes in released schedules should be entered in red to flag all pertinent data. All changes must be entered on all outstanding copies of the schedules to insure that the concerned departments are informed. For example, marketing/sales must know the availability of product, and the finance department must be aware of potential changes in receipts for planning.

3
WORK SIMPLIFICATION

Today, any enterprise, whether private or government, is competitive. Effective management must not only be policy—it must also be practice. Thus, many philosophies have been developed to provide a basic framework within which resources may be effectively directed to obtain optimum production at lowest cost.

The purpose of this chapter is to present a management philosophy and to provide assistance to those manufacturing engineers who may be engaged in identifying the basic procedural problems of their organizations, systematically analyzing those problems, and designing and installing realistic improvements.

This chapter contains a description of the basic techniques of work simplification and an explanation of how to apply them. However, it is not intended that work simplification should be limited by the contents of this manual. There is no monopoly on management know-how. Enlightened managers will have fresh viewpoints and ideas, which will serve to complement existing work-simplification programs.

PRINCIPLES OF WORK SIMPLIFICATION

There have been many approaches to work simplification, ranging from simple to very sophisticated procedures. Basically, work simplification is the organized application of common sense to the finding of easier and better ways of performing work. The accepted premise must be that there is a better way to do any task.

Whether a work-simplification program is implemented by in-house talent or by outside management consultants, the same basic fundamentals apply. To obtain maximum benefits, an organized, uniform approach must be used in applying the basic techniques. A proper relationship must also exist between the human and technical aspects of improvements or even the best improvements will not be fully accepted.

Work simplification as presented herein includes five basic techniques: work-distribution analysis, process analysis, motion analysis, work count, and layout studies. These techniques provide an organized approach to improving operations. All five techniques are related but each plays a different role in analyzing procedural problems and work situations. One, several, or all of the techniques may be used depending on the type of organization being studied.

Regardless of the techniques used, a uniform method is necessary for all work-simplification efforts. Best results will be achieved if a step-by-step plan is followed. The following steps constitute a firm foundation for work simplification.

Selecting the Job to Be Improved

Selecting the job to be improved requires careful consideration. Improvements should be made first where returns will be the greatest. The consideration of a job to improve may primarily involve economics, trouble spots, or unsatisfactory conditions. Selecting a job with which the observer is very familiar may be ideal for the first attempt at applying work-simplification techniques. The following list provides clues to assist in selecting the job to be improved.

1. Greatest cost: work that involves the largest expenditure of funds, man-hours, and use of equipment.
2. Greatest workload: the largest volume of work being done by the activity.
3. Repetitiveness of work: work that will last over a long period of time.
4. Number of persons assigned: work that requires large numbers of personnel to perform similar tasks.
5. Bottlenecks: work not flowing smoothly.
6. Schedules not met: failure to meet deadlines resulting in backlog of work or overtime.
7. Excessive waste: work that results in waste of energy, materials, and time.
8. Unsafe work practices: work that results in numerous accidents or that is dangerous.
9. Fatiguing: work that involves great physical effort or that is being done with frequent rest periods.

10. Unpleasant: work that is undesirable because of extreme conditions such as dust, noise, fumes, or temperature.

There are three parts to every job: "make ready," "do," and "put away." Reduction of the "make ready" and "put away" time is important because this effort quite often does not contribute to the end result. Elimination of the "do" phase automatically eliminates the "make ready" and "put away" time. A reduction of the "do" time, although insignificant, becomes important when the volume of work is large.

Recording the Job Details

The next step is to record, in order of occurrence, each detail of the job you have selected to improve. Recording the order of occurrence is important since it permits analysis of each detail in proper sequence. Do not omit seemingly unimportant details. These details may not be important individually, but collectively they can play an important role in developing the new method. The various charts that are used to assemble the job details are discussed within this chapter.

Analyzing the Job Details

The purpose of analyzing the job details is to aid in the development of the best method. Each detail is analyzed by asking a series of questions, which, while appearing to be extremely simple, normally produce improvements. The most important question, *Why* is it necessary?, is asked over and over. Very often, the answer to this will result in the elimination of details or portions of the job being analyzed.

Next, several questions are asked.

The answers furnish hints or clues to developing a better method, as shown in the following list:

Questions	*Answers*
Why is it necessary?	Lead to elimination.
What is the purpose of each detail?	
Where should each detail be performed?	Lead to combining details when practical or changing sequence to
When should each detail be done?	improve method.
Who should perform each detail?	
How should each detail be accomplished?	Lead to simplifying work when possible.

Remember, when asking these questions, the analysis must deal with facts and reasons, not opinions and excuses, or the conclusions will not be valid. To find a solution, the problem must be examined objectively. The attitude with which this examination is made is the key to the total procedure of simplifying work.

Developing the Improvements

Improvements may be developed in many ways. A particular task or portion of a job that is found to be unnecessary should be eliminated. Related tasks in a process should be combined. Changing the sequence of details in an operation may eliminate or reduce transportations and/or delays.

Installing the Improvements

Personnel in the activity being studied should participate in the development of new methods, for they possess a vast reservoir of untapped ideas. It is also advisable to have interested personnel review proposed methods, as they may be aware of related problems that others might overlook. A trial or test period for major changes is necessary to determine how much of the new method is workable. Human factors and technical requirements should also be considered when implementing proposed methods.

TECHNIQUES OF WORK SIMPLIFICATION

Many formal and informal techniques are being used to improve work methods. This section describes five simple work-simplification techniques, which provide the framework for a successful work-simplification effort. These techniques have been tested on a large scale both in government and private industry. They are best adapted to meet the needs of supervisors and are appropriate with operating procedures. The work-simplification techniques are:

1. Work-distribution analysis
2. Process analysis
3. Motion analysis
4. Work count
5. Layout studies.

Work-Distribution Analysis

The purpose of work-distribution analysis is to provide the optimum relationship between the workload assigned to an organization and the resources allotted to perform the work by making the most equitable distribution of tasks. Through use of this technique, it is easy to determine what work is being done, how much time is spent on it, and who is doing it. Work-distribution analysis, which is frequently used when groups of people are involved, helps to determine the contribution of each individual. After such contributions are determined in detail, an analysis is made and improvements are developed. This analysis is accomplished through use of the work-distribution chart and allied forms.

Preparation of Work-Distribution Chart

The division of work in an organization can be better analyzed by recording in one place all the activity of a unit and the contribution that each member of the unit makes in accomplishing each activity. A chart reflecting this breakout should be prepared in the following manner:

1. Each individual in the organization, including the supervisor, prepares Form A, a task list of individual jobs (Figs. 3-1 through 3-5), listing in detail each significant duty and type of work, the average number of hours per week spent on each task, and the volume of work for each when appropriate. The task list should not be a repetition of an occupational specialty or job classification statement. It should be a clear, complete portrayal of what each individual actually does during the period covered by the task list, usually 1 week. Ambiguous phrases should be avoided. Specific tasks must be described (for example, "Prepare Form A," "Test Model 100"). Each task list is reviewed by the supervisor for completeness and correctness. Sample forms are shown in Figs. 3-1 through 3-5. In these samples, tasks have been numbered to assist in completing the work-distribution chart. (Normally, the tasks would be numbered after completing the operations list.)

2. Form B, an operations list, is prepared by the supervisor, analyst, or observer making the appraisal. It lists the basic functions that are accomplished by the organization and includes all the major jobs that are performed to complete the assigned mission. The functions or major jobs are numbered and matched with the tasks on the task list, and these tasks are annotated accordingly. A sample operations list is shown in Fig. 3-6.

3. The third step is preparation of Form C, the work-distribution chart (Fig. 3-7), for which Forms A and B provide the basis. The operations or major jobs of the organization are listed under the Operation/Process column of the chart in order of relative importance. Several lines should be left blank between each item listed. Beginning with the supervisor of the organization, list all employees from left to right in order of grade. The employee's name, working title, and grade are entered in appropriate blocks. Take the task list of each individual and match the tasks with the proper entries in the Operation/Process column. Tasks that don't fit can be listed opposite "Miscellaneous." The hours each individual

TASK LIST OF INDIVIDUAL JOBS *(For Work Distribution Chart)*				
NAME Grey	WORKING TITLE Supervisor		GRADE	
TASK NO.	DESCRIPTION OF TASK		WORK UNIT VOLUME	HOURS PER WEEK
1	Proofread catalog card copy		250	10
2	Inventory and maintain records			7
6	Assign work and give instructions			3
9	Attend briefings on printouts			3
8	Editing			3
5	Prepare manual			4
7	Answer phone			6
6	Train employees			4
		TOTAL HOURS OF WORK		40
DATE	FUNCTION Proofreading Section		CERTIFIED BY *(Signature and grade of supervisor)*	

Fig. 3-1. Form A, task list of individual jobs.

TASK LIST OF INDIVIDUAL JOBS (*For Work Distribution Chart*)

TASK NO.	DESCRIPTION OF TASK	WORK UNIT VOLUME	HOURS PER WEEK
NAME Jones	WORKING TITLE Proofreader	GRADE	
3	Make final review of tab copy	300	15
1	Proofread catalog card copy	240	12
7	Answer phone		5
2	Inventory and maintain records		5
6	Assign work and give instructions		3
		TOTAL HOURS OF WORK	40
DATE	FUNCTION Proofreading Section	CERTIFIED BY (*Signature and grade of supervisor*)	

Fig. 3-2. Form A, task list of individual jobs.

TASK LIST OF INDIVIDUAL JOBS (*For Work Distribution Chart*)				
NAME White	WORKING TITLE Proofreader		GRADE	
TASK NO.	DESCRIPTION OF TASK		WORK UNIT VOLUME	HOURS PER WEEK
1	Proofread catalog card copy		300	14
3	Make final review of tab copy		360	18
4	Proofread special notices for tab		60	4
7	Answer phone			4
			TOTAL HOURS OF WORK	40
DATE	FUNCTION Proofreading Section	CERTIFIED BY (*Signature and grade of supervisor*)		

Fig. 3-3. Form A, task list of individual jobs.

TASK LIST OF INDIVIDUAL JOBS (For Work Distribution Chart)				
NAME Smith	WORKING TITLE Proofreader		GRADE	
TASK NO.	DESCRIPTION OF TASK		WORK UNIT VOLUME	HOURS PER WEEK
1	Proofread catalog card copy and indicate errors		300	20
2	Count documents and maintain records			2
3	Make final security review of tab copy		150	5
4	Proofread special notices for tab and indicate mistakes		75	5
6	Assign work and give instructions to proofreaders			8
			TOTAL HOURS OF WORK	40
DATE	FUNCTION Proofreading Section		CERTIFIED BY (Signature and grade of supervisor) Gray	

Fig. 3-4. Form A, task list of individual jobs.

TASK LIST OF INDIVIDUAL JOBS *(For Work Distribution Chart)*				
NAME Green	WORKING TITLE Proofreader		GRADE	
TASK NO.	DESCRIPTION OF TASK		WORK UNIT VOLUME	HOURS PER WEEK
2	Count incoming and outgoing documents and maintain records			10
1	Read catalog card copy and indicate errors		296	10
3	Make final security review of tab copy		240	10
4	Proofread special notices for tab and indicate errors		150	8
7	Answer phone – take messages			2
		TOTAL HOURS OF WORK		
DATE	FUNCTION Proofreading Section	CERTIFIED BY *(Signature and grade of supervisor)* Gray		

Fig. 3-5. Form A, task list of individual jobs.

OPR NO.	OPERATION LIST *(For work distribution chart)*	DATE
	DESCRIPTION OF OPERATION	WEEKLY VOLUME
1	Proofread catalog card copy	1800
2	Inventory documents and maintain records	
3	Review tab copy	1200
4	Proofread special notices for tab	400
5	Special projects	
6	Administration and supervision	
7	Answer telephone	
8	Edit tab copy	
9	Miscellaneous	
FUNCTION Proofreading Section	CERTIFIED BY *(Signature and grade of supervisor)* Gray	

Fig. 3.6. Form B, operations list.

WORK DISTRIBUTION

BY:
CURRENT:
PROPOSED:

FUNCTION CHARTED: Proofreading Section
APPROVED: M. Rose, Branch Chief
DATE:
REVISION:

Oper No.	Activity	Work Count	Hrs Wk	Gray — Supervisor: TASK	Work Count	Hrs Wk	Jones — Proofreader: TASK	Work Count	Hrs Wk	White — Proofreader: TASK	Work Count	Hrs Wk	Brown — Proofreader: TASK	Work Count	Hrs Wk	Smith — Proofreader: TASK	Work Count	Hrs Wk	Green — Proofreader: TASK	Work Count	Hrs Wk
1	Proofread catalog card copy	1911	91	Read copy and indicate errors	250	10	Read copy and indicate errors	240	12	Read copy and indicate errors	300	14	Read copy and indicate errors	525	25	Read copy and indicate errors	300	20	Read copy and indicate errors	296	10
2	Inventory documents & maintain records		24	Count incoming & outgoing documents and maintain records.		7	Count incoming & outgoing documents and maintain records.		5							Count incoming & outgoing documents and maintain records.		2	Count incoming & outgoing documents and maintain records.		10
3	Review tab copy	1160	58				Make final security review	300	15	Make final security review	360	18	Make final security review	100	10	Make final security review	150	5	Make final security review	240	10
4	Proofread special notices for tab	360	22							Read and indicate errors	60	4	Read and indicate errors	75	5	Read and indicate errors	75	5	Read and indicate errors	150	8
5	Special projects		4	Prepare manual		4															
6	Administration & supervision		14	Assign work & give instructions		3	Assign work & give instructions		3							Assign work and give instructions		8			
	Train employees		4	Train employees		4															
7	Answer phone		17	Answer questions and take messages		6	Answer questions and take messages		5	Answer questions and take messages		4							Answer questions and take messages		2
8	Edit tab copy		3	Perform initial edit on documents		3															
9	Miscellaneous		3	Get briefings on computer printouts		3															
	TOTAL (Man-hours)		240			40			40			40			40			40			40

ANALYSIS: WHAT TAKES THE MOST TIME?...IS THERE MISDIRECTED EFFORT?...ARE SKILLS USED PROPERLY?...ARE THERE TOO MANY UNRELATED TASKS?...ARE TASKS SPREAD TOO THINLY?...IS WORK DISTRIBUTED EVENLY?

Fig. 3-7. Form C, work-distribution chart.

spends on each task are entered in the Hours per Week column under his or her name. The volume of work also is entered when appropriate. The hours per week spent on each task are totaled and entered in the Hours per Week column beside each Operation/Process entry. The volume of work is totaled, also when appropriate. The employees' hours are totaled. Total man-hours will equal total hours for all employees. Thus, all the time in a typical week is accounted for. The volume of work data cannot be totaled at the bottom of the chart as work units will not be the same for all tasks.

Analysis of the Work-Distribution Chart

With the completion of the work-distribution chart depicting the present situation, a clear picture is presented of the work distribution in the organization. Pertinent facts have been collected, classified, and organized in usable form to facilitate easy and effective study of the work distribution. (See Fig. 3-7.)

Analysis of the present work distribution is the "payoff point." It should be done in a systematic manner as follows:*

1. Analyze operations as a whole.
2. Analyze each operation independently.
3. Analyze each person's effort independently.

The following questions will assist in examining the details.

1. Which operations take the most time? Should these operations take the largest number of man-hours? It is reasonable to assume that the largest total time should be spent on the major operation or function of the organization. The analyst or observer may find that the operation he or she has listed as first in importance to the work of the organization is not the operation upon which the most time is spent. A critical look at the Operation/Activities column and the Total Hours column may reveal several situations that warrant further study to see if the time spent is justified. If the operation is a continuing one and consists of several steps, a more detailed analysis should be made by preparing a process chart. Circle such man-hour totals for further process charting (see Fig. 3-7, Form C).

2. Is there misdirected effort? Is the organization spending too much time on relatively unimportant operations or unnecessary work? Clues to misdirected effort frequently are found in miscellaneous or administrative categories. The misdirected effort of any one individual may be insignificant; however, it often

---

*See Fig. 3-8 for a sample proposed work-distribution chart developed after analysis of present work-distribution chart. (This analysis has been put on a separate chart for the purpose of illustration. Normally, however, analysis would be made on the first work-distribution chart.)

WORK DISTRIBUTION

BY:
CURRENT:
PROPOSED:

FUNCTION CHARTED: Proofreading Section   APPROVED: M. Rose, Branch Chief   DATE: REVISION:

Opr No.	Activity	(MOST TIME STUDY FURTHER) Work Count	Hrs Wk	Gray — Supervisor TASK	Work Count	Hrs Wk	Jones — Proofreader TASK	Work Count	Hrs Wk	White — Proofreader TASK	Work Count	Hrs Wk	Brown — Proofreader TASK	Work Count	Hrs Wk	Smith — Proofreader TASK	Work Count	Hrs Wk	Green — Proofreader TASK	Work Count	Hrs Wk
1	Proofread catalog card copy	1911	91	Read copy and indicate errors	250	10	Read copy and indicate errors	240	10	Read copy and indicate errors	300	12	Read copy and indicate errors	525	25	Read copy and indicate errors	300	20	Read copy and indicate errors	296	10
2	Inventory documents & maintain records			Count incoming & outgoing documents and maintain records			Count incoming & outgoing documents and maintain records		5				Count incoming & outgoing documents and maintain records			Count incoming & outgoing documents and maintain records		2	Count incoming & outgoing documents and maintain records		2
3	Review tab copy	1160	58				Make final security review	300	15	Make final security review	360	18	Make final security review	100	10	Make final security review	150	5	Make final security review	240	5
4	Proofread special notices for tab	360	22							Read and indicate errors	60	4	Read and indicate errors	75	5	Read and indicate errors	75	5	Read and indicate errors	150	8
5	Special projects			Prepare manual		4															
6	Administration & supervision			Assign work & give instructions		14	Assign work & give instructions		3							Assign work and give instructions		8			
	(Train employees)			Train employees		4															
7	Answer phone			Answer questions and take messages		6	Answer questions and take messages		4	Answer questions and take messages		5							Answer questions and take messages		2
8	Edit tab copy			Perform initial edit on documents		3															
9	Miscellaneous			Get briefings on computer printouts		3															
	TOTAL (Man-hours)		240			40			40			40			40			40			40

Annotations: MOST TIME STUDY FURTHER · SKILLS PROPERLY USED? · COMBINE RELATED? · SPREAD TOO THIN? · SKILLS PROPERLY USED? · SPREAD TOO THIN? · NOTE WHO MAINTAINS OFFICE SUPPLIES?

ANALYSIS: WHAT TAKES THE MOST TIME?...IS THERE MISDIRECTED EFFORT?...ARE SKILLS USED PROPERLY?...ARE THERE TOO MANY UNRELATED TASKS?...ARE TASKS SPREAD TOO THINLY?...IS WORK DISTRIBUTED EVENLY?

Fig. 3-8. Sample analysis of Form C.

becomes a sizable total when several people are involved. When individuals are involved in tasks not contributing directly to the mission of the organization, misdirected effort generally can be discovered.

3. Are skills being used properly? Is everyone being utilized in the best possible manner, or are special skills and abilities being wasted? It is wasteful and demoralizing to have personnel working below their skills. In addition, having personnel working too far above their skills is also undesirable. In a small organization, it may not be possible to have employees spend time only on work equal to their abilities. However, in striving to attain the ideal, the waste of skills will be reduced.

4. Are individuals performing too many unrelated tasks? A large number of tasks recorded in any one Task column on the work-distribution chart may indicate that the individual is involved in such a variety of tasks that there is lost motion. Few individuals can perform a variety of tasks equally well. Greater efficiency generally results if workers are assigned related tasks.

5. Are tasks spread too thinly? In addition to the "jack-of-all-trades" situation, the work-distribution chart reveals tasks that are "everybody's job and nobody's job." To find instances where tasks are spread too thinly, look across the chart from left to right. If everyone has some part in an activity, it may be a danger signal indicating duplication of effort. One person working steadily at a task is more productive than several individuals working the same number of man-hours. The assignment of a task to one person pinpoints responsibility.

6. Is work distributed evenly? Some work situations may pass all the tests discussed so far and still seem out of balance. This may be due to unequal distribution of tasks or of employees. The result may be a situation where one employee is usually working overtime while another employee, of the same grade and with similar duties, is finished at the end of the workday. The relative importance of tasks assigned to persons engaged in similar activities should be measured. Urgent and important tasks should be spread as evenly as possible to insure that all work is completed according to schedule.

The answers to the preceding questions combined with the experience of the analyst and/or supervisor usually will suggest improvement. After each question has been asked and areas that require improvement have been indicated, consideration should be given to possible eliminations, additions, and rearrangement of tasks. Once a decision has been made regarding sound improvements, a proposed work-distribution chart should be made. (See Fig. 3-7.) This chart will show what tasks and operations should be done, who should do them, how much time should be spent, and, when appropriate, how much should be done.

The improved work-distribution chart will help make sure that improvements are well thought out. It may be used as a proposal document when high-level authority is needed to make changes. It will be a useful plan of operations that can have many uses—such as orientation of new employees, helping to justify personnel requirements, and providing a source document of pertinent facts about the organization.

Remember, if continuing benefits are to be obtained, work-distribution analysis must be made on a recurring basis. The best situation remains the best only for a relatively short time. Procedure changes can be expected, and improvements are mandatory. Whenever either occurs to a significant degree, the work-distribution chart should be reviewed.

Process Analysis

The second tool or technique in the work-simplification program is process analysis. Through process analysis, it can be determined how work is done as well as who does it, and where and when it is done. Process charting can contribute to the analysis of procedures in the following situations:

1. When a major change of personnel, procedures, or volume of work is taking place
2. When a procedural problem arises
3. When making a periodic review of operating methods
4. When establishing a new organization

There are many types of process charts that can be used for specific purposes: operations process, flow process, multiple activity (man and machine), right- and left-hand, and simultaneous-motion cycle charts (cycle charts). The chart described herein is applicable to procedures involving materials and/or personnel. Use of other process charts is encouraged, as appropriate, after the analyst/observer becomes familiar with the use of Form D, flow-process chart (Fig. 3-9).

The flow-process chart is a graphic presentation of steps in a process. It is used primarily to aid in a systematic analysis of a procedure, to find ways to improve and simplify work. The chart will help in improving the manner in which steps are carried out, and aid in eliminating unnecessary travel or transportation.

Flow-process charts also are used very effectively in briefings and in group discussions. They enable the speaker to point out relationships of all the essential characteristics of a system, procedure, or process. They afford an opportunity to pinpoint and compare advantages and disadvantages of different alternatives leading to and supporting the conclusions and recommendations. Use of the "before" and "after," or "present" and "proposed," charts effectively demonstrates improvements already made and those proposed.

WORK MEASUREMENT FLOW PROCESS AND STANDARD SUMMARY CHART		REFERENCE/FILE NUMBER		DATE 6 Jan 65			

WORK MEASUREMENT
FLOW PROCESS AND STANDARD SUMMARY CHART

PROCESS Scrambling Eggs	SUMMARY						
	ACTIONS	PRESENT		PROPOSED		DIFFERENCE	
"X" APPLICABLE BLOCK PRESENT		NO.	TIME	NO.	TIME	NO.	TIME
☒ MAN ☐ MATERIAL	OPERATIONS	22	145	15	124	7	21
CHART BEGINS Get Fry Pan / CHART ENDS Salt & Pepper Eggs	TRANSPORTS	23	56	11	28	12	28
	INSPECTIONS	1	2	1	2	0	0
CHARTED BY S. Spade	DELAYS	2	110	1	74	1	36
	STORAGES						
ORGANIZATION / WORK CENTER	TOTAL	112	313	28 56	228		85

DETAILS OF METHOD	POSITION TITLE	OPERATION	TRANSPORT	INSPECTION	DELAY	STORAGE	WORK UNIT STANDARD SUMMARY			
							REFERENCE Distance Quan.	ELEMENT TIME Seconds	FREQUENCY	BASE TIME
Get fry pan and put on range	"	O⇨☐D▽					1	8		
Walk to refrigerator	"	O⇨☐D▽					6	3		
Get butter	"	O⇨☐D▽					1	6		
Butter to range	"	O⇨☐D▽					6	3		
Walk to cupboard	"	O⇨☐D▽					4	2		
Get spoon	"	O⇨☐D▽					1	6		
Spoon to range	"	O⇨☐D▽					4	2		
Butter to fry pan	"	O⇨☐D▽						10		
Walk to refrigerator	"	O⇨☐D▽						3		
Get eggs	"	O⇨☐D▽					2	10		
Eggs to range	"	O⇨☐D▽					6	3		
Walk to cupboard	"	O⇨☐D▽					4	2		
Get measuring cup	"	O⇨☐D▽					1	6		
Measuring cup to range	"	O⇨☐D▽					4	2		
Walk to refrigerator	"	O⇨☐D▽					6	3		
Get milk	"	O⇨☐D▽					1	4		
Milk to range	"	O⇨☐D▽					6	3		
Turn on range burner	"	O⇨☐D▽						4		

CONTINUED ON REVERSE SIDE

METHOD APPROVED (Signature of Supervisor)	TOTAL BASE TIME	313
	PF&D ALLOWANCE _____%	
	STANDARD TIME	

SHEET_____ OF _____ SHEET(S)

Fig. 3-9. Form D, flow-process chart.

WORK MEASUREMENT **FLOW PROCESS AND STANDARD SUMMARY CHART**			REFERENCE/FILE NUMBER				DATE		

PROCESS

"X" APPLICABLE BLOCK PRESENT
☐ MAN ☐ MATERIAL

CHART BEGINS	CHART ENDS

CHARTED BY

ORGANIZATION	WORK CENTER

SUMMARY					
ACTIONS	PRESENT		PROPOSED		DIFFERENCE
	NO.	TIME	NO.	TIME	NO.	TIME
OPERATIONS						
TRANSPORTS						
INSPECTIONS						
DELAYS						
STORAGES						
TOTAL						

WORK UNIT

DETAILS OF METHOD	POSITION TITLE	OPERATION TRANSPORT INSPECTION DELAY STORAGE	STANDARD SUMMARY			
			REFERENCE	ELEMENT TIME	FREQUENCY	BASE TIME
Wait for butter to melt	"	○⇨☐D▽	Distance Quan.	20		
Pick up and break	"	○⇨☐D▽	2	10		
Inspect eggs	"	○⇨☐D▽		2		
Measure & pour milk in fry pan	"	○⇨☐D▽		8		
Walk to cupboard	"	○⇨☐D▽	4	2		
Get turner	"	○⇨☐D▽		6		
Turner to range	"	○⇨☐D▽	4	2		
Stir eggs	"	○⇨☐D▽		15		
Wait for eggs to cook	"	○⇨☐D▽		90		
Turn of burner	"	○⇨☐D▽		4		
Walk to cupboard	"	○⇨☐D▽	4	2		
Get plate	"	○⇨☐D▽		6		
Plate to range	"	○⇨☐D▽	4	2		
Eggs to plate	"	○⇨☐D▽		10		
Get eggshells	"	○⇨☐D▽		4		
Eggshells to trash	"	○⇨☐D▽	4	2		
Walk to range	"	○⇨☐D▽	4	2		
Get milk and butter	"	○⇨☐D▽		2		

CONTINUED ON REVERSE SIDE

METHOD APPROVED (*Signature of Supervisor*)

TOTAL BASE TIME	313
PF&D ALLOWANCE _____ %	
STANDARD TIME	Figure 1

Fig. 3-9. (*Continued*).

WORK MEASUREMENT
FLOW PROCESS AND STANDARD SUMMARY CHART

	REFERENCE/FILE NUMBER			DATE	

PROCESS

	SUMMARY					
ACTIONS	PRESENT		PROPOSED		DIFFERENCE	
	NO.	TIME	NO.	TIME	NO.	TIME
OPERATIONS						
TRANSPORTS						
INSPECTIONS						
DELAYS						
STORAGES						
TOTAL						

"X" APPLICABLE BLOCK PRESENT
☐ MAN   ☐ MATERIAL

CHART BEGINS   CHART ENDS

CHARTED BY

ORGANIZATION   WORK CENTER

WORK UNIT

DETAILS OF METHOD	POSITION TITLE	OPERATION	TRANSPORT	INSPECTION	DELAY	STORAGE	STANDARD SUMMARY			
							REFERENCE	ELEMENT TIME	FREQUENCY	BASE TIME
Milk & butter to refrigerator	"	○ ⇨ ☐ D ▽					Distance Quan. 6	Seconds 3		
Put milk & butter in refrigerator	"	○ ⇨ ☐ D ▽						6		
Walk to range	"	○ ⇨ ☐ D ▽					6	3		
Get spoon, turner and measuring cup	"	○ ⇨ ☐ D ▽					3	6		
Spoon, turner and measuring cup to dishpan	"	○ ⇨ ☐ D ▽					4	2		
Walk to range	"	○ ⇨ ☐ D ▽					4	2		
Get frying pan	"	○ ⇨ ☐ D ▽					1	2		
Frying pan to dishpan	"	○ ⇨ ☐ D ▽					4	2		
Walk to range	"	○ ⇨ ☐ D ▽					4	2		
Get plate of eggs	"	○ ⇨ ☐ D ▽					1	2		
Eggs to table	"	○ ⇨ ☐ D ▽					8	4		
Salt & pepper eggs	"	○ ⇨ ☐ D ▽						10		
		○ ⇨ ☐ D ▽								
		○ ⇨ ☐ D ▽								
		○ ⇨ ☐ D ▽								
		○ ⇨ ☐ D ▽								
		○ ⇨ ☐ D ▽								
		○ ⇨ ☐ D ▽								

CONTINUED ON REVERSE SIDE

METHOD APPROVED (*Signature of Supervisor*)

TOTAL BASE TIME	313
PF&D ALLOWANCE _____%	
STANDARD TIME	

SHEET __3__ OF __3__ SHEET(S)

Fig. 3-9. (*Continued*).

Preparation of Form D

Important details of doing work may be overlooked if observations of what is done are made without recording the details. The flow-process chart helps make a record of the step-by-step procedure in doing work. The chart also gives a graphic picture of the work through symbols representing various elements of work. In making a flow-process chart of the present method, the following eight steps will be used:

1. *State the process being studied.* The work-distribution chart as previously mentioned will provide potential areas for study. The analyst may wish to select for the initial effort a process that he or she is very familiar with. The process should be stated clearly and precisely. During the study, do not deviate from that process.

2. *Determine the subject to be followed.* It is necessary to determine whether the chart will follow a person, activity, or material. After the decision has been made, the chart must not deviate from the subject being followed. Each step listed must be on the chosen subject and must actually follow the work being done.

3. *Determine starting and ending points.* Be specific in identifying the place or time that the process chart begins and ends. Don't try to cover too much with one chart, but be sure to give complete coverage between the start and stop points. If the process is long and complicated, break the analysis into smaller areas of less coverage by making several charts.

4. *Briefly describe each step in the process.* Make the description brief and pertinent to a single step of the process. Include every detail, in the order in which it occurs, and verify information with the people actually doing the work. Unless you include every specific detail on the chart, the chart will be of little value; accurate analysis and good improvements depend on a detailed picture of current practice.

5. *Classify each step by using symbols.* When completing a flow-process chart, use symbols to identify the type of work that is being done in each step on the chart. The symbols make it possible to show graphically the sequence of work being done and help to quickly spot needed improvements. For this purpose, each step is classified by symbol into one of five elements as follows:

   a. An *operation* occurs when an object is intentionally changed in any of its physical or chemical characteristics, is assembled or disassembled, or is arranged or prepared for another operation, transportation, inspection, or storage. An operation occurs also when information is given or received or when planning or calculating takes place. Examples of operations are:

   (1) Load cartons on pallet.

(2) Drill hole.

(3) Register mail.

(4) Answer.

The symbol for operation is ○.

b. *Transportation* occurs when an object is moved from one place to another, except when such movements are part of the operation or are caused by the operator at the work station during an operation or an inspection. Examples of transportation activities are:

(1) Trucking supplies to a warehouse.

(2) Carrying mail to another desk.

(3) Walking between work stations.

(4) Movement of cartons on power conveyor.

The symbol for transportation is ⇒.

c. An *inspection* occurs when an item is examined, verified, received, or checked for quality or quantity but not changed. Examples of inspection are:

(1) Proofread a letter.

(2) Test-fire a weapon.

(3) Check weight of supplies.

(4) Count items to verify quantity.

The symbol for inspection is □.

d. A *delay* occurs in a process when conditions do not permit or require immediate performance of the next planned action. A delay symbol should not be used when an object is changed, created, or added to, even though a person does not handle the object for a period of time. Examples of delay are:

(1) Letter in outgoing box, awaiting messenger.

(2) Breakdown in assembly line.

(3) Holding an item to prepare a complete shipment.

(4) Document waiting for signature.

The symbol for delay is D.

e. A *storage* occurs when an object is retained in one place protected against unauthorized removal. Examples of storage are:

(1) Correspondence in a permanent file.

(2) Supplies placed in a warehouse.

(3) Suspense copy in file.

The symbol for storage is ▽.

6. *Record time and distance*. Enter distances in feet for all transportations. Distances will be entered in Frequency block of Form D. Remember that the moving of items within a work location or about a desk usually is not considered transportation. Normally, time will be entered in minutes for each step.

7. *Record quantity*. Make an entry of item quantity where applicable. Entry will be made in the Frequency block of Form D.

8. *Summarize*. Count the number of operations, transportations, inspections, delays, and storages, and enter each in the Summary portion of the flow-process chart under Present. The total time for each of the elements is entered, and the total distance traveled for the study is noted in the lower half of the Total column. A completed flow-process chart of the Present situation is shown as an example in Fig. 3-9.

As previously stated, work simplification is part of a total integrated management process involving several other management techniques. For purposes of work-simplification study, the Work Center block and Standard Summary data (with the exception of Base Time) do not need to be completed *per se*. Also, the Personal Fatigue and Delay Allowance and the Standard Time blocks may be left blank. These blocks are for use by methods-and-standards personnel in making work measurement studies.

During this phase of the study, the analyst/observer also can obtain valuable information from those interviewed about improvements that may be made and how. Some of the ideas may appear impractical. However, others may lead to a better way of carrying out the intent of the procedure. At every opportunity, the analyst/observer should cultivate and encourage the full cooperation of all concerned.

Analysis of Flow-Process Chart

The finished flow-process chart shows the procedure as it now exists. The analyst/observer should make a systematic analysis of the data on the chart. Improvements will result from eliminating, combining, rearranging, and simplifying the details. Thus, every detail should be questioned. For example,

1. *Why* is it necessary? Can the entire process or some of the steps be eliminated?
2. *What* is actually done? Are all the steps included? What should be done?
3. *Where* should the step be done? Can it be done better in another place?
4. *When* should this be done? Is it done in the right sequence? Can it be combined or simplified by moving it back or ahead?
5. *Who* should do it? Is the right person handling it, or can someone else do it better?

6. *How* is the job being done? Can it be improved with different equipment of different layout?

Persistent questioning with an open mind may uncover weaknesses that can be corrected and that may lead to the development of better methods. As the details are questioned, notes should be made on the chart indicating possible improvements. The Reference block of the Standard Summary portion of chart may be used for this purpose.

After the analyst/observer has arrived at certain conclusions and recommendations as a result of the analysis, a new flow-process chart should be prepared showing the proposed procedure and steps that incorporate the recommendations. It is essential that the analyst/observer work in complete cooperation with those concerned throughout the study. The analyst/observer should review and discuss thoroughly with interested supervisors the whole proposed procedure, particularly the changes. Such a review will make possible additional improvements as well as the correction of any erroneous information. Areas of disagreement should be restudied and further discussions held to obtain general concurrence. This coordinated effort will speed up agreements and make the installation of the revised procedure easier, because it will be the product of the joint efforts of those concerned.

The Summary portion entries from the Present flow-process chart and summaries of the Proposed chart will be entered on the Summary portion of the Proposed chart. The Difference block of the Proposed chart will reflect the improvements that have been made.

A sample Proposed flow-process chart is presented as Fig. 3-10.

Motion Analysis

Motion analysis is a work-simplification technique that concentrates on the basic elements of a job. Most work is the result of many and separate motions of the human body. Wasted time and effort have been mentioned as reasons why work methods must be improved. One of the greatest wastes is in using human resources. New ways to save time must be found. When taken together, even small savings are significant in contributing to more effective accomplishment of assigned missions at lower cost. The two predominant reasons for interest in motion analysis are:

1. To reduce worker effort and fatigue, thus making work easier and workers more comfortable.
2. To reduce the time required to perform a complete cycle of work.

An individual does not have to be a trained methods engineer to apply the fundamentals of motion economy to the work done in an organization. Having the proper attitude and a desire to find easier ways of getting work done are the prime requisites. These include, for example, being alert to the possible substitution of mechanical and electronic methods for manual methods, and examining the possibility of relocating equipment or facilities to reduce worker effort in using them.

Principles

Many dynamic individuals in the industrial engineering field have devoted their lives to the subject of motion economy. As a result, today we have a set of principles to guide us in obtaining the best results for effort spent. These principles are generally applicable to repetitive operations. Most operations involving manual effort could be performed with less effort and greater efficiency if these principles were observed. There is a definite physical area in which materials and supplies should be placed. Within this area, objects can be manipulated and work accomplished without excessive effort. In some cases, it may be impractical to apply every principle of motion economy because of physical limitations; however, the larger the number of fundamentals applied, the greater the improvement will be.

There is a *maximum* efficient work area for an individual, either in a standing or sitting position. This area is determined by extending each arm from the shoulder and making a circular path with the shoulders as pivotal points. The overlapping area formed by the arcs of the arms constitute a zone beyond which the hands cannot work closely together without moving the body.

The *normal* efficient work area is determined in the same manner. However, only the forearms are extended, the upper arms are relaxed, and the elbows are held close to the body. The overlapping area is the zone in which both hands may work together most conveniently. The maximum and normal work areas are illustrated in Fig. 3-11. Measurements are only general. Details are worked out to fit the individual situation.

The principles of motion economy are:

1. Both hands should begin movement simultaneously.
2. Both hands should complete their movement at the same time.
3. Both hands should not be idle at the same time except during rest periods.
4. Motions of the arms should be in opposite and symmetrical direction and should be made simultaneously.
5. Hesitation should be studied and eliminated wherever possible.

WORK MEASUREMENT
FLOW PROCESS AND STANDARD SUMMARY CHART

| REFERENCE/FILE NUMBER | | | | DATE | |

PROCESS Scrambling eggs in iron fry pan

"X" APPLICABLE BLOCK
☒ MAN ☐ MATERIAL

CHART BEGINS Get fry pan
CHART ENDS Salt & pepper eggs

CHARTED BY S. Matisoff

ORGANIZATION WORK CENTER

ACTIONS	SUMMARY					
	PRESENT		PROPOSED		DIFFERENCE	
	NO.	TIME	NO.	TIME	NO.	TIME
OPERATIONS			15	124		
TRANSPORTS			11	28		
INSPECTIONS			1	2		
DELAYS			1	74		
STORAGES						
TOTAL			28	228		
			56			

WORK UNIT

DETAILS OF METHOD	POSITION TITLE	OPERATION TRANSPORT INSPECTION DELAY STORAGE	STANDARD SUMMARY			
			REFERENCE	ELEMENT TIME	FREQUENCY	BASE TIME
Get fry pan & put on range		○⇨☐D▽	Distance Quan. 1	Seconds 8		
Turn on range burner		○⇨☐D▽		3		
Walk to refrigerator		○⇨☐D▽	6	3		
Get eggs, butter & milk		○⇨☐D▽	3	15		
Walk to range		○⇨☐D▽	6	3		
Walk to cupboard		○⇨☐D▽	4	2		
Get measuring cup, spoon, turner & plate		○⇨☐D▽	3	15		
Walk to range		○⇨☐D▽	4	2		
Butter to fry pan		○⇨☐D▽		10		
Pick up & break eggs		○⇨☐D▽	2	10		
Inspect eggs		○⇨☐D▽		2		
Measure & pour milk in pan		○⇨☐D▽		8		
Stir eggs		○⇨☐D▽		15		
Wait for eggs to cook		○⇨☐D▽		90 (74)		
Get eggshells		○⇨☐D▽	Next 6 steps	4		
Eggshells to trash		○⇨☐D▽	4 done while eggs	2		
Walk to range		○⇨☐D▽	4 are cooking.	2		
Get butter & milk		○⇨☐D▽	(90−16=74)	2		

CONTINUED ON REVERSE SIDE

METHOD APPROVED (*Signature of Supervisor*)

TOTAL BASE TIME	
PF&D ALLOWANCE _____%	
STANDARD TIME	

Fig. 3-10. Sample proposed flow-process chart.

WORK MEASUREMENT
FLOW PROCESS AND STANDARD SUMMARY CHART

REFERENCE/FILE NUMBER | DATE

PROCESS

"X" APPLICABLE BLOCK
☐ MAN   ☐ MATERIAL

CHART BEGINS | CHART ENDS

CHARTED BY

ORGANIZATION | WORK CENTER

SUMMARY

ACTIONS	PRESENT		PROPOSED		DIFFERENCE	
	NO.	TIME	NO.	TIME	NO.	TIME
OPERATIONS						
TRANSPORTS						
INSPECTIONS						
DELAYS						
STORAGES						
TOTAL			50 feet			

WORK UNIT

STANDARD SUMMARY

DETAILS OF METHOD	POSITION TITLE	OPERATION TRANSPORT INSPECTION DELAY STORAGE	REFERENCE Distance	Quan.	ELEMENT TIME	FREQUENCY	BASE TIME
Butter & milk to refrig.	"	○⇨☐D▽	6		3		
Walk to range	"	○⇨☐D▽	6		3		
Turn off range	"	○⇨☐D▽			4		
Put scrambled eggs on plate	"	○⇨☐D▽		2	10		
Place spoon, cup, turner in fry pan	"	○⇨☐D▽		3	8		
Fry pan to dishpan	"	○⇨☐D▽	4	1	2		
Walk to range	"	○⇨☐D▽	4		2		
Get plate of eggs	"	○⇨☐D▽		1	2		
Eggs to table	"	○⇨☐D▽	8		4		
Salt & pepper eggs	"	○⇨☐D▽			10		
		○⇨☐D▽					
(Note: Time is in seconds only		○⇨☐D▽					
for illustrative purposes. Time may be in minutes or other		○⇨☐D▽					
logical increment of time.)		○⇨☐D▽					
		○⇨☐D▽					
		○⇨☐D▽					
		○⇨☐D▽					
		○⇨☐D▽					

CONTINUED ON REVERSE SIDE

METHOD APPROVED (*Signature of Supervisor*)

TOTAL BASE TIME	
PF&D ALLOWANCE _____%	
STANDARD TIME	

SHEET __2__ OF __2__ SHEET(S)

Fig. 3-10. (*Continued*).

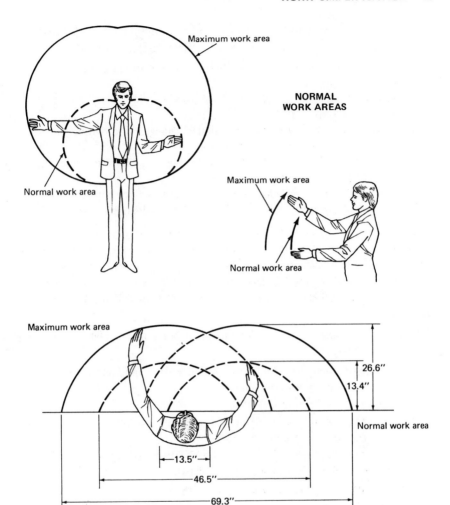

Fig. 3-11. Maximum and normal work areas.

6. Hand motions should be as simple as possible so that they may be performed in a shorter time. The classifications of motions in order of economy are:

   a. Finger motions only
   b. Finger and wrist motions
   c. Finger, wrist, and forearm motions
   d. Finger, wrist, forearm, and upper arm motions

e. Finger, wrist, forearm, upper arm, and body motions. (A change in position is required to perform this classification of motion.)

7. Material and equipment should be located as nearly as possible within the normal grasp area.

8. Sliding, rolling, and pushing rather than carrying is usually quicker in transporting small objects.

9. Straight line motions requiring sudden changes in direction are not as desirable as continuous curved motions.

10. The sequence of motions should be arranged so that it increases the possibility of rhythm becoming automatic.

11. Work equipment and materials should be put in place before work begins; prepositioning reduces searching, finding, and selecting actions.

A basic step in motion study is learning to recognize the basic elements an individual performs in completing a particular task. Unnecessary motions can be eliminated, and more effective motion sequences can be substituted. The following partial list of significant basic elements, which are classified according to the purpose of the motion(s) involved, is presented as an aid in improving work methods.

1. *Transport empty* is the act of moving a body member that is empty in reaching for an object.

   a. Rule: Reduce the number and distances to a minimum. Provide a return load whenever possible.

   b. Suggestions for improvement:

      (1) Can a conveyor chute be used to supply object?

      (2) Can a gravity system be used?

      (3) Can transport be handled more effectively in large units?

      (4) Can one hand be used instead of two?

2. *Grasp* is the act of getting control of anything.

   a. Rule: Study the number of times grasp occurs in a cycle. Provide suitable hoppers, trays, etc., to facilitate the grasp action.

   b. Suggestions for improvement:

      (1) Can a combination tool be used?

      (2) Can a mechanical grasp be used?

      (3) Is it possible to grasp more than one object at a time?

      (4) Can a conveyor be used?

      (5) Can objects be slid, pushed, or rolled instead of grasped?

      (6) Can tools be prepositioned?

      (7) Is the article transferred from one hand to the other?

3. *Transport loaded* is the act of moving an object.

  a. Rules: Reduce the number and distance of transport loaded to a minimum. If 50 percent or more of the time needed to perform an operation is spent in moving articles or tools, or in transporting over an identical motion pattern, a conveyor of some sort usually can be designed. Provide a return load whenever possible.

  b. Suggestions for improvement: See section on transport empty.

4. *Release load* is the act of relinquishing control of an object. Release load as a motion cycle takes place from the time any part of the transportation medium leaves an object until all parts have lost control.

  a. Rules: Study the number of times release load occurs in a cycle. Ordinarily, once should be enough for one item. Study the possibility of facilitating release load by the proper design of finished material containers and tool holders.

  b. Suggestions for improvement:

    (1) Can drop delivery be used?
    (2) Can multiple units be released?
    (3) Can a mechanical or air ejector be used?
    (4) Can a conveyor be used?

5. *Preparation* is the preparation of the articles transported for the next motion to be performed or positioning the articles or transportation medium so they are in approximate position for use in the next motion.

  a. Rule: Study the possibility of rearranging the materials or tools so that they may be grasped properly to permit the prepositioning to be performed in transit.

  b. Suggestions for improvement:

    (1) Can the object be prepositioned in transit?
    (2) Can tools be suspended?
    (3) Can tools be stored in proper location for use?
    (4) Can a hopper be used to keep parts in constant relation to each other?

6. *Position* is the act of placing an object in an exact and predetermined position.

  a. Rule: Study the possibility of using fixed mechanical guide.

  b. Suggestions for improvement:

    (1) Can a guide be used?
    (2) Can a stop be used?

(3) Can a funnel be used?

(4) Can a template be used?

7. *Hold* is the act of retaining and maintaining control of the object grasped, or of holding object while work is being performed on the object.

   a. Rule: Do not peform a hold with a human hand where a mechanical device can be used.

   b. Suggestions for improvement:

     (1) Can clamp, jog, shelf, trough, clip, adhesive, vacuum, mechanical device, conveyor, friction, etc., be used to eliminate the necessity for holding?

Application of Motion Analysis

In applying the fundamentals of motion economy and improving basic motion sequences, the basic elements of each task are identified and analyzed. The number and type of basic elements will be contingent upon the work being done. In an assembly-type operation, the basic elements probably will be small and detailed. In an office situation, the elements possibly will be more gross. Once the elements have been identified, the fundamentals of motion economy are applied and improvements are made wherever possible.

In addition to asking the basic questions outlined previously, more specialized questions may provide additional clues to improvements. For example,

1. Are the simplest motions being used?
2. Are desks, files, or workbenches arranged to permit efficient work flow?
3. Is work stored within easy reach?
4. Is there too much walking for tools, materials, or finished products?
5. Are the right tools or equipment being used?
6. Can mechanical equipment replace manual work?
7. Can jigs or fixtures be used?
8. Is the most efficient means of transportation used?
9. Has the comfort of all workers been considered?

Motion analysis above the elementary level may require the assistance of a trained methods-and-standards technician. Each primary-level field activity should have a trained methods-and-standards staff to provide consultant service and assistance as required.

Work Count

Work count shows the volume of work. As a work-simplification technique, work count is used to determine the impact of the volume of work on the methods and procedures under study. Counting or measuring work assists in improving processes, adjusting work assignments, evaluating performance, eliminating bottlenecks, and getting optimum production at lowest cost. Work count is particularly helpful in scheduling work, assigning work, and distributing work fairly and efficiently.

Work-count data may be collected using Form E, work sampling data collection and/or computation (Fig. 3-12). The work-distribution chart provides space for inserting appropriate work-count data for each task. Through use of the data on this form, a task can be analyzed with respect to individual effort, number of man-hours involved, and relationship of output to the appropriate peformance standard.

Work Units

The basic step in work count lies in determining "significant" work to count. A work unit is an item of work selected to express quantitatively the work accomplished in a specific area. Normally, you can find an end product that can be counted or measured; however, that product may or may not be a meaningful work unit. Results of the productive effort must be related to the individuals or group making the effort. In addition, it is essential that a work unit have the following characteristics:

1. The work unit must be readily identifiable by both analyst/observers and operating personnel.
2. The work unit must be countable both by definition and as a practical matter.
3. The work unit must reflect output or volume of work completed.

Work count can be applied to both industrial- and office-type operations. In industrial-type operations, such as an engine repair shop, the work unit might be "generators repaired" or "starters repaired," for the respective sections. At the work-center or repair-shop level, the individual work units may be pyramided to a work-center work unit of "engines repaired." In office or clerical situations, "letters," "postings," "vouchers edited," and so on, might be selected as work units.

A list should be made of all possible work units in an organization. All possibilities should be considered, and only significant ones should be selected. Once established, the work count should be made as accurately as possible over a period of time. The work count when analyzed will provide a way to develop improved procedures and evaluate worker effectiveness. In addition, the work unit, if properly selected, forms the basis for work measurement and local cost accounts. At the local activity level, specific work units such as pieces, documents, or pallets can be used. These work units would be pyramided and summarized to the appropriate command-level performance indicator.

WORK SAMPLING DATA COLLECTION AND/OR COMPUTATION

ORGANIZATION: A-B-C-D-E Processing Unit
WORK CENTER
DATE PREPARED
OBSERVER
REFERENCE/FILE NUMBER
WORK MEASUREMENT SUPERVISOR

TIME SCHEDULE		NAME OF INDIVIDUAL	A's Processed	B's Processed	C's Processed	D's Processed	E's Processed	WORK UNITS		HOURS WORKED	GROUP/INDIVIDUAL PERFORMANCE RATING	LEVELED BASE HOURS	PF & D ALLOWANCE (%)	STANDARD TIME	NONPRODUCTIVE	OUT OF AREA
07:00	12:00															
07:10	12:10															
07:20	12:20															
07:30	12:30															
07:40	12:40															
07:50	12:50															
08:00	13:00	Parker	HHT HHT		HHT HHT											
08:10	13:10	Helms		HHT HHT		HHT //										
08:20	13:20	Martin		HHT HHT	HHT HHT											
08:30	13:30	Starkey		HHT HHT		HHT HHT										
08:40	13:40	Fleming			HHT HHT											
08:50	13:50															
09:00	14:00	Davies	HHT HHT	HHT HHT	HHT HHT											
09:10	14:10															
09:20	14:20	McCoy		HHT HHT		HHT HHT										
09:30	14:30	Cooper			HHT HHT											
09:40	14:40															
09:50	14:50	Horton	HHT /				HHT /									
10:00	15:00															
10:10	15:10	Blout			HHT ////											
10:20	15:20															
10:30	15:30															
10:40	15:40															
10:50	15:50															
11:00	16:00															
11:10	16:10															
11:20	16:20															
11:30	16:30	TOTALS														
11:40	16:40															
11:50	16:50	PRODUCTION COUNT	22	40	67	32	11									

48

Fig. 3-12. Form E, work sampling data collection and/or computation.

Method of Counting

The method of counting depends upon the type of operation being measured. Automatic methods that are available should be used whenever possible. For example, the number of vouchers prepared in a given time might be determined through the use of serially numbered forms. It is ideal when an impartial source determines the work volume. When the volume is small, a simple tally or manual count may suffice.

Application of Work Count

In the work-simplification effort to improve processes, work count will be a valuable technique if applied in the following ways:

1. To schedule work. Each step in a process does not require the same amount of time, effort, and skill. Work count assists in determining the time required for each step so that a realistic schedule can be made.
2. To relate tasks. When several individuals in an organization are performing unrelated tasks, work count will aid in analyzing these tasks to determine if they can be combined and assigned to fewer individuals, resulting in a better, coordinated effort.
3. To divide work. Work count is used to identify the type of action required by various kinds of work and to isolate exceptions for special treatment, thereby reducing idle time.
4. To locate bottlenecks. Optimum production results when work flows smoothly through the various steps in a process. An excessive number of units or lack of proper number of units passing through a critical point may provide a clue to making adjustments in the process, which will improve the operation.
5. To demonstrate personnel needs. Using work count, increased or revised work volumes may be demonstrated with facts and figures that will be useful in adjusting personnel requirements.
6. To stimulate interest. Individuals are interested in the accomplishment of their organization in relation to other organizations engaged in similar work. A work count of production effort can be used to arouse interest and maintain a competitive spirit as well as determine relative effectiveness of personnel engaged in similar work.

Layout Studies

Layout studies are used in connection with all the other techniques; however, they especially complement process analysis and motion analysis. The objectives of layout studies are:

1. To improve production.
2. To provide the greatest physical ease to the largest number of individuals.
3. To provide as short and straight a distance as possible in the processing and travel of material, documents, personnel, etc.

Layout studies can lead to significant improvements by themselves but they usually are more valuable when used in conjunction with other techniques. Layout studies are made through the use of a layout chart, which is a plan or sketch of the physical facilities of an office, shop, warehouse, or area upon which the flow of work is traced.

Principles of Physical Layout

Physical layout is a compromise between many factors. Quite often, it is complicated because several different tasks must be performed in the same work area with the same facilities. An ideal layout cannot always be achieved because of the type and amount of space and equipment required.

Physical layout is an inescapable problem. The question is not, Shall we have a layout? It is, How good is the layout we have? While an ideal layout may be impossible, the layout may be improved through systematic analysis and application of the following fundamentals:

1. Work flow should follow straight lines with a minimum of backtracking and cross travel.
2. Persons having the most frequent contact should be located near each other.
3. Files, cabinets, and other records and materials should be located in close proximity to those who use them most.
4. Surplus facilities should be released to provide space for other requirements.
5. The allocation of space should be in keeping with the requirements of the work.
6. The arrangement of personnel and equipment should facilitate supervision.
7. The layout should be adapted to the maintaining of proper safety and security practices.
8. Persons using the same equipment should be grouped together.
9. Persons receiving visitors or required to maintain outside contacts should be located near entrances.
10. The capacities and characteristics of the building should be studied to make certain that heavy or bulky equipment such as safes and lifts can be accommodated.

Of course, a physical facility may never ideally meet these criteria. However, any physical layout must be planned if it is to be a good one. Remember, the

best layout is that which integrates the personnel, materials, machinery, supporting activities, and any other considerations in a way that results in the best compromise. In essence, we plan the ideal layout, and then, from that, we plan the
practical layout.

Layout Charts

The size and type of the layout chart used will depend on the size and complexity
of the facility under study. Floor plans and maps, templates and three-dimen-
sional models, and photographs may be used. Drawings and diagrams are basic.

Office Layout

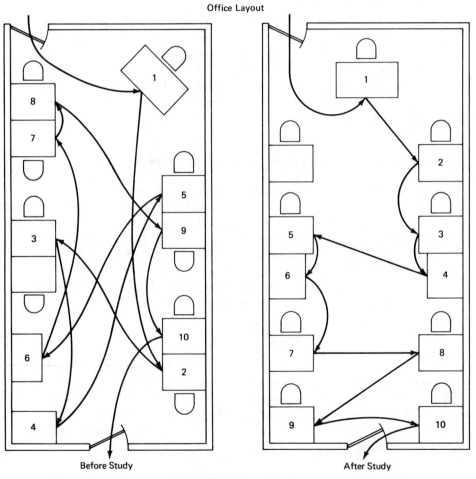

Before Study   After Study

Fig. 3-13. Layout chart.

Sketches and worksheet-type drawings are readily made, easily altered, and inexpensive. Templates are very valuable to the layout analyst. Photographs provide excellent portrayals of "before" and "after" situations. In selecting the visual aids, keep in mind the following desirable traits:

1. Clarity—enough for the person using it and enough for others to understand.
2. Flexibility—enough to suggest and show many different plans in a short time.
3. Economy—in terms of initial outlay and effectiveness in its use.
4. Availability—on hand when needed.

The amount of detail on each chart is a matter of individual judgment. Normally, two charts are prepared, one showing the present situation and the other the proposed. Putting the present situation on a chart will help in analyzing and improving the arrangement of facilities. The proposed chart will give a clear plan when proposals are implemented. It may also be useful in presenting "before" and "after" pictures to obtain approval of improvements. Figure 3-13 illustrates an example of a layout chart.

INSTALLATION AND FOLLOW-UP OF IMPROVEMENTS

The idea of finding better ways of doing work is not new; it has been carried on by successful managers for years. Progress could have been greater and success more rapid if all of the sound improvements that were developed had been properly implemented. Work-simplification effort cannot be considered complete until the sound proposals that have been developed are put into operation.

Any proposed change in existing procedures requires consideration of two basic problems: human factors and technical requirements. Careful consideration of these can make the selling, installation, and follow-up of proposals easier and more effective.

Human Factors

Individuals involved in a work area are affected each time the details or method of performing a job are modified or changed. Workers do not like to believe that they have been doing their jobs in the wrong way, nor do they like to change methods they have been using for a long time. When new processes or improvements are proposed, resistance to change or resentment of criticism may develop among the individuals who are affected or concerned.

Resentment of criticism can be overcome if personnel are improvement-conscious. It should be explained to all concerned that new proposals are not intended as criticism of past or existing methods. In addition, the proposals should emphasize the benefits and improvements that will result if they are adopted.

Resistance to change will be reduced if concerned personnel participate in the development of proposals and feel free to submit ideas of their own. Normally, people will cooperate in implementing proposals when they originate or assist in developing them. The beneficial factors must be emphasized, and negative factors should be minimized. If the beneficial factors regarding the proposal outweigh the detrimental ones, it will be much easier to gain acceptance of the proposal.

Technical Requirements

A goal for any improvement is better utilization of existing resources, including buildings, equipment, and personnel. A careful review should be made of the technical requirements involved in proposals to ascertain that the change will result in concrete savings, increased efficiency, or more effective operations. Improvements must be of more value than the cost to implement the proposal. Being able to demonstrate that proposals are workable and practical is an essential element in selling and installing improvements. The flow-process chart, work count, and layout studies provide means for making this determination.

Installing the New Method

Benefits of work simplification will not be realized until the improvements are actually implemented. In many cases, proposals for improvement will require approval. To get approval, it must be proved that a better way of doing work has been found.

If possible, proposed changes should be tested and validated in actual use under controlled conditions. Where testing would disrupt production, this may not be possible. Interest personnel in trying the new method. Point out that the reason for any change is to make work easier. Don't allow the changeover to become stymied, and stay receptive to still further improvement that may become apparent. Make sure that credit and praise is given to those who have contributed ideas or made changes easier by their cooperation.

Reports

Form F, methods improvement project summary, is used to indicate action taken or recommended resulting from application of work-simplification techniques (see Fig. 3-14). Improvements resulting in personnel savings, cost avoidance, or cost reductions will be reported accordingly.

METHODS IMPROVEMENT PROJECT SUMMARY			
1. ORGANIZATION Storage Division	2. WORK CENTER Bin Issue	3. SUBJECT Packing for Shipment	4. REFERENCE/FILE NO.

5. PRESENT METHOD

Packing supplies are stored in a central supply area. Each employee goes to the central supply area and selects and transports appropriate containers to the working areas where the supplies are packed for shipment.

6. PROPOSED METHOD

Packing tables have been fabricated. A supply of containers is maintained at packing station. Packers do not have to leave packing area to get supplies.

(NOTE: FILL IN DATA AS APPROPRIATE. REVERSE SIDE OF THIS FORM MAY BE USED FOR FOLLOW-UP DATA.)

7. ECONOMIC ANALYSIS DATA	PRESENT	PROPOSED
A. LABOR RATE (*$/Hr*)	$ 2.46	$ 2.46
B. ANNUAL VOLUME (*No. of Work Units*)	5000	5000
C. LABOR STANDARD (*Hrs/Unit*)	.10	.08
D. COST		
(1). LABOR COST PER WORK UNIT	$ .25	$ .20
(2). MATERIAL COST PER WORK UNIT	$ –	$ –
(3). MISCELLANEOUS COST PER WORK UNIT	$ –	$ –
E. TOTAL COST PER WORK UNIT	$ .25	$ .20
F. MANHOURS PER YEAR	5000	4000
G. LABOR COST PER YEAR	$ 12500	$ 10000
H. MATERIAL COST PER YEAR	$ –	$ –
I. FLOOR SPACE (*Sq. Ft*)	–	–
J. COST OF FLOOR SPACE	$ –	$ –
K. OTHER COST (*Specify*)	$ –	$ –
8. TOTAL OF LINES 7G, 7H, 7J, AND 7K	$ 12500	$ 10000
9. TOTAL GROSS SAVINGS (*First Year*)		$ 2500
10. IMPLEMENTATION COST (*$*)		
A. LABOR COST		$ 200
B. EQUIPMENT AND MATERIAL COST		$ 300
C. ENGINEERING COST		$ –
D. BUILDING MODIFICATION		$ –
E. UTILITIES		$ –
F. OTHER COST (*Specify*)		$ –
11. TOTAL IMPLEMENTATION COST		$ 500
12. COST OF LIBERATED INVESTMENT		$ –
13. TOTAL NET SAVINGS (*First Year*)		
A. DOLLARS		$ 2000
B. MANPOWER SPACES		.5 (w-4)
14. AMORTIZATION PERIOD		
CONTINUED ON REVERSE SIDE		

Fig. 3-14. Form F, methods improvement project summary.

15. INTANGIBLE ANALYSIS

16. FOLLOWUP ACTION

DATE								
% IMPLEMENTATION								

17. REASON FOR INCOMPLETE IMPLEMENTATION

18. REMARKS

19. ANALYST (Signature and Date)	20. METHODS/STANDARDS SUPERVISOR (Signature and Date)	21. SUPERVISOR (Signature and Date)

Fig. 3-14. (Continued).

Review

Periodic reviews are essential to insure that organizations do not return to former methods of operations. The full benefits of the improvement will not accrue unless the improvements are continued. All work papers should be filed and kept in a readily accessible place to assist in the review. The type and frequency of review depend on the amount of change in the process and the degree of acceptance by operating personnel. An objective review and evaluation of improvements is essential to insure that improvements are working and providing benefits and/or savings.

4
MANUFACTURING ENGINEERING METHODS

Manufacturing Engineering is that branch of professional engineering which requires the education and experience necessary to understand and apply engineering procedures to the manufacturing processes and production methods of industrial products. It requires the ability to *plan* manufacturing, to *research* and to *develop* tools, processes, machines, and equipment, and to *integrate* the facilities and systems for producing quality products with the optimal capital expenditure.

Manufacturing Engineering embraces the activities in the planning and selection of the methods of manufacturing, the development of the production equipment, the research and development to improve the efficiency of established techniques, and the development of new techniques. These activities include but are not limited to:

1. Facilities planning, including processes, plant layout, and equipment layout.
2. Tool and equipment selection, design, and development.
3. Value analysis and cost control relating to manufacturing methods and procedures.
4. Feasibility studies for the manufacture of new or different products with respect to the possible integration of new items into existing facilities.
5. Review of product plans and specifications, and the possible changes to provide for more efficient production.
6. Research and development of new manufacturing methods, techniques,

tools, and equipment to improve product quality and reduce manufacturing costs.

7. Coordination and control of production within a plant and between separate plants.
8. Maintenance of production control to assure compliance with scheduling.
9. Recognition of current and potential problems and the implementation of corrective action to eliminate them.
10. Economic studies related to the feasibility of acquiring new machinery, tools, and equipment.

Manufacturing Engineering is divided into four basic functional areas:

1. *Manufacturing Planning:* the preliminary engineering work relative to the establishment of a manufacturing system for the production of a product. The selection and specification of the necessary facilities, equipment, and tools, as well as the plant layout to provide the most efficient operation.
2. *Manufacturing Operations:* the engineering work involved in the routine functioning of an existing plant or facility to provide efficient and economical production output to quality standards. The improvement of existing layouts, procedures, tooling, and product plans and specifications.
3. *Manufacturing Research:* the pursuit of new and better materials, methods, tools, techniques, and procedures to improve manufacturing processes and reduce costs. The development of new concepts and innovative uses of existing items.
4. *Manufacturing Control:* the management of manufacturing operations to assure compliance to required schedules. Coordination of all manufacturing departments and support departments such as purchasing, materials, etc.

The subject of this chapter is the sub-discipline of manufacturing engineering known as *Methods Engineering*. Methods engineering is a major contribution to the manufacturing engineering department. We will review the techniques and procedures employed by the methods engineer beginning at the point that the engineering department releases the product documentation to the production department. See Fig. 4-1.

MANUFACTURING FLOW TRAVELER

Working from the released engineering drawings, the methods engineer will generate a step-by-step procedure for the assembly of each part or assembly. This procedure will become the "Manufacturing Flow/Traveler" (normally referred

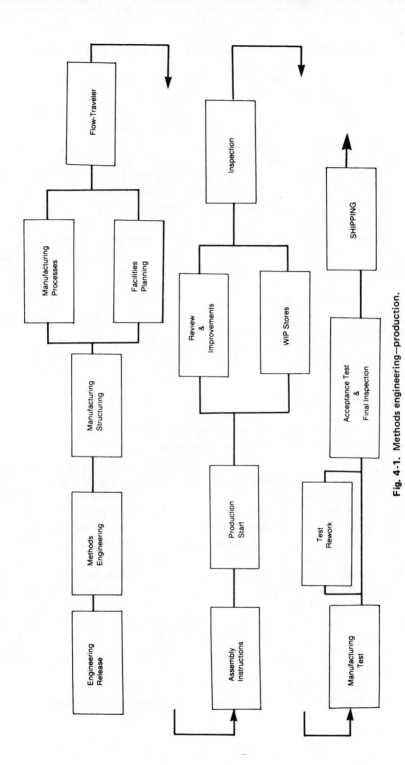

Fig. 4-1. Methods engineering—production.

to as the "traveler"). See Figs. 4-2 and 4-3. A separate traveler will go through the manufacturing cycle with each part or assembly as a means of tracking. The traveler provides for the buy-off (acceptance) of each part or assembly at every level of inspection. The traveler is the only record of manufacturing a product that indicates the sequence of operations, the identity of the operator, and the inspection of each operation.

The methods engineer must carefully review the assembly steps of every product to prevent any disassembly of parts installed out of sequence. When parts must be removed to allow clearance for the installation of another part that should have been previously installed, it is costly in terms of labor and it may cause damage to the original part. Each part and component in an assembly should be reviewed for preparation operations. For example, the identification marking of a printed circuit board must be done prior to installation of the components to provide a flat surface for the silkscreen. Each of the preparation operations should be organized into a logical sequence aimed at moving the parts and materials down the production line to prevent a holding situation of any parts prior to the assembly operation. The printed circuit board should not be completed and issued to the line when it may be held up waiting for the components to be prepped.

TASK BREAKDOWN

Working from the released drawing from the engineering department and starting with the top assembly, the methods engineer will structure each part, subassembly and assembly into manageable units for manufacturing. This is accomplished by generating a planning chart. See. Fig. 4-4. The next step is to record in order of occurrence each detail of the job. This permits analysis of each detail in proper sequence. Do not omit seemingly unimportant details. Collectively these details can play an important role in developing the methods. From the planning chart we can decide which parts to manufacture in house and which parts to obtain from outside vendors. These are the "Make or Buy" decisions. The planning chart also provides a means of establishing the labor breakdowns by discipline (i.e., the parts that will be made in the machine shop, mechanical assembly, electrical assembly and so forth). The planning chart defines each subunit or part of a product for each specialized discipline within the manufacturing department. The card cage assembly is a product of the mechanical assembly group whereas a printed circuit assembly is a product of the printed circuit line.

Each block on the planning chart identifies a traveler requirement and, based on the complexity of the part or assembly, an assembly instruction.

MANUFACTURING FLOW/TRAVELER

	ASSY DWG NO.	DASH NO.	REV.	ASSY NOMENCLATURE	CONTROL	SYSTEM
CONFIGURATION	MFG. ENGINEER	DATE		NEXT ASSY NO.	OPER.	NEXT OPER.	PAGE OF

SCHEMATIC NO.	DWG NO.	REV
WIRE LIST NO.	DWG NO.	REV
A	DWG NO.	REV
B	DWG NO.	REV
C	DWG NO.	REV
D	DWG NO.	REV

CHG NO.	EO NUMBER	LTR CHG	DISP	EFF	MFG SERIAL NO FROM THRU	MFG ENGINEER	DATE	OPER	WAS CONDITION OR REASON FOR CHANGE	QA	DATE

Fig. 4-2. Manufacturing flow/traveler.

Control	Revision	Job Number	Serial No.	Next Assembly	
Manufacturing Engineer	Date	Quality Engineer	Date	Test Engineer	Date
Assembly Drawing	Revision	Assembly Nomenclature	Systems No.	Final Acceptance	

Flow No.	Oper. No.	Flow Description	Std. Hrs.	Fixt./Tool Process No.	Operator Inspector	Date	Notes

Fig. 4-3. Manufacturing flow/traveler.

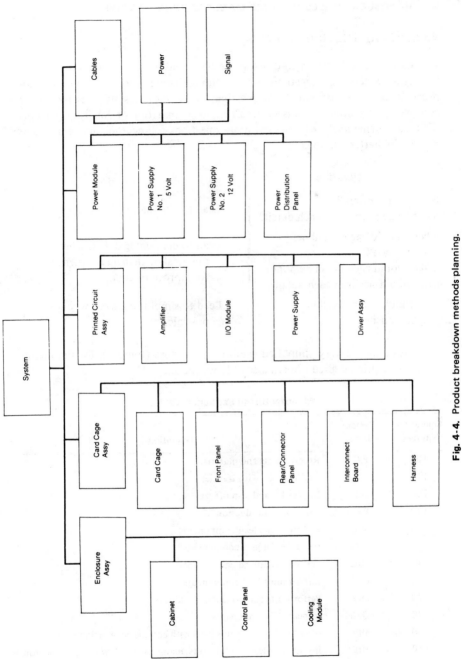

Fig. 4-4. Product breakdown methods planning.

ANALYZING THE JOB DETAILS

We analyze the job details to develop the best method. Each detail is analyzed by asking a series of questions, which, though simple, normally produce improvements. The most important question, "Why is it necessary?", is asked over and over. Very often, the answer results in the elimination of details or portions of the job being analyzed. Several more questions follow; their answers help in developing better methods.

Question	Answer
Why is it necessary? *What* is the purpose of each detail?	Lead to elimination.
Where should each detail be performed? *When* should each detail be done? *Who* should perform each detail?	Lead to combining details when practical or changing sequence to improve method.
How should each detail be accomplished?	Lead to simplifying work when possible.

For example, a very simplified methods procedure (traveler), for the manufacture of a printed circuit board assembly would be:

Manufacturing Sequence Table.

Operation Number	Person/ Dept.	Operation
100	P.C.	Receive parts and materials
110	P.C.	Assemble kit with traveler
700	Q.C.	Inspect kit and sign-off traveler
120	P.C.	Issue kit to production
200	Mfg.	Preform axial lead components
710	Q.C.	Inspect axial lead components
210	Mfg.	Perform multilead components
720	Q.C.	Inspect multilead components
220	Mfg.	Perform I/C components, Machine
730	Q.C.	Inspect I/C components
300	Mfg.	Install resistors & capacitors with automatic insertion
310	Mfg.	Install multilead components; manual with 50 Watt soldering iron
320	Mfg.	Install I/C components Manual with 50 Watt soldering iron

Manufacturing Sequence Table. (*Continued*)

Operation Number	Person/ Dept.	Operation
330	Mfg.	Clean assembly-vapor degrease Observe all safety precautions
740	Q.C.	Inspect assembly
340	Mfg.	Inspection rework
750	Q.C.	Inspect all rework & buy-off assembly
400	Mfg.	Package printed circuit assembly in sealed plastic bags
130	P.C.	Route to WIP stores

The operation numbers are arbitrarily selected to indicate the type of person or department responsible for each operation.

100,110 Indicates operations performed by Production Control Department

200,210 Indicates operations performed by the section in the Manufacturing Department that preps the components.

300,310 Indicates operations performed by operators on the assembly line.

400,410 Indicates operations performed by the section in the Manufacturing Department that performs materials handling functions.

500,510 Machine Shop, Sheet Metal Shop, etc.

700,710 Quality Control (Inspection)

Abbreviations used above are: Q.C. Quality Control
P.C. Production Control
Mfg. Manufacturing

Figure 4-5 shows the step-by-step procedure in a graph form preferred by many engineers.

On completion of the traveler, the methods engineer has all of the information needed to review the existing facilities and equipment. It is his/her responsibility to provide the tools and equipment to manufacture the parts, assemblies and units in his/her area. The methods engineer is responsible for the action required to make the production line capable of performing the task required in the most efficient manner.

ASSEMBLY INSTRUCTIONS

Normally the people working on the assembly line or production line are not technically trained personnel. It is the methods engineer's job to train the manu-

66

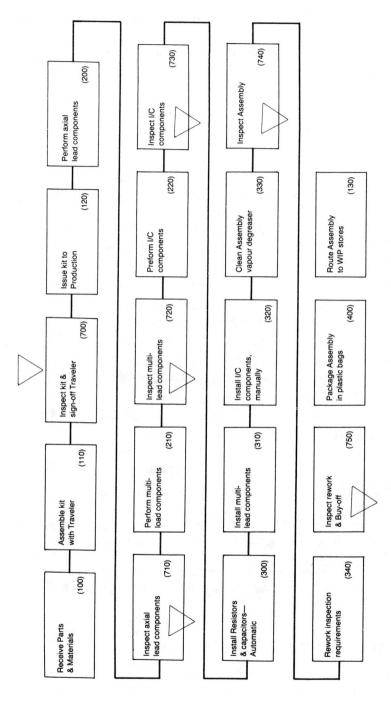

Fig. 4-5. Product flow/chart printed circuit assembly.

facturing personnel and to translate the engineering data into a useable format. To do this he uses assembly instructions and aids. The assembly instructions are a critical factor in achieving cost effectivity, productivity and reliability.

The assembly instructions (AI) range from simple hand sketches to more exotic step-by-step drawings. The cost of the assembly instructions is based on the value and length of the manufacturing program. In manufacturing millions of automobiles the most elaborate assembly aids can be justified. Yet in manufacturing twenty mainframe computers a more lower cost method should be employed. With the exception of the high volume electronic companies (radios, watches, T.V., etc.), most organizations can use the conventional method of xeroxing copies of portions of the engineering drawing with the methods engineer's notes and additions.

An example assembly instruction, including the flow traveler, is shown in Figs. 4-6 to 4-12.

Definitions

Production Release.   The release of engineering drawings that define the product or part to be manufactured.

Product Breakdown.   The methods engineer's organization of the parts or assembly to be manufactured.

Facilities Planning.   The methods engineer's task of layout and equipment requirements for the manufacture of a specific product.

Equipment.   The equipment required for the manufacture of a specific product.

Line Analysis.   The methods engineer's organization of the labor disciplines available and required for the manufacture of a specific product.

Methods.   The step-by-step procedure to be followed in the manufacture of a specific product.

Assembly aids.   Pictorial presentations and/or actual hardware displayed for production personnel to see how parts and assemblies are manufactured.

Wip stores.   Storage of complete parts and assembly until next operation (Work-in-Progress).

Control	Revision		Job Number		Serial No.		Next Assembly	
Manufacturing Engineer	date		Quality Engineer	date		Systems No.	Test Engineer	
		Rev.	Assembly Drawing		Assembly Nomenclature		date	

Left section

FLOW NO.	FLOW DESCRIPTION	STD HRS	FIXT/TOOL NO. INSP/TEST PROC	OPER.	DATE
01	Load Cores-P/N7140		1021		
02	Identify frames by Fire		PG-of A.I.		
	Lot Number				
[01]	Inspect				
03	Install temperature		14167		
	Sensor				
	Clean and Prime		514488		
04	Bond Core Mats		751448		
[02]	Inspect				
	Near Side				
05	Install even numbered		51448		
	"Y" (BIT) lines				
06	"Y" (BIT) lines				
07	Install Folded "Y" lines (BIT)				
	Lines				
08					
09	Install odd numbered				
	"X" word lines				

Right section — Final Acceptance

FLOW NO.	FLOW DESCRIPTION	STD HRS	FIXT/TOOL NO. INSP/TEST PROC	OPER.	DATE
10	Install even numbered				
11	"Y" (BIT) Lines				
12					
13					
14	Install odd numbered				
	"X" (Word) Lines				
[03]	Inspect				
15	Install sense line				
16	Install sense line				
	(Far side)				
[04]	Inspect				
	Near Side				
17	Solder X-A Leads		75141		
18					
19					
20	Solder Y-B Leads		751416		

Fig. 4-6. Manufacturing flow/traveler.

Control	Revision				
Manufacturing Engineer	date			Serial No.	
Assembly Drawing	Rev.	Job Number			
		Quality Engineer			Next Assembly
		Assembly Nomenclature		date	Systems No.
				Test Engineer	
					Final Acceptance

FLOW NO.	FLOW DESCRIPTION	STD HRS	FIXT/TOOL NO. INSP/TEST PROC	OPER.	DATE
21	Solder Y-C Leads				
22	Solder Y-D Leads				
23	Solder X-A Leads				
24					
25					
26					
27	Solder Y-C Leads				
28	Solder Y-D Leads				
29					
30	Route & Solder Sense Lines Far Side				
31	Clean				
Q5	Inspect				
T1	CTC Test				
32	Bond Flat Packs		514167		
33	Solder Flat Packs		4164		
34	Clean				
Q6	Inspect				
T2	AMB Test 25° C(CTC)		Eng'rg		

FLOW NO.	FLOW DESCRIPTION	STD HRS	FIXT/TOOL NO. INSP/TEST PROC	OPER.	DATE
35	Conformal Coat		4489		
Q7	Inspect				
T3	Test		Eng'rg		
QB	Inspect				
TO	ENGINEERING				

Fig. 4-7. Manufacturing flow/traveler.

ASSEMBLY INSTRUCTIONS

CHG NO.	ITEM NO.	PART NUMBER	DESCRIPTION/ COLOR CODE	QTY	FLOW NO.	SKETCH
						NOTE: ALL MATERIALS AND EQUIPMENT MUST BE CLEANED BEFORE STARTING THE CORE LOADING OPERATION AND MUST BE HANDLED CAREFULLY TO INSURE CLEANLINESS THROUGHOUT THE PROCEDURE.
						CLEANING CAVITY PLATE
			Ultrasonic		01	PLACE THE CAVITY PLATE INTO THE ULTRASONIC CLEANER WITH THE CAVITY SURFACE FACING DOWN. TURN ON THE CLEANER TO MED. (MARKED ON SWITCH), AND CLEAN FOR 5 MINUTES. REMOVE THE CAVITY PLATE AND ALLOW IT TO DRAIN OVER THE ULTRASONIC TANK. RINSE THOROUGHLY WITH D.I. WATER.
			Cleaner			USING THE BASKET, LOWER THE CAVITY PLATE INTO THE VAPOR DEGREASER WITH THE CAVITY SURFACE FACING DOWN FOR APPROX. ONE MIN.—THEN WHILE HOLDING THE CAVITY PLATE BY ONE EDGE, RINSE THE PLATE ON BOTH SIDES WITH THE NOZZLE— ALLOW TO DRAIN OVER THE TANK.
			Vapor			
			Degreaser			RINSE THOROUGHLY IN FOAMING AMMONIA IN A SHALLOW TRAY USING A MEDIUM SOFT BRUSH—OVEN DRY AT 150° FOR 30 MINUTES. DO NOT TOUCH THE CAVITY SURFACE.
			Brush, Camel			
			Hair			CLEANING THE CORES
			Ammonia			USING A SMALL INSTRUMENT TRAY, PLACE A KIMWIPE OVER THE BOTTOM TO PROTECT THE CORES FROM THE HARD SURFACE AND ADD APPROX. 1½ IN. OF SOLVENT. CAREFULLY PLACE THE CORES INTO THE TRAY USING A PENCIL MAGNET. USING A SMALL SOFT BRUSH, BRUSH THE CORES BACK AND FORTH, AND IN A CIRCULAR MOTION, DRAIN THE SOLVENT AND REPEAT THE OPERATION. A-MIN
			Trichord		02	

Fig. 4-8.

ASSEMBLY INSTRUCTIONS

SKETCH

USE PLENUM COVER & CAVITY PLAST. ASSY. #1021

CORE LOADING

1. ALL CORE MATS SHALL BE MADE FROM CONTROLLED FIRE LOTS — CORE P/N 714

2. FIRE LOTS MUST NOT BE MIXED IN INDIVIDUAL MATS.

3. SEE PAGE _____ OF THIS A.I. FOR FIRE LOT DISTRIBUTION IN THE COMPLETE ARRAY.

4. CHECK FIRE LOT NO. ON CONTAINER BEFORE STARTING.

5. START VIBRATOR & VACUUM "PUMP".

6. TILT PLENUM CHAMBER SLIGHTLY FORWARD — POUR APPROX. 50,000 CORES OVER CAVITIES IN THE PLATE, HOLDING THE CONTAINER ABOUT ½ TO ¾ OF AN INCH ABOVE THE PLATE.

7. ROCK THE PLENUM CHAMBER FROM SIDE TO SIDE & BACK & FORTH TO CAUSE THE CORES TO MOVE OVER THE CAVITIES IN THE PLATE. ADJUST THE VIBRATOR, ON THE FRONT OF THE SYNTRON UNIT & THE VACUUM, BY TURNING THE HANDLES, UNTIL ALL OF THE CAVITIES ARE FILLED. CORES THAT BUNCH UP SHALL BE MOVED BY TOUCHING THEM WITH A SMALL CAMEL HAIR BRUSH—LEAVE THE VIBRATOR & THE VACUUM ON, UNTIL ALL CAVITIES ARE FILLED. USE A MAGNETIZED NEEDLE TO FILL INDIVIDUAL CAVITIES.

8. USE THE LUXOR MAGNIFIER LAMP TO DETERMINE THAT ALL CAVITIES ARE FILLED AND ALL LOOSE CORES ARE OUT OF THE MAT AREA.

TOOLS REQUIRED.

MAGNETIC TWEEZER — (ORBIT INDUSTRIES)

CAMEL HAIR BRUSH #1

MAGNETIZED NEEDLE IN PIN VISE

KIMWIPES

ALCOHOL IN DISPENSE

LOCATING FRAMES #10219

THIS CAVITY PLATE IS MADE OF SOFT COPPER

IF CAVITIES BECOME CLOGGED, CALL YOUR SUPERVISOR — PROBING THE CAVITIES WITH METAL OBJECTS WILL DESTROY IT. SEE PAGE _____ FOR CLEANING INSTRUCTIONS.

Fig. 4-9.

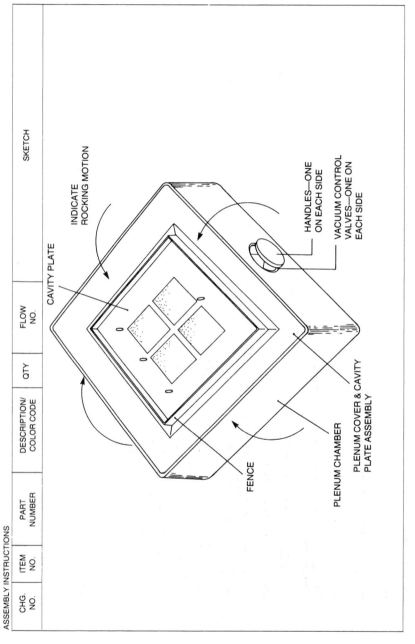

Fig. 4-10.

ASSEMBLY INSTRUCTIONS

SKETCH

SENSE WIRING — GENERAL NOTES

1. THE SENSE WIRES USED IN THIS MEMORY COME FROM THE VENDOR WITH A 5" NEEDLE ATTACHED.

2. A SHORT NEEDLE IS REQUIRED TO MAKE THE CROSS OVER BETWEEN THE MATS.

3. FIG. 16 SHOWS HOW THE NEEDLE IS REMOVED AFTER EACH PASS THRU A LINE OF CORES.

4. START WIRING IN THE USUAL MANNER—AS THE NEEDLE POINT COMES OUT OF THE LINE OF CORES BETWEEN THE MATS, PLACE THE POINT OF A TOOTHPICK OR ORANGE STICK UNDER THE POINT OF THE NEEDLE—AS YOU CONTINUE TO PUSH THE NEEDLE, GUIDE (DO NOT LIFT) THE POINT OF THE NEEDLE OVER THE CORES IN THE NEXT MAT.

5. DRESS THE WIRE INTO THE REQUIRED POSITION BY PULLING THE WIRE GENTLY AND AS CLOSE TO THE BOARD OR FIXTURE AS POSSIBLE.

6. BE VERY CAREFUL NOT TO PULL UP ON THE WIRE AS THIS WILL LIFT THE CORES OFF THE BACKING TAPE.

7. SELECT THE CORRECT ROW OF CORES IN THE NEXT MAT AND CONTINUE AS ABOVE.

Fig. 4-11.

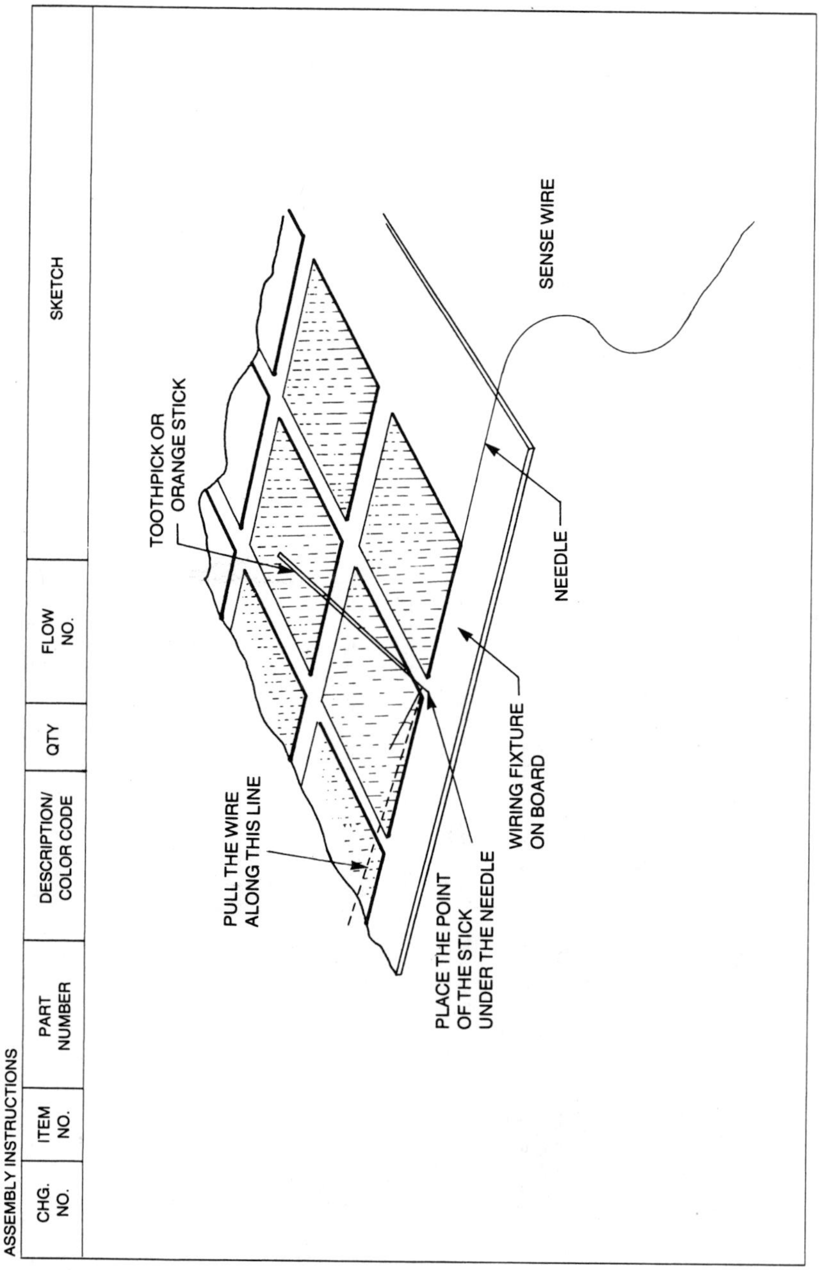

ASSEMBLY INSTRUCTIONS

CHG. NO.	ITEM NO.	PART NUMBER	DESCRIPTION/ COLOR CODE	QTY	FLOW NO.	SKETCH

TOOTHPICK OR ORANGE STICK

PULL THE WIRE ALONG THIS LINE

PLACE THE POINT OF THE STICK UNDER THE NEEDLE

WIRING FIXTURE ON BOARD

NEEDLE

SENSE WIRE

Fig. 4-12.

5
MANUFACTURING STANDARDS FOR SETTING LABOR COSTS

INTRODUCTION

The purpose of this section is to explain the organization of the chapter and to provide information for using the data presented to establish consistent and useful standards.

Organization of Chapter 5

This chapter is divided into four sections: Introduction, Allowances, Electromechanical Manufacturing, and Mechanical Manufacturing. Data related to each section of this chapter are grouped to allow an easier search of the chapter for desired information. For example, all information related to the mounting of parts with hardware (screw, nuts, washers) is presented in the section on electromechanical manufacturing.

Each section is divided into three parts:

1. *Data summary tables.* The data summary tables present the data given in this chapter in their most useful form. They are intended to provide a quick reference to the time required to perform a task without having to develop a standard from elemental time values each time a standard is needed.

2. *Limits of application.*   The limits of application are provided to explain within what limits the data summarized in the tables are valid and applicable.
3. *Source data.* The source data are an incremental breakdown of each item listed in the tables. This section gives the elemental time standards used, and shows how these elemental values were combined to arrive at the standard data given in the tables.

Use of Chapter 5

This chapter may be used to set the following time standards on any parts assembly:

1. Unit standard
2. Budget standard
3. Bid standard.

The relationships among these standards are:

Bid standard = Budget standard + Bid allowances

Budget standard = Unit standard + Budget allowances

Unit standard = Base (unit) hours + Manufacturing allowances

Only the base (unit) hours can be developed directly from the predetermined standard data presented herein.   The allowances referred to can be found on pp. 77-82; they must be applied before a useful standard can result.

1. *Developing the base (unit) hours.*   The base (unit) hours for a job are developed directly from the data in the manual.   The general steps to be followed are:

   a. Analyze the job thoroughly.
   b. Break the job down into the tasks required to accomplish the work involved.
   c. Record the necessary tasks in the order of their accomplishment.
   d. Determine and record the frequency (number of times) that each task must be performed until the whole job is completed.
   e. Determine and record the time value of each elemental task involved in the job. (Note that each time value presented in the manual is in terms of minutes.)
   f. Multiply the time value for each task times the frequency of occurrence.
   g. Sum the results of these calculations to determine the total number of minutes that are required to complete the job.

    h. Convert the total number of minutes to the number of hours required.
    i. The result of step (h) gives the base (unit) hours.
2. *Developing the Standard(s).* To obtain useful standards, allowances must be made for the various influences that will prevent the operator from performing the work at a rate identical to the base (unit) hours. Examples of these influences are personal time, initial job learning, and poor workmanship. The allowances necessary to convert the base (unit) hours to a good standard are presented in the next section, Allowances.

ALLOWANCES

As explained previously, more than one standard would be developed for each assembly or part to be produced. The difference between these standards will be the allowances that are progressively applied to them. The purpose of this section is to define the allowances that will be used and to explain how they will be applied in the development of the standards.

A simplified list of the categories of allowances that will be used is:

1. Manufacturing allowances
2. Budget allowances
3. Bid allowances.

The general formulas for using allowances are:

1. Unit standard $= \sum\limits_{i=1}^{3}$ (Base standard + Manufacturing allowances)$_i$

2. Budget standard $= \sum\limits_{i=1}^{3}$ (Unit standard + Budget allowances)$_i$

3. Bid standard $= \sum\limits_{i=1}^{3}$ (Budget standard + Bid allowances)$_i$

where

    $i = 1$ represents standards for production activities
    $i = 2$ represents standards for inspection activities
    $i = 3$ represents standards for test activities.

Manufacturing Allowances

The manufacturing allowance is an allowance necessary to compensate for lost time due to (1) operator personal needs, (2) general working conditions, and

(3) the inherent difficulty of the unit being processed when the work should be beyond the "learning" period. Two allowances will be used to compensate for these factors: the personal, fatigue, and delay (P.F.&D.) allowance and the complexity factor.

1. *P.F.&D. allowance.* The P.F.&D. allowance will be applied to the base standard to compensate for scheduled and unscheduled breaks, operator fatigue, normal unavoidable delay during processing, and daily operator setup and cleanup time.

2. *Complexity allowance.* The complexity allowance will be applied to compensate for the difficulty required to build the particular unit in question. This allowance will be based on the nature of the work involved and will vary depending upon the base time (+P.F.&D.) required to produce the unit in question. The complexity factor increases as (Base time + P.F.&D.) increases.

Budget Allowances

The budget allowance is an allowance necessary to compensate for operator lost time due to both setup and repair operations. Budget allowances are applied above and beyond the operator's production-time target (the unit standard). They give a true approximation of the average work content per unit involved in producing a lot throughout the entire duration of the job. Two allowances will be applied as budget allowances:

1. *Learning allowance.* The learning allowance will be used to compensate for the "operator learning" process during start-up operations of the job. This allowance will vary dependent upon the type of work involved and the number of units to be produced.

2. *Touch-up and/or rework allowance.* This allowance is used to compensate for lost time due to repair operations on in-process work. This allowance will be applied based on the type of work involved.

Bid Allowances

The bid allowance is an allowance necessary to level the standard with respect to the expected operator job performance, and to compensate for nonoperator time expected to be accrued to the unit during processing. Examples of bid allowances are:

1. *Allowance for expected job performance.* This allowance is used to level the work content of the job with respect to the past performance of operators on similar types of work. If operators have demonstrated less than 100 percent

average performance in the past (as compared with predetermined standards), then it is reasonable to assume that the same will hold true for future jobs. Compensation for this fact must be made in advance.

2. *Allowances for nonoperator time—Direct-support activities.* Generally, it will be necessary for personnel other than operators to devote some portion of their time to support manufacture of each particular unit required. Where this time can be directly attributed to the support needs of particular units, then an allowance can be used to allocate the average nonoperator time that will be accrued against each unit to be produced. Examples of allowances for direct-support activities are:

a. Setup allowances. Time required to set up work stations, tools, and so on by group leaders.
b. Auxiliary allowances. Supervisory time that can be directly charged or allocated to a particular unit (group leaders and forepersons).
c. Expediting/Material handling. Support time necessary for transporting in-process materials.
d. Manufacturing engineering. Support time required to develop work aids and fixtures, and to provide additional miscellaneous floor support for a particular unit.
e. Cost control. Support time required to develop standards and cost controls that can be directly attributed to the needs of a particular unit.
f. Production control. Support time required to establish production orders and controls in order to satisfy the demands for the job in question.

Manufacturing Allowances (Production)

Table 5-1. Station Allowance: Personal, Fatigue, and Delay.

Category	Description	Min/Day
Personal	Scheduled breaks, two, 10 min each	20
	Unscheduled breaks	15
Fatigue		12
Delay[a]	Set-up	5
	In-process	15
	Clean-up	5
Total P.F.&D. time		72
P.F.&D. allowance $= \dfrac{72 \text{ min}}{480 \text{ min}} =$		15%

[a]Unavoidable delay *only.*

Table 5-2. Complexity Allowance.

Base Hours + P.F. & D.	Complexity Allowance (%)
0–0.25	0
0.25–0.50	5
0.50–1.00	10
1.00–2.00	15
2.00–4.00	20
4.00–8.00	25
Over 8.00	30

Table 5-3. Total Manufacturing Allowance.

Base Hours for Unit	P.F. & D. Factor	Complexity Factor	Total Factor for Manufacturing Allowance
0.00–0.24	1.15	1.00	1.15
0.24–0.44	1.15	1.05	1.21
0.44–0.87	1.15	1.10	1.27
0.87–1.74	1.15	1.15	1.32
1.74–3.48	1.15	1.20	1.38
3.48–6.96	1.15	1.25	1.44
Over 6.96	1.15	1.30	1.50

Budget Allowances (Production)

Table 5-4. Learning Allowance Rates.[a]

Production Area	Learning Rate (%)
A. Printed-Circuit and	90
terminal boards	90
B. Components	90
C. Metal fabrication	90
D. Harness	85
E. Wire preparation	95
F. Module sub-assembly	85

[a]To determine the learning allowance:

1. Determine the lot size to be produced and the production area where the unit will be made.
2. Knowing the production area where the unit will be made, determine the applicable learning rate from Table 5-4.
3. Enter Table 5-5 at the point reflecting the number of units in the lot size and determine the learning factor in the column representing the applicable learning rate.
4. The learning factor is used as a multiplier to convert the unit standard to (unit standard + learning allowance).

Table 5-5. Production Learning Factors.

No. Units	Cumulative Average Learning Factors[a]		
	85%	90%	95%
1	5.33	2.94	1.68
2	4.50	2.65	1.60
3	4.12	2.50	1.54
4	3.79	2.39	1.51
5	3.60	2.30	1.49
6	3.43	2.23	1.47
7	3.33	2.18	1.45
8	3.20	2.13	1.43
9	3.10	2.10	1.41
10	3.04	2.05	1.40
20	2.58	1.84	1.33
30	2.33	1.73	1.30
40	2.18	1.66	1.27
50	2.06	1.60	1.24
60	1.98	1.55	1.22
70	1.90	1.51	1.21
80	1.85	1.48	1.20
90	1.80	1.46	1.19
100	1.75	1.43	1.18
150	1.60	1.34	1.14
200	1.49	1.28	1.10
250	1.40	1.24	1.09
300	1.34	1.20	1.08
350	1.29	1.18	1.08
400	1.25	1.15	1.07
450	1.22	1.13	1.05
500	1.19	1.11	1.04
600	1.14	1.08	1.03
700	1.10	1.05	1.02
800	1.06	1.03	1.01
900	1.03	1.02	1.00
1000	1.00	1.00	1.00

[a]These factors taken from learning curves (Fig. 5-1) where standard (1.000) should be reached on the 400th unit during production.

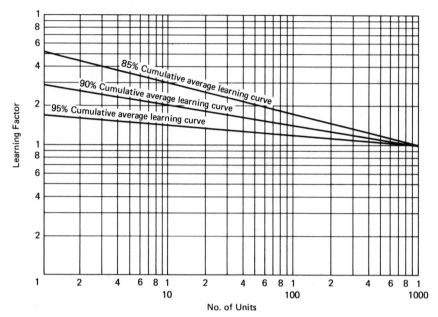

Fig. 5-1. Cumulative average learning curves.

Table 5-6. Touch-up and/or Rework Allowances.

Production Area	Allowance for Touch-up and/or Rework
A	27.5%
B	
C	
D	
E	
F	

BASIC ELEMENTS OF WORK

Standard Elements

Table 5-7. Basic Elements of Work.

Item No.	Description	Min/Occurrence 6"	12"	18"
1	Reach to object by itself and grasp object	.008	.010	.013
2	Reach to jumbled objects and grasp one object	.013	.015	.018
3	Reach to jumbled objects and grasp handful of objects	.024	.026	.029
4	Move object(s) to indefinite location	.005	.008	.010
5	Move object to exact location (within 0.5 in.)	.006	.009	.012
6	Move object to exact location and position in symmetrical hole (loose fit)	.009	.013	.016
7	Move object to exact location and position in nonsymmetrical hole (loose fit)	.012	.015	.018
8	Move object to exact location and position in symmetrical hole (close fit)	.016	.019	.022
9	Move object to exact location and position in nonsymmetrical hole (close fit)	.018	.021	.024

	Optional Additions (min/occurrence)	
1	Grasp additional jumbled object after grasping first object	.012
2	Regrasp hard-to-handle object to maintain control	.003
3	Apply pressure	.010

Table 5-8. Common Gets of Parts and Wires.

Object		Source	Min/Get 6"	12"	18"	24"
1	Part	Workbench	.011	.017	.022	.027
2	Part or wire	Jumbled in stackbin or wire rack	–	.024	.029	.034

Table 5-9. Get Part from Shelf Behind Operator (Includes Chassis).

Object		Source	Min/Get
1	Part	Jumbled in totepan	.200
2	Part	Shelf	.213

Table 5-10. Get Tool from Workbench and Aside Tool to Workbench.

	Tool	Min/Tool
1	Pliers, cutters, or wire strippers	.025
2	Soldering iron and solder	.062

Table 5-11. Mount Parts with Screws and/or Nuts with Lock Washers.

A. Minutes per Part Held with One Screw and/or One Nut with Lock Washer.[a]

Threads/In.

Rundown Length	(1) One Screw						(2) One Nut									(3) One Nut and One Screw								
	20-24		28		32-36		20-24			28			32-36			20-24			28			32-36		
	A	C	A	C	A	C	A	B	C	A	B	C	A	B	C	A	B	C	A	B	C	A	B	C
3/16	.26	.28	.27	.29	.28	.30	.24	.30	.35	.25	.31	.36	.25	.32	.37	.34	.42	.46	.35	.43	.47	.35	.44	.48
1/4	.28	.30	.30	.32	.31	.33	.25	.32	.37	.26	.34	.39	.26	.35	.40	.35	.44	.48	.36	.46	.50	.36	.47	.50
3/8	.32	.34	.35	.37	.37	.39	.27	.36	.41	.28	.39	.44	.29	.41	.46	.37	.48	.52	.38	.51	.55	.39	.53	.57
1/2	.37	.39	.39	.41	.42	.44	.29	.41	.46	.30	.43	.48	.32	.46	.51	.39	.53	.57	.40	.55	.59	.42	.58	.61

B. Minutes per Additional Screw and/or Nut with Lock Washer.

Rundown Length	(1) One Screw						(2) One Nut									(3) One Nut and One Screw								
	20-24		28		32-36		20-24			28			32-36			20-24			28			32-36		
	A	C	A	C	A	C	A	B	C	A	B	C	A	B	C	A	B	C	A	B	C	A	B	C
3/16	.20	.22	.21	.23	.22	.24	.18	.24	.29	.19	.25	.30	.19	.26	.31	.25	.33	.37	.26	.34	.38	.26	.35	.39
1/4	.22	.24	.24	.26	.25	.27	.19	.26	.31	.20	.28	.33	.20	.29	.34	.26	.35	.39	.27	.37	.41	.27	.38	.42
3/8	.26	.28	.29	.31	.31	.33	.21	.30	.35	.22	.33	.38	.23	.35	.40	.28	.39	.43	.29	.42	.46	.30	.44	.48
1/2	.31	.33	.33	.35	.36	.38	.23	.35	.40	.24	.37	.42	.26	.40	.45	.30	.44	.48	.31	.46	.50	.33	.49	.53

C. Additional Minutes per Nut for Rundown Performed with Wrench.[b]

No. Reposition of Wrench/Rev

	4		8		
	28	32-36	20-24	28	32-36
20-24					

D. Additional Minutes per Flat Washer or Spacer.

Min/Part

Condition	To Screw	To Nut
Unrestricted	.05	.05
Restricted	–	.07
Inaccessible	.05	.09

[a] Note: A = unrestricted, B = restricted, C = inaccessible.
[b] Applies to restricted and inaccessible conditions only.

Table 5-12. Threaded Fastener Installation Times.

A. Get Component and Position in Chassis
(Includes One Get and Aside of Each Required Tool).

Hardware Used	Tool(s) Used	Min/Part
Screw(s)	Screwdriver	.07
Nut(s)	Spintite	.07
Screw(s) and nut(s)	Screwdriver and spintite	.09

B. Install Hardware with Lock Washers.

No. Screws or Nuts	Min/Quantity of Screws/Nuts Installed								
	(1) Screws		(2) Nuts			(3) Screws and Nuts		
	A	C	A	B	C	A	B	C
1	.14	.16	.15	.18	.23	.22	.27	.31
2	.28	.32	.30	.36	.46	.44	.54	.62
3	.42	.48	.45	.54	.69	.66	.81	.93
4	.56	.64	.60	.72	.92	.88	1.08	1.24
5	.70	.80	.75	.90	1.15	1.10	1.35	1.55
6	.84	.96	.90	1.08	1.38	1.32	1.62	1.86
7	.98	1.12	1.05	1.26	1.61	1.54	1.89	2.17
8	1.12	1.28	1.20	1.44	1.84	1.76	1.16	2.48
9	1.26	1.44	1.35	1.62	2.07	1.98	1.43	2.79
10	1.40	1.60	1.50	1.80	2.30	2.20	2.70	3.10

Note: A = Unrestricted, B = Restricted, C = Increasing.

C. Install Hardware without Lock Washers.

No. Screws or Nuts	Min/Quantity of Screw(s) or Nut(s)								
	(1) Screws		(2) Nuts			(3) Screws and Nuts		
	A	C	A	B	C	A	B	C
1	.09	.11	.10	.11	.14	.16	.20	.22
2	.18	.22	.20	.22	.28	.32	.40	.44
3	.27	.33	.30	.33	.42	.48	.60	.66
4	.36	.44	.40	.44	.56	.64	.80	.88
5	.45	.55	.50	.55	.70	.80	1.00	1.10
6	.54	.66	.60	.66	.84	.96	1.20	1.32
7	.63	.77	.70	.77	.98	1.12	1.40	1.54
8	.72	.88	.80	.88	1.12	1.28	1.60	1.76
9	.81	.99	.90	.99	1.26	1.44	1.80	1.98
10	.90	1.10	1.00	1.10	1.40	1.60	2.00	2.20

Note: A = Unrestricted, B = Restricted, C = Increasing.

Table 5-12. (Continued)

D. Rundown Nut with Fingers, Normally *Unrestricted* (No Resistance to Rundown).

Rundown Length	Min/Nut Rundown									
	Threads/In.									
	20	24	28	32	36	40	44	48	52	56
$\frac{1}{16}$	.01	.01	.01	.01	.02	.02	.02	.02	.02	.02
$\frac{1}{8}$	.02	.02	.02	.03	.03	.03	.04	.04	.04	.05
$\frac{3}{16}$	.03	.03	.04	.04	.05	.05	.06	.06	.07	.07
$\frac{1}{4}$	.03	.04	.05	.05	.06	.07	.07	.08	.09	.09
$\frac{3}{8}$	.05	.06	.07	.08	.09	.10	.11	.12	.13	.14
$\frac{1}{2}$	.07	.08	.09	.11	.12	.13	.15	.16	.17	.19

E. Rundown Screw with Screwdriver or Nut with Spintite Normally Restricted or Inaccessible (No Resistance to Rundown).

Rundown Length	Min/Screw or Nut Rundown									
	Threads/In.									
	20	24	28	32	36	40	44	48	52	56
$\frac{1}{16}$	.02	.02	.03	.03	.03	.04	.04	.04	.05	.05
$\frac{1}{8}$	.04	.04	.05	.06	.06	.07	.08	.08	.09	.10
$\frac{3}{16}$	.05	.06	.07	.08	.09	.10	.12	.12	.14	.15
$\frac{1}{4}$	.07	.08	.10	.11	.12	.14	.15	.17	.18	.19
$\frac{3}{8}$	.10	.12	.15	.17	.19	.21	.23	.25	.27	.29
$\frac{1}{2}$	.14	.17	.19	.22	.25	.28	.30	.33	.36	.39

F. Rundown Nut or Screw with Spintite or Screwdriver Using Wrist Turns (Minor Resistance).

Rundown Length	Min/Rundown Length									
	Threads/In.									
	20	24	28	32	36	40	44	48	52	56
$\frac{1}{16}$	.04	.05	.06	.06	.07	.08	.09	.10	.11	.11
$\frac{1}{8}$	.08	.10	.11	.13	.14	.16	.18	.19	.21	.22
$\frac{3}{16}$	.12	.14	.17	.19	.22	.24	.26	.29	.31	.33
$\frac{1}{4}$	.16	.19	.22	.25	.29	.32	.35	.38	.41	.45
$\frac{3}{8}$	.24	.29	.33	.38	.43	.48	.57	.57	.62	.67
$\frac{1}{2}$	.32	.38	.45	.51	.58	.64	.70	.76	.83	.89

Table 5-12. (Continued)

G. Rundown Nut with Spintite or Screw with Screwdriver against Heavy Resistance.

Rundown Length	Min/Nut Rundown									
	Threads/In.									
	20	24	28	32	36	40	44	48	52	56
$\frac{1}{16}$	.05	.06	.07	.08	.09	.10	.11	.12	.13	.14
$\frac{1}{8}$	.10	.12	.14	.15	.17	.19	.21	.23	.25	.27
$\frac{3}{16}$	.15	.17	.20	.23	.26	.29	.32	.35	.38	.40
$\frac{1}{4}$	.19	.23	.27	.31	.35	.39	.42	.46	.50	.54
$\frac{3}{8}$	.29	.35	.41	.46	.52	.58	.64	.69	.75	.81
$\frac{1}{2}$	.39	.46	.54	.62	.69	.77	.85	.92	1.00	1.08

H. Rundown Nut with Open-End or Box Wrench.[a]

Rundown Length	Min/Nut Rundown									
	Threads/In.									
	20	24	28	32	36	40	44	48	52	56
$\frac{1}{16}$	.15	.17	.20	.23	.26	.28	.32	.34	.37	.40
$\frac{1}{8}$	.28	.34	.39	.49	.51	.56	.62	.68	.73	.79
$\frac{3}{16}$	.43	.51	.60	.68	.77	.84	.94	1.02	1.11	1.18
$\frac{1}{4}$	.56	.68	.79	.90	1.02	1.13	1.24	1.35	1.47	1.58
$\frac{3}{8}$	.85	1.03	1.19	1.35	1.52	1.69	1.86	2.03	2.20	2.37
$\frac{1}{2}$	1.13	1.35	1.58	1.81	2.03	2.26	2.48	2.71	2.93	3.16

[a]Reposition wrench four times per revolution.

I. Rundown Nut with Open-End or Box Wrench.[a]

Rundown Length	Min/Nut Rundown									
	Threads/In.									
	20	24	28	32	36	40	44	48	52	56
$\frac{1}{16}$	.25	.29	.35	.39	.45	.48	.54	.58	.64	.68
$\frac{1}{8}$	.48	.58	.68	.77	.87	.97	1.06	1.16	1.26	1.35
$\frac{3}{16}$	.74	.77	1.02	1.16	1.31	1.45	1.61	1.74	1.89	2.03
$\frac{1}{4}$	.97	1.16	1.35	1.55	1.74	1.93	2.12	2.32	2.52	2.71
$\frac{3}{8}$	1.45	1.74	2.03	2.32	2.61	2.89	3.18	3.48	3.76	4.06
$\frac{1}{2}$	1.93	2.32	2.71	3.08	3.48	3.86	4.25	4.64	5.03	5.41

[a]Reposition wrench eight times per revolution.

Table 5-12. *(Continued)*

J. Rundown Nut with Pneumatic Driver.[a]

	Min/Rundown									
	Threads/In.									
Rundown Length	20	24	28	32	36	40	44	48	52	56
$\frac{3}{8}$	.01	.01	.01	.01	.01	.02	.02	.02	.02	.02
$\frac{1}{2}$	.01	.01	.02	.02	.02	.02	.02	.03	.03	.03
$\frac{5}{8}$	.01	.02	.02	.02	.03	.03	.03	.04	.04	.04
$\frac{3}{4}$	.02	.02	.02	.03	.03	.04	.04	.04	.05	.05
$\frac{7}{8}$	.02	.02	.03	.03	.04	.04	.05	.05	.06	.06
1	.02	.03	.03	.04	.04	.05	.05	.06	.07	.07

[a]Average rpm: 1200 with no load, 800 loaded.

K. Rundown Nut with Ratchet.

	Min/Rundown									
	Threads/In.									
Rundown Length	20	24	28	32	36	40	44	48	52	56
$\frac{1}{16}$	.07	.08	.10	.11	.12	.14	.15	.16	.18	.19
$\frac{1}{8}$	.14	.16	.19	.22	.24	.27	.30	.32	.35	.38
$\frac{3}{16}$	.21	.24	.29	.33	.37	.41	.45	.49	.53	.57
$\frac{1}{4}$	.27	.32	.38	.43	.49	.54	.59	.65	.70	.76
$\frac{3}{8}$	.41	.49	.57	.65	.75	.82	.89	.97	1.05	1.13
$\frac{1}{2}$	.54	.65	.76	.87	.97	1.08	1.19	1.29	1.40	1.51

L. Mount Extra Washers or Spacers.[a]

	Min/Part	
Condition	To Screw	To Nut
Unrestricted	0.05	0.05
Restricted	–	0.07
Inaccessible	0.05	0.09

[a]Optional addition.

Table 5-13. Mount Parts with Screw(s) and/or Nut(s) and Liquid Stake.

A. Min/Part Held by One Screw and/or One Nut.

Rundown Length	Screw			Nut			Screw and Nut		
	Threads/In.								
	20 or 24	28	32 or 36	20 or 24	28	32 or 36	20 or 24	28	32 or 36
$\frac{3}{16}$	.28	.29	.30	.24	.25	.25	.36	.37	.37
$\frac{1}{4}$	.30	.32	.33	.25	.26	.26	.37	.38	.38
$\frac{3}{8}$	.32	.35	.37	.27	.28	.29	.39	.40	.41
$\frac{1}{2}$	.39	.41	.44	.29	.30	.33	.41	.42	.44
$\frac{3}{16}$	.22	.23	.24	.18	.19	.19	.30	.31	.31
$\frac{1}{4}$	.24	.26	.27	.19	.20	.20	.31	.32	.32
$\frac{3}{8}$	.26	.29	.30	.21	.22	.23	.33	.34	.35
$\frac{1}{2}$	.33	.35	.38	.23	.24	.27	.35	.36	.38

Table 5-14. Mount Parts with Elastic Stop Nuts (Including Rundown).[a]

A. Min/Part Held by One Screw and/or Stop Nut.

Rundown Length	Screw		Nut			Screw and Nut		
	A[b]	C	A	B	C	A	B	C
$\frac{3}{16}$	.38	.40	.39	.40	.43	.48	.52	.54
$\frac{1}{4}$	.46	.48	.47	.48	.51	.56	.60	.62
$\frac{3}{8}$	.61	.63	.62	.63	.66	.71	.75	.77
$\frac{1}{2}$	.77	.79	.78	.79	.82	.87	.91	.93
$\frac{3}{16}$	.32	.34	.33	.34	.37	.39	.43	.45
$\frac{1}{4}$	.40	.42	.41	.42	.45	.47	.51	.53
$\frac{3}{8}$	.55	.57	.56	.57	.60	.62	.66	.68
$\frac{1}{2}$	.71	.73	.72	.73	.76	.78	.82	.84

[a]32 threads/in.
[b]A = unrestricted, B = restricted, C = inaccessible.

Special Standards for Mounting Parts with Hardware

Component Description	Installation Procedure	Decimal Min
Diode	Mount diode to chassis (6 washers, 1 nut): 1. Position 3 washers to diode. 2. Position diode in mounting hole. 3. Position remaining 3 washers on diode. 4. Install nut. Note: 1 washer is actually a solder lug.	0.62
Wafer switch	Mount wafer switch to panel (4 washers, 1 nut): 1. Position washer (key) to switch. 2. Position switch to mounting hole. 3. Position remaining 3 washers to switch. 4. Install nut.	0.65
Resistor	Mount resistor to panel or chassis with 2 screws, 2 lock washers, and 2 locknuts 1. Position screw and lockwash (quantity: 2). 2. Position resistor. 3. Install nut.	0.95
Connector	Mount connector to chassis or panel (2 screws, 2 lock washers, 2 flat washers, 2 locknuts): 1. Position screw and lock washer. 2. Unpackage connector and lock washer. 3. Position flat washers. 4. Install locknuts.	1.60

Special Standards for Mounting Parts with Hardware (*Continued*)

Component Description	Installation Procedure	Decimal Min
Variable resistor	Mount variable resistor (2 washers, 2 nuts): 1. Install nut and washer on resistor. 2. Position variable resistor in mounting hole. 3. Install remaining washer and nut.	0.47
Double throw switch	Mount double throw switch (2 lock washers 2 nuts): 1. Install nut and washer on switch. 2. Position switch and install second washer and nut.	0.49
Fuseholder	Mount fuseholder with washer and nut: 1. Get fuseholder. 2. Remove nut threaded on fuseholder. 3. Position fuseholder to mounting hole. 4. Get washer and position. 5. Install nut.	0.26
Input jack	Mount input jack with washer and nut: 1. Get input jack. 2. Remove nut threaded on input jack. 3. Position input jack to mounting hole. 4. Get washer and position. 5. Install nut.	0.31
Alarm light	Mount alarm light with washer and nut: 1. Get alarm light. 2. Remove nut threaded on alarm light. 3. Position alarm light to mounting hole. 4. Get washer and position. 5. Install nut. 6. Install bulb and cover.	0.35 0.25

Special Standards for Mounting Parts with Hardware (Continued)

Component Description	Installation Procedure	Decimal Min
Knob	Mount knob on stud with 2 setscrews.	1.25
Fuse	Install fuse into fuseholder with threaded cap; install fuse into clip.	0.10 0.06
Fuseholder	Mount fuseholder with 2 screws, 2 flat-washers, and 2 locknuts.	1.25
Nameplate	Install nameplate on panel (with 4 screws): 1. Get nameplate and position. 2. Get screws and install.	0.66
Nameplate	Install nameplate on panel (with 4 screws and 4 lock washers). 1. Get nameplate and position. 2. Get lock washers and screws, and install.	1.13

Special Standards for Mounting Parts with Hardware (*Continued*)

Component Description	Installation Procedure	Decimal Min
Nameplate	Install nameplate on panel (with 4 screws, 4 flat washers, and 4 locknuts): 1. Get nameplate and position. 2. Get screws and install. 3. Get flat washers and install. 4. Get nuts and install.	2.25
Handle	Install handle to panel or chassis (with 2 screws, 2 lock washers): 1. Get screws and lockwashers, and position. 2. Get flat washers and position. 3. Get handle, unwrap, and position. 4. Run down screws. 5. 2 additional flat washers add:	0.72 0.10
Spiral wrap	Install spiral wrap on panel handle: 1. 8 wraps. 2. 10 wraps. 3. 13 wraps.	 1.96 2.42 3.11
Front panel	Mount front panel to assembly fixture with captive screws: 1. Get panel from shelf. 2. Unwrap panel and discard wrapping. 3. Position panel to upright assembly fixture. 4. Run down 4 captive panel screws with fingers to secure panel to fixture. 5. For each additional captive screw, add:	 0.50 0.10

Special Standards for Mounting Parts with Hardware (*Continued*)

Component Description	Installation Procedure	Decimal Min
	Remove front panel fastened to assembly fixture with captive screws: 1. Remove captive screws from fixture by loosening with fingers (quantity: 4). 2. Remove panel from fixture. 3. Aside front panel to shelf. 4. For each additional captive screw, add:	 0.50 0.08
Meter	Mount meter onto front panel with 3 each screw, flat washer, lock washer, and nut: 1. Get meter from bin. 2. Get 3 each screw, flat washer, lock washer, nut. 3. Assemble flat washers to screws. 4. Position meter to panel. 5. Install 3 each hardware combinations. 6. Get spintite and screwdriver. 7. Rundown hardware. 8. Aside tools.	 1.90
Handle	Mount handle to panel or chassis with 2 each ferrule, screw, flat washer: 1. Get 2 each ferrules, screws, flat washers. 2. Assemble flat washers to screws. 3. Apply loctite to screws. 4. Get handle from bin. 5. Get ferrules and position onto handle. 6. Position handle to panel. 7. Install screw/flat washer with fingers. 8. Get screwdriver. 9. Run down screws. 10. Aside tool.	 0.90

Special Standards for Mounting Parts with Hardware (*Continued*)

Component Description	Installation Procedure	Decimal Min
Spiral wrap	Install spiral wrap on panel handle:	
	1. 8 wraps.	0.15
	2. 9 wraps.	0.17
	3. 10 wraps.	0.19
Lug	Mount lug (screw, lock washer, locknut):	0.61
	1. Install screw and lock washer.	
	2. Position lug.	
	3. Install nut.	
Cable clamp	Mount cable clamp (screw, lock washer, clamp retainer, locknut):	0.78
	1. Position clamp on cable.	
	2. Install screw and lock washer.	
	3. Position clamp.	
	4. Position retainer.	
	5. Install nut.	
Connector and cap	Mount connector and cap (4 screws, 4 lock washers, 4 flat washers, 4 plastic washers, and 1 gasket):	2.90
	1. Position gasket to connector.	
	2. Position connector in mounting hole.	
	3. Install screws (3).	
	4. Position washers (9, 3 each).	
	5. Install nuts (3).	
	6. Install screw and cap.	
	7. Install remaining 3 washers and nut.	
Cable strap	Mount (plastic) molded cable strap (with screw, 2 flat washers, lock washer, and nut):	0.70
	1. Get screw and flat washer.	
	2. Assemble flat washer to screw.	
	3. Insert screw with flat washer through hole in upright panel.	
	4. Get cable strap and assemble to screw projecting through backside of panel.	

Special Standards for Mounting Parts with Hardware (*Continued*)

Component Description	Installation Procedure	Decimal Min
	5. Get flat washer, lock washer, and nut.	
	6. Assemble flat washer, lock washer, and nut to projecting screw and over cable strap.	
	7. Get screw driver and spintite.	
	8. Run down nut while holding screw.	
	9. Aside tools to work surface.	
Variable resistor (small)	Mount small variable resistor to panel with 1 each flat washer, lock washer, and nut:	
	1. Get variable resistor from bin.	
	2. Get lock washer, flat washer, and nut.	
	3. Position variable resistor to panel.	
	4. Install flat washer, lock washer, and nut with fingers.	
	5. Get spintite.	
	6. Run down hardware.	
	7. Aside tool.	0.41
Input jack	Mount input jack (panel connector) to front panel with lug, lock washer, and nut:	
	1. Get input jack from bin.	
	2. Get lug, lock washer, and nut.	
	3. Position jack to panel.	
	4. Install lug, lock washer, and nut.	
	5. Position lug.	
	6. Get spintite.	
	7. Run down hardware.	
	8. Aside tool.	0.33
Wafer switch (small)	Mount small wafer switch to panel with flat washer, lock washer, and nut:	
	1. Get wafer switch from bin.	
	2. Get flat washer, lock washer, and nut.	
	3. Position switch to panel.	
	4. Install lock washer, flat washer, and nut.	
	5. Get spintite.	
	6. Run down hardware.	
	7. Aside tool.	0.50

Special Standards for Mounting Parts with Hardware (*Continued*)

Component Description	Installation Procedure	Decimal Min
Chain mount	Mount connector-cap chain mount to panel with screw. 1. Get chain mount from bin. 2. Get screw. 3. Apply loctite to screw. 4. Assemble chain mount to screw. 5. Get screwdriver. 6. Install screw/chain mount to panel. 7. Aside tool. 8. For each additional chain mount on screw, add:	 0.33 0.08
Miniature toggle switch	Mount miniature toggle switch (with O-ring and dust-cover/nut): 1. Get switch. 2. Get O-ring and install on switch shaft next to switch body. 3. Assemble switch to panel backside with switch shaft projecting through switch mounting hole. 4. Get dust-cover/nut. 5. Install dust-cover/nut over switch arm and next to panel. 6. Get spintite. 7. Run down dust-cover/nut with spintite. 8. Aside spintite.	0.45

Table 5-15. Basic Elements of Wire Preparation for Insulated, Stranded, or Solid Wire.

Description	Min/Wire		Hourly Rate
	First	Additional	
1. Dereel, measure, and cut wire to length	1.14	0.12	400
2. Strip insulation to length (both ends, includes all tool-handling)	0.20	0.12	400
3. Twist strands of insulated stranded wire (hand or machine)	0.18	0.09	550
4. Tin wire (14–28 gauge) with soldering iron	0.18	0.12	400
5. Tin wire (14–28 gauge) in solder pot	0.14	0.06	800

97

Table 5-16. Hand Preparation of Insulated, Stranded, or Solid Wire
(Dereel, Measure, Cut, Strip, and/or Twist and Tin
Insulated, Stranded, or Solid Wire).

Description	Min/Wire		Hourly Rate
	First	Additional	
1. Dereel, measure, cut, strip, twist, and tin with iron (insulated stranded wire, 14–28 gauge)	1.70	0.45	133
2. Dereel, measure, cut, strip, twist, and tin in solder pot (insulated stranded wire, 14–28 gauge)	1.64	0.39	154
3. Measure, cut, strip, twist, and tin with iron (precut insulated stranded wire, 14–28 gauge)	0.75	0.45	133
4. Measure, cut strip, twist, and tin in solder pot (precut insulated stranded wire, 14–28 gauge)	0.69	0.39	154
5. Dereel, measure, and cut to length solid insulated wire (14–28 gauge)	1.14	0.12	500
6. Measure and cut to length solid insulated wire (14–28 gauge)	0.19	0.12	500
7. Measure, cut, and strip solid insulated wire (14–28 gauge)	0.39	0.24	250
8. Dereel, measure, cut, and strip solid insulated wire (14–28 gauge)	1.34	0.24	250
9. Measure, cut, strip, twist, and tin with iron on board or in chassis (insulated stranded wire, 14–28 gauge) *per wire end*	0.30	0.23	250

Table 5-17. Basic Elements of Wire Preparation for Shielded Wire,
Subminiature Coaxial Cable, and Coaxial Cable.

Description	Min/End	Min/Wire	Hourly Rate
1. Hand-cut coax wire from reel		0.23	260
2. Measure and mark strip length of outer jacket with dykes		0.13	460
3. Strip outer jacket and trim with dykes (1 in.; per each additional in.)	0.19	0.33	180
4. Extract 1 in. of center conductor from shield with scribe and pigtail 1 in. of shield (per additional in.)	0.38 0.04	0.73 0.08	82
5. Extract 1 in. of center conductor from shield with scribe and terminate shield without combing or pigtailing, includes get-and-aside of cutters (per additional in.)	0.43 0.03	0.80 0.06	75
6. Comb-out shield (includes get and aside of comb)	0.50	0.98	61
7. Terminate shield without combing or pigtailing (includes get-and-aside of cutters)	0.13		
8. Strip inner jacket or dielectric from center conductor (includes get-and-aside strippers)	0.08	0.16	375
9. Wrap ground wire around shield and solder (includes get-and-aside of pliers, ground wire, iron, and solder)	0.50	1.00	60

Table 5-18. Hand Preparation of Shielded Wire, Subminiature Coaxial Cable, and Coaxial Cable. These standards are presented with no inclusions for manufacturing allowances.

Item No.	Task Description	Min/ Unit	Hourly Rate
1	Single coaxial: dereel, hand-cut, hand-prep, pigtail one end	2.37	25
2	Single coaxial: dereel, hand-cut, hand-prep, wrap one groundwire	2.99	20
3	Single coaxial: dereel, hand-cut, hand-prep, wrap two groundwires	3.49	17
4	Single coaxial: dereel, hand-cut, hand-prep, wrap three groundwires	3.99	15
5	Coaxial pair: dereel, hand-cut, hand-prep, pigtail one end, join shields to make pair	5.55	11
6	Coaxial pair: dereel, hand-cut, hand-prep, wrap two groundwires, three groundwire terminations	6.02	10
7	Coaxial pair: dereel, hand-cut, hand-prep, wrap three groundwires, four groundwire terminations	6.56	9

These standards are presented with no inclusions for manufacturing allowances.

Item No.	Task Description	Min/Unit	Hourly Rate
8	Coaxial pair: same as No. 7 except no pigtailing. Wrap three groundwires, four groundwire terminations.	6.94	9
9			
10		9.50	6
11	13.84 13.34		
12	9.39 9.89 9.50		
13		13.34	4.5
14	Group of six coaxial tied together: 9 groundwires with 14 ground wire terminations; pigtail two ends. Dereel, hand-cut, hand-prep.	20.98	3

Table 5-19. Sleeving Preparation.

Description	Min/Sleeve		Hourly Rate
	First	Additional	
1. Dereel, measure, and hand-cut sleeving (0–2 in.)	1.08	0.05	1200
2. Measure and hand-cut sleeving (0–2 in.)	0.53	0.05	1200
3. Measure and hand-cut marked sleeving (0–2 in.)	0.11	0.04	1500
4. Measure and handcut marked sleeving (2–6 in.)	0.53	0.08	750
5. Machine-cut sleeving (0–2 in.)	0.40	0.005	7000
6. Heat-stamp sleeving or markers	5.00	0.05	1200

Sleeving Installation Times

Table 5-20. Basic Elements of Sleeving.

Item No.	Description	Decimal Min
1	Get sleeve, move sleeve to wire, position sleeve on wire end.	0.04
2	Additional time for positioning sleeve with tight fit on wire end.	0.02
3	Arrange two or more wires for a single sleeving.	0.03
4	Additional time to position one sleeve on two or more wire ends (allow for each wire over one).	0.02
5	Additional time to position sleeve on tagged wire (allow per wire).	0.02
6	Slide sleeve on terminal (by hand).	0.03

Table 5-21. Sleeving—One End Only.

Distance Sleeve Is Moved On First Wire	Sleeve Length (In.) Min/Sleeve[a]																			
	1	2	3	4	5	6	7	8	9	10	11	12	13	14	15	16	17	18	19	20
1																				
2	.05	.06	.06	.07	.07	.08	.08	.09	.10	.10	.11	.11	.12	.13	.13	.14	.14	.15	.16	.16
3																				
4	.06																			
5		.07	.07																	
6																				
7				.08																
8					.08															
9						.09														
10							.09													
11						.09			.10	.11										
12			.08						.11											
13	.07								.11											
14												.12								
15													.12							
16														.13						
17													.13		.14					
18											.12				.14					
19					.10											.15				
20				.09													.15			
21																		.16		
22		.08															.16			
23	.08															.15			.17	.17
24													.14							
25																				
26												.13								
27											.12									
28																				.18
29					.11															
30				.10														.17		
31																				
32			.09													.16				
33	.08														.15					
34																				
35													.14							
36																				

[a]Time values represent the time required to:
1. Get one sleeve of given length.
2. Move the sleeve to the end of a wire.
3. Position the sleeve on the end of the wire (up to 1 in. from the end).
4. Slide the sleeve down the wire and into position at some given length from the end of the wire.

Table 5-22. Optional Additions to Sleeving.

Sleeve group of wires in which the length varies more than 4 in.

1. Divide wires into groups in which the length does not vary more than 4 in.
2. Select "sleeving time" for each group from Table 4-21.
3. Subtract "get, move, and position sleeve" time (from Table 4-20) for each additional group above.
4. Allow "arrange two or more wires for a single sleeve" (Table 4-20) for each group.
5. Allow "place sleeving over two or more wires" for each wire above one (Table 4-20).

Example: Install 1-in. sleeving on 50 wires. Wires are 36 in. long and are to be sleeved in groups as follows: 20 @ 10 in., 10 @ 5 in., 10 @ 20 in., and 10 @ 8 in.

Solution: 1. Four Groups
2. 0.06 for 5 in, 8 in., 10 in., 0.07 for 20 in. = 0.25
3. 0.04 + 0.04 + 0.04   = 0.12
4. 0.03 + 0.03 + 0.03 + 0.03   = 0.12
5. (50 − 1) (0.02)   = 0.98
             1.47

Basic Cable Assembly Times

Table 5-23. Basic Elements of Cabling.

Item No.	Description	Decimal Min
1	Get connector from bin, unpackage, aside package position connector in fixture	0.32
2	Get soldering iron, get solder and aside	0.09
3	Fill connector tubelet with solder	0.07
4	Get wire from rack, position wire in fixture spring, position wire end at connector tubelet	0.08
5	Get group of wires, position group in fixture, remove group from fixture	0.13
6	Grasp wire end with pliers, heat connector tubelet with soldering iron, insert wire end into heated tubelet	0.12
7	Check solder joint (visual)	0.04
8	Apply cleaner to brush; clean solder joint.	0.06
9	Remove connector from fixture.	0.17
10	Remove wires from fixture spring.	0.02
11	Get group of wires (or cable end), align wire ends, get tape, wrap tape around wire ends, aside group.	0.23
12	Get group of taped wires and route through conduit or sleeving.	(See p. 000)
13	Install marker.	(See p. 000)
14	Shrink marker into position on cable with hair drier (up to 3-in. marker).	0.32

Table 5-24. Optional Additions to Cabling Operations.

Item No.	Description	Decimal Min
1	Tie tag on cable, remove tag.	0.18
2	Place cable in bag; staple bag.	0.24
3	Disassemble MS connector.	0.15
4	Install washer, ring, or bushing (solid) on cable:	
	a. Loose tolerance	0.08
	b. Close tolerance.	0.10
5	Install insulating bushing for MS connector on group of wires, where each wire must be routed through bushing individually . . . per wire.	0.11

Table 5-25. Special Cabling Standards.

Item No.	Description of Task	Decimal Min	Hourly Rate
1	Install BNC–type connector on coax a. Wire-prep coax end b. Install connector	8.00	7.5

Cable Harness Fabrication Times

Table 5-26. Get Wire(s) from Wire Rack, Lay in Cable on Board, and Secure Wire Ends in Springs.

Wire Length (In.)	One Wire	Wires Laid Simultaneously to Same Breakout[a]												
		2	3	4	5	6	7	8	9	10	11	12	13	14
Through 5	.08	.14	.18	.22	.26	.31	.35	.39	.43	.47	.51	.56	.60	.64
6-10	.10	.15	.20	.24	.28	.32	.36	.41	.45	.49	.53	.57	.61	.66
11-15	.11	.16	.21	.25	.29	.33	.37	.41	.46	.50	.54	.58	.62	.66
16-20	.12	.17	.21	.26	.30	.34	.38	.42	.47	.51	.55	.59	.63	.67
21-25	.13	.18	.22	.27	.31	.35	.39	.43	.48	.52	.56	.60	.64	.68
26-30	.14	.19	.24	.28	.32	.36	.40	.44	.49	.53	.57	.61	.65	.70
31-35	.14	.20	.24	.28	.33	.37	.41	.45	.49	.54	.58	.62	.66	.70
36-40	.16	.21	.25	.30	.34	.38	.42	.46	.51	.55	.59	.63	.67	.72

[a]When laying various length wires simultaneously, use the longest wire to select wire length. When wire length exceeds 40 in., add 0.01 min for each additional 6 in. in length.

Table 5-27. A. Select Wire(s) from Connector, Lay Wire(s) in Cable,
Secure Second End in Springs.

Wire Length (In.)	One Wire	Wires Laid Simultaneously to Same Breakout								
		2	3	4	5	6	7	8	9	10
Through 5	.05	.10	.14	.19	.23	.28	.32	.37	.41	.46
6-10	.07	.11	.16	.20	.25	.29	.34	.38	.43	.47
11-15	.08	.12	.17	.21	.26	.30	.35	.39	.44	.48
16-20	.09	.13	.18	.22	.27	.31	.36	.40	.45	.49
21-25	.10	.14	.19	.23	.28	.32	.37	.41	.46	.50
26-30	.11	.15	.20	.24	.29	.33	.38	.42	.47	.51
31-35	.12	.16	.21	.25	.30	.34	.39	.43	.48	.52
36-40	.13	.17	.22	.26	.31	.35	.40	.44	.49	.53

B. If Wires Can Be Laid Only One at a Time, and Are Selected from
Connector with Wires over 10 and up to and Including 20,
Use Time Value under One Wire Below.

Wire Length (In.)	One Wire	Wires Laid Simultaneously to Same Breakout[a]										Over 20 Wires
		11	12	13	14	15	16	17	18	19	20	One Wire
Through 5	.14	.60	.73	.86	.99	1.12	1.25	1.38	1.51	1.64	1.77	.21
6-10	.15	.62	.75	.88	1.01	1.14	1.27	1.40	1.53	1.66	1.79	.23
11-15	.16	.64	.77	.90	1.03	1.16	1.29	1.42	1.55	1.68	1.81	.24
16-20	.17	.66	.79	.92	1.05	1.18	1.31	1.44	1.57	1.70	1.83	.25
21-25	.18	.68	.81	.94	1.07	1.20	1.33	1.47	1.60	1.73	1.86	.26
26-30	.19	.70	.83	.96	1.09	1.22	1.35	1.48	1.61	1.74	1.87	.27
31-35	.20	.72	.85	.98	1.11	1.24	1.37	1.50	1.63	1.76	1.89	.28
36-40	.21	.74	.87	1.00	1.13	1.26	1.39	1.52	1.65	1.78	1.91	.29

[a]When laying various length wires simultaneously, use the longest wire to select wire length.
When wire length exceeds 40 in., add .01 min for each additional 6 in. in length.

Table 5-28. Optional Additions to Harnessing Operations.

	Decimal Min
1. Form bend in wire(s) while laying cable—each bend. (Allow only once per bend when laying multiple wire simultaneously.)	0.01
2. Form wire(s) around each cable button.	0.01
3. Route wire through aperture.	
a. Easy (no force required) up to 5 in. of wire.	0.03
b. Difficult (force required) up to 5 in. of wire.	0.06
c. Each additional 5 in. of wire.	0.01
4. Get connector from stock bin, mount on cable board, and remove upon completion of cable.	
a. Install in mating connector; remove.	0.11
b. Secure with toggle clamp; remove.	0.08
c. Place on (2) pins; secure with (2) Teflon washers; remove.	0.16
d. Secure with movable spring clip, remove.	0.08
e. Over pins, between pins, or in slotted fixtures, remove.	0.06
f. Secure with (1) thumb screw in fixture; remove.	0.11
g. Secure with (2) screws or (2) nuts; remove.	0.59
h. Add when connector is on shelf in front.	0.03
5. Get and mount cable button on cable board and remove upon completion of cable.	
a. Place on pins and secure with Teflon washer; remove.	0.11
b. Secure on standoff with screw; remove.	0.27
6. Remove wires from each spring.	0.02
7. Remove cable, and place on shelf.	0.06
8. E Z Coding	
a. Identify E Z Code while laying wire; add to lay time per wire coded.	0.02
b. Prepare E Z Code card for coding.	0.11
c. Apply E Z Code to wire from card, one wire end.	0.08
d. Apply E Z Code to wire from card, both wire ends.	0.16
e. Apply E Z Code on wire from tape roll, one wire end.	0.07
f. Apply E Z Code on wire from tape roll, both wire ends.	0.16
g. Remove E Z Code; pull off with fingers.	0.04
h. Remove E Z Code; unwrap with soldering aid.	0.06
9. Apply "tie wrap" to cable, first "tie wrap."	0.21
a. Each additional "tie wrap."	0.18
10. Cut temporary tie from cable with scissors.	0.09
11. Wrap tape around cable.	
a. Tape applied as protective coat, first wrap.	0.20
b. Tape applied as protective coat, each additional wrap.	0.04
c. Tape applied as cable tie.	0.24

Table 5-28. (*Continued*)

	Decimal Min
12. Apply spiral wrap	
a. Using tool:	
(1) First wrap	0.26
(2) Additional wrap	0.04
b. Hand method:	
(1) First wrap	0.35
(2) Additional inches.	0.23
13. Trim wire to length on board, first wire.	0.05
a. Trim wire to length on board, each additional wire.	0.03
14. Tin wire, first wire	0.10
a. Tin wire, each additional wire.	0.04
15. Apply shield clip	
a. Prepare and solder cable shield clip to wire.	0.24
b. Install wire and clip on pigtail shields.	0.60
16. Solderless lug application	
a. Crimp solderless lug to wire end, first lug and wire.	0.15
b. Crimp solderless lug to wire end, each additional lug and wire.	0.13
c. Crimp solderless lug to wire end, each additional wire in lug.	0.04
17. Crimp solder lug (10-to-14-gauge wire) (does not include strip, twist, or tin wire)	0.36
18. Tie cable, cable knot	
a. Unrestricted: more than $\frac{3}{4}$ in. to finger obstruction.	0.31
b. Semirestricted: $\frac{3}{8}$ to $\frac{3}{4}$ in. to finger obstruction.	0.47
c. Very restricted: less than $\frac{3}{8}$ in. to finger obstruction.	0.83
d. Get and aside iron.	0.06
19. Lace cable–shuttle method	
a. First lace stitch.	1.16
b. Each additional single stitch.	0.13
20. Lace cables–hand method	
a. Lace: single stitch–unrestricted	0.11
semirestricted	0.21
very restricted	0.28
b. Lace: double stitch–unrestricted	0.24
semirestricted	0.38
very restricted	0.67

**Table 5-29.   Additional Selecting Time for White Stamped Wires
Laid from Connector.**

No. Wires/Connector	Decimal Min	No. Wires/Connector
1	–	26
2	.04	27
3	.08	28
4	.15	29
5	.23	30
6	.34	31
7	.46	32
8	.61	33
9	.77	34
10	.95	35
11	1.16	36
12	1.38	37
13	1.62	38
14	1.88	39
15	2.16	40
16	2.46	41
17	2.78	42
18	3.12	43
19	3.48	44
20	3.86	45
21	4.25	46
22	4.59	47
23	5.16	48
24	5.56	49
25	6.04	50

Normal time for connectors with more than 50 wires should be calculated from the following:

$$\text{Normal time} = \frac{N(N+1)}{4}(.039) - (.012N) \text{ decimal min}$$

where N = No. wires in connector. (Note: The normal time covers the total select time for all wires from N wires in the connector. This time is to be added to the times for laying wires from a connector.)

Wire/Terminal Terminations

Table 5-30. Wire Connections to Turret and Pierced Terminals.
A. Individual Method.

| | Min/Connection | | | | | |
| | Turret Condition | | | Pierced Condition | | |
Type of Connection	1	2	3	1	2	3
1. Buss jumper (first end with "J" hook)	0.12	0.16	–	0.10	0.14	–
2. Buss jumper (extended end)	0.25	0.32	0.42	0.23	0.30	0.40
3. Stranded jumper (either end)	0.23	0.30	0.39	0.21	0.28	0.37
4. Stranded cable wire (selected from 1–10 wires)	0.25	0.32	0.42	0.23	0.30	0.39
5. Stranded cable wire (selected from 11–20 wires)	0.30	0.38	0.47	0.28	0.35	0.45
6. Stranded cable wire (selected from more than 21 wires)	0.38	0.45	0.54	0.36	0.43	0.52

B. Group Method.

Type of Connection	1	2	3	1	2	3
1. Buss jumper (first end with "J" hook)	0.12	0.16	–	0.10	0.14	–
2. Buss jumper (extended end)	0.20	0.27	0.37	0.18	0.25	0.35
3. Stranded jumper (either end)	0.18	0.25	0.34	0.16	0.23	0.32
4. Stranded cable wire (selected from 1–10 wires)	0.20	0.27	0.37	0.18	0.25	0.34
5. Stranded cable wire (selected from 11–20 wires)	0.25	0.33	0.42	0.23	0.30	0.40
6. Stranded cable wire (selected from more than 21 wires)	0.33	0.40	0.49	0.31	0.38	0.47

C. Get-and-Aside Tools to Connect Wire.

Description of Operation	Min/Occurrence
1. Initial get-and-aside of pliers to wrap wire	0.03
2. Internal get-and-aside of cutters to cut and crimp	0.05

Table 5-31. Component Lead Connections to Turret and Pierced Terminals.

	Condition		
	1	2	3
Method of Connection	Min/Two Lead Components Connected		
1. Individual method of connect (includes all tool handling)	0.48	0.68	0.88
2. Group method of connect (includes all but internal tool handling)	0.38	0.58	0.78

Table 5-32. Extended: End Wire Connections to
Printed-Circuit–Board Tubelets.

	Min/Wire End(s) Connected									
	Individual Method	Group Method								
		Quantity of Wire Ends Connected								
	1	2	3	4	5	6	7	8	9	10
1. Connect only	0.19	0.33	0.47	0.61	0.75	0.89	1.03	1.17	1.31	1.45
2. Connect and solder	0.29	0.47	0.65	0.83	1.01	1.19	1.37	1.55	1.73	1.91

Table 5-33. A. Soldering.

| | Min/Connection Classification of Terminal | | | | | |
| | 9–14 Gauge Turret Slotted Hook | | 14–19 Gauge Turret Slotted Hook | | Pierced | |
Wire Gauge	Restricted	H Restricted	Restricted	H Restricted	Restricted	H Restricted
28	0.06	0.07	0.04	0.05	0.04	0.05
26	0.06	0.07	0.04	0.05	0.04	0.05
24	0.06	0.07	0.04	0.05	0.04	0.05
23	0.06	0.07	0.04	0.05	0.04	0.05
22	0.06	0.07	0.04	0.05	0.04	0.05
21	0.07	0.08	0.05	0.06	0.05	0.06
20	0.07[a]	0.07[a]	0.05	0.06	0.05	0.06
19	0.07[a]	0.07[a]	0.05	0.06	0.05	0.06
18	0.07[a]	0.07[a]	0.06[a]	0.07[a]	0.05	0.06
17	0.07[a]	0.07[a]	0.06[a]	0.07[a]	0.05[a]	0.06[a]
16	0.07[a]	0.07[a]	0.06[a]	0.07[a]	0.05[a]	0.06[a]
15	0.07[a]	0.07[a]	0.07	0.08	0.05[a]	0.06[a]
14	0.10	0.11	0.07	0.08	0.06	0.07
13	0.10	0.11	0.08	0.09	0.07	0.08
12	0.10	0.11	0.09	0.10	0.09	0.10
11	0.10	0.11	0.09	0.10	0.09	0.10
10	0.11	0.12	0.11	0.12	0.11	0.12

[a]Denotes standard time based on 100-watt iron beyond this point.

B. Conversion Tables—Wire Gauge Combinations.

Gauge Combinations	Resultant Gauge
$\frac{2}{26}$	23
$\frac{3}{26}$	21
$\frac{4}{26}$	20
$\frac{2}{22}$	19
$\frac{3}{22}$	17
$\frac{4}{22}$	16
$\frac{2}{20}$	17
$\frac{3}{20}$	15
$\frac{4}{20}$	14
$\frac{2}{18}$	15

Table 5-33. (*Continued*)

B. Conversion Tables—Wire Gauge Combinations.

Gauge Combinations	Resultant Gauge
$\frac{3}{18}$	13
$\frac{2}{16}$	13
$\frac{2}{26} + \frac{1}{22}$	20
$\frac{2}{26} + \frac{1}{20}$	18
$\frac{2}{26} + \frac{1}{18}$	17
$\frac{2}{26} + \frac{1}{16}$	15
$\frac{2}{22} + \frac{1}{20}$	16
$\frac{2}{22} + \frac{1}{18}$	16
$\frac{2}{22} + \frac{1}{16}$	14
$\frac{2}{20} + \frac{1}{18}$	15
$\frac{2}{20} + \frac{1}{16}$	13

C. Cross-Sectional Areas.

Wire Gauge	Area (In.$^2 \times 10^{-4}$)
10	81.7
11	65.0
12	51.5
13	40.7
14	32.2
15	25.5
16	20.4
17	15.9
18	12.6
19	10.2
20	8.0
21	6.2
22	4.9
23	4.2
24	3.1
25	2.5
26	2.0
27	1.5
28	1.2
29	0.9
30	0.8

Table 5-34. Optional Additions to Wire Connecting and Soldering.

Item No.	Description	Min/Occurrence
1	Push wire into place and crimp in preparation for second wire on same turret terminal (per additional wire on the same turret terminal).	0.05
2	Make bend in wire or lead with pliers.	0.02
3	Wrap bus wire around terminal with pencil-type tool.	0.09
4	Prepare pencil-type tool for No. 3 above. Allow once per group of terminal wrapped.	0.19
5	Shorten lead.	0.02
6	Pull lead through pierced terminal with pliers.	0.03
7	Open hole with scribe.	0.03
8	Additional allowance per wire obtained from wire rack over 12 in. long.	0.02
9	Get and aside soldering iron.	0.06

Touch-up and Repair of Printed-Circuit–Board Assemblies.

Table 5-35. Base Handling Time.[a]

Task Description	Decimal Min
1. Board and component pre-inspection (before wave soldering)	4.00
a. Get board and position in fixture.	
b. Inspect board for component defects.	
(1) Missing component	
(2) Damaged component	
(3) Wrong component	
(4) Improperly mounted component	
(a) Loose component	
(b) Misoriented component	
(c) Poor wrap	
(d) Leads too long	
(e) Improper insulation length	
(f) High component.	
c. Correct defects.	
d. Aside board.	
2. Board touch-up (after wave soldering)	3.00
a. Get board and position.	
b. Inspect board for component defects	
(1) Missing component	
(2) Damaged component	
(3) Improperly mounted component	
(a) Loose component	

Table 5-35. (*Continued*)

Task Description	Decimal Min
(b) High component	
(c) Leads too long.	
c. Inspect board for solder joint defects	
(1) Excess solder	
(2) Voids	
(3) No (or insufficient) solder	
(4) Cold solder	
(5) Fractured solder	
(6) Exposed copper.	
d. Inspect board for overall appearance	
(1) Dirty board (foreign matter and debris)	
(2) Damaged finish	
e. Aside board.	
3. Repair of board (after inspection)	
a. Get board and position.	
b. Locate defects noted by inspection.	
c. Aside board.	

[a]Base handling times given for touch-up and repair operations do not include any time required for the correction of defects.

Table 5-36. Correction of Board Defects.

Task Description	Decimal Min Touch-up
1. Correction of component defects (per component)	
a. High component	1.00
b. Long lead (per lead)	0.20
c. Install missing component	
(1) 2-Lead component	
(2) 3-Lead component	
d. Remove wrong, damaged, or misoriented component	
(1) 2-Lead component	
(2) 3-Lead component	
2. Correction of solder-joint defects (per joint)	
a. Excess solder	1.50
b. Voids	0.17
c. No (or insufficient) solder	0.17
d. Cold solder	1.50
e. Fractured solder	0.22
f. Exposed copper	0.17
g. Poor solder flow	0.17
h. Poor wetting	0.17

Table 5-37. Printed-Circuit-Board Assembly.

Contract No.		Equipment description		By	Tbl	Date
Assembly No.		Assembly description			Page	Of

Operation Description	Component Description	Task Description	Qty	Std. Time	Ext'd Min
Component Preparation	(1) Capacitor	(a) Clean leads		0.12	
		(b) Tin leads		0.05	
		(c) Hand-form leads		0.17	
	(2) Diode, coil,	(a) Machine-form leads		0.05	
	Resistor	(b) Hand-form leads		0.10	
	(3) I.C., Flatpack	Preform (14) leads with arbor press		0.55	
	(4) Transistor	Precut (3) leads (transistor/socket combination)		0.25	
	(5) Sleeving	Install sleeving on component		0.20	
	(6) Board	Clean board prior to assembly		1.38	
	(7)				
	(8)				
Bench Assembly	(1) Board	Get and aside, position and remove from fixture		0.27	
	(2) Buss wire	Hand-prep, sleeve, connect (2) ends		1.00	

Table 5-37. (*Continued*)

Contract No.		Equipment description		By	Tbl	Date	
Assembly No.		Assembly description				Page	Of

Operation Description	Description	Task Description	Qty	Std. Time	Ext'd Min
	(3) Capacitor	Install		0.10	
	(4) Coil	Install		0.15	
	(5) Connector	Install and hand-solder		5.80	
	(6) Diode	Install		0.12	
	(7) Hardware	Install screw, nut, lock washer combination		0.55	
	(8) Integrated circuit	(a) Install dual-in-line (14 leads)		0.35	
		(b) Install flatpack (14) leads)		0.85	
	(9) Resistor	Install		0.12	
	(10) Resis. var.	Install, tape or glue to board		0.18	
	(11) Test point	(a) Install 2-lead		0.16	
		(b) Install 1-lead		0.18	
	(12) Transistor	(a) Install with pad		0.18	
		(b) Install with socket		0.70	
	(13) Bonding	Bond component to board		1.00	

(14) Solder resist	Apply solder resist over hardware	0.25

Finishing Operations

(1) Bend and cut	Bend and cut lead on back of board (per lead)	0.02
(2) Tool-handling	Get and aside iron and solder	0.12
(3) Hand-solder	Hand-solder lead to board (per lead)	0.15
(4) Skinpack	Wrap and unwrap board	1.18
(5) Wave solder	Wave-solder board	0.75
(6) Cleaning	Clean board in degreaser	0.25
(7) Inspect leads	Check lead length with gauge	0.15
(8) Mark and tag	Mark and tag board	0.66
(9) Ejector	Install	0.50
(10) Integrated circuit	Push up from backside of board with tool	0.08

Total Min

Total Hr

+ Touch-up time @ 0 0009 Hr/lead
+ Conformal coat (A/R)

117

6

STANDARD MANUFACTURING PROCESSES

Chapter 6 is a guideline for a standard of excellence for workmanship and assembly standards covering all practices that justify standardization at this time.

The standards contained herein were compiled through extensive study of the latest assembly methods and the reliability requirements of quality assurance. They shall prevail over all production equipment and be adhered to unless instructed otherwise on the applicable engineering drawings.

It is our desire to provide manufacturing personnel with a comprehensive workmanship manual that will aid in the production of reliable products. This chapter should be read in its entirety and referred to by those concerned in order to gain complete understanding of workmanship requirements.

GENERAL PRACTICE

The purpose of this chapter is fourfold:

1. To provide a guide for all manufacturing employees.
2. To control personal conduct in a manner acceptable to all those concerned.
3. To promote an atmosphere of cooperation and improve productive work habits.
4. To create a better understanding of safety and housekeeping regulations.

Personal Conduct

Falsifying of personnel or company records is reason for dismissal.

Punching another's time card or allowing another to punch yours is reason for dismissal.

Be at work station on time.

Do not report for work while under the influence of alcohol or drugs.

Personal conversation is distracting and should be avoided.

Use pay phones for personal calls. Limit to 3 minutes.

Parts or prints should not be in your toolbox.

Do not attempt to hide or conceal errors or mistakes. Seek help from lead or supervisor.

Contributing to unsanitary or obscene conditions on company premises is reason for disciplinary action.

Do not sell or solicit contributions for any purpose on company premises.

Do not remove property belonging to the company or to other persons from the premises.

Notify your supervisor when unable to attend assigned working hours.

Do not leave work station without notifying your supervisor or lead.

Housekeeping

It is difficult, if not impossible, to produce a clean, efficient, electronic unit in a dirty, cluttered, unorganized facility. Therefore, we should strive to maintain good housekeeping in order to promote clean, safe, and productive working conditions. We ask your assistance to:

Keep work stations in a neat, orderly manner.

1. Only tools necessary to perform present job should be out on bench; all others should be in toolbox.
2. Parts, components, and hardware in properly marked containers.
3. Prints conveniently but neatly arranged.

Keep personal belongings, handbags, lunches, sweaters, and coats in lockers or baskets provided, not on workbench or back of chair.

Keep coffee cups, food, and food wrappers away from work stations.

Maintain lunch areas, lounge facilities, and rest rooms in a clean, comfortable, and sanitary condition.

Eliminate cigarette burns on benches and fixtures. Do not throw cigarettes or other litter on floor.

Avoid cluttering spare parts, boxes, packing material, hardware, and similar goods on workbenches or floor.

When asked to participate in special housekeeping programs, remember that this is your home away from home. Take pride in it.

Work Habits

It is our intention to help all employees develop safe, productive, and harmonious work habits. The following suggestions are to aid in this development:

Keep personal differences to a minimum.

It is each individual's responsibility to uphold the decisions of and to carry out the instructions as given by the lead or supervisor.

At times it may be necessary to share tools or work areas. This shall be done courteously.

Always return tools to tool crib promptly when you finish using them. Do not hoard tools.

Be considerate of your fellow employees.

Be sure you understand what to do.

Be sure you have the proper materials.

Be sure you know how to perform the job.

Be sure your workmanship is the best possible.

Check your work to be absolutely sure that every task is complete; when you are not sure, *ask*.

Safety

Report all industrial injuries promptly. Take the time to be safe, and you may save the time it takes to recover from careless and painful injuries. The time you save may be your own.

Wear safe and appropriate clothing and footwear.

1. Sturdy leather shoes with closed heel and toe, and low wide heels.
2. No loose hair or clothing around power tools.
3. *Safety glasses* are standard apparel in areas where drilling, grinding, dip soldering, or any other hazardous job is being done.

Heat guns and soldering irons or pots must be used with care.

To avoid eye injuries from flying lead ends, dikes should have open face toward board or chassis and a cupped hand over leads while clipping. Be careful, as leads may bounce. Eye protection must be worn.

Keep hands away from moving machinery. Leave adjustment and repair to qualified personnel.

Do not distract others while operating power equipment, nor be distracted while operating such equipment yourself.

Improperly stacked material and cluttered aisles and floor areas present a safety hazard and are forbidden.

The use of *long* air hoses and power cords shall be avoided whenever possible. When long cords or hoses are required, they shall be taped or fixed such that they are not a safety hazard.

Never lean back on chairs or stools.

Do not engage in horseplay.

MARKING AND IDENTIFICATION

The purpose of this section is twofold:

1. To provide tested and approved methods of identifying units, parts, and material and to avoid loss, errors, and rework due to improper or illegible markings.
2. To promote a greater understanding of inspection stamps and their meaning, in regard to status, operation, or condition of units during assembly.

This section is to be used by all manufacturing personnel as a guide in the absence of clear, or specific instructions on engineering drawings. In the event of conflict between this section and the requirements of engineering drawings, the drawing shall take precedence.

Marking

All marking, unless otherwise specified on engineering drawings, shall be in accordance with this standard.

Marking shall be accomplished with rubber stamps, code markers, or other engineering-specified methods only.

Hand stamping, unless otherwise noted, shall be performed with 0.125-inch letters in a manner that is visibly clear, straight in line, and evenly impressioned. When necessary to add or change markings, stamping shall be comparable to original in size and type.

Stamping ink will be of an approved type and of a color that is in contrast to the color of the item to which it is being applied.

When required, a protective coating of clear varnish or spray shall be applied in a suitable manner to avoid contaminating areas to be soldered or damaging finished surfaces.

Markings should be visible after assembly, except on exterior finished surfaces.

Markings shall read from left to right, from bottom to top, unless otherwise noted on drawing.

Markings must not adversely affect the soldering of any circuit pattern.

Acceptable (Enlarged for clarification)	Stamp was at angle S
	Stamp not fully inked G
A-3082-02	Stamp overloaded with ink T
	Stamp moved during stamping E
Not Acceptable	Stamp not vertically aligned D
a-3082/02	
A-3082-02	
A-3082-02	

Identification

Parts or components not individually identified shall be kept segregated in suitable containers marked with part numbers.

When no other means is practical, an I.D. tag with all pertinent information shall be attached to part. Manner of attachment will not cause damage to part or adjacent units, nor mar finish surfaces.

Parts and component identification, or value markings, when stamped on chassis or panel, shall be visible after assembly unless space provision or orientation prohibits.

Assemblies and subassemblies (small units) shall be identified with their part number or drawing number per engineering drawings.

Each cable shall be identified with its part number and should have connector designations stamped on or near connectors.

Manufacturing personnel will stamp their assembly number on each unit when it is completed.

Inspection stamps are a practical method of identifying the status or condition of a unit or assembly. The following are given as a reference to the meaning of each stamp.

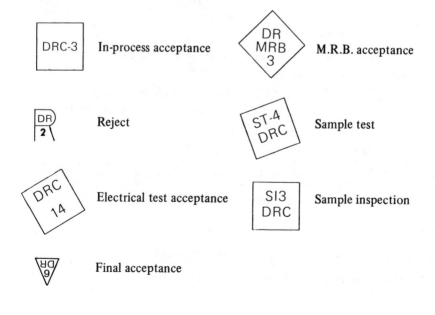

FABRICATION

The purposes of this section are:

1. To provide a convenient reference to quality standards governing the fabrication of mechanical assemblies.
2. To promote a better understanding of engineering drawings and the operation and care of tools in general use throughout the electronics industry.
3. To document the requirements for various types of hardware and control
  . the installation thereof.
4. To provide a method of assuring that the construction and appearance of products are in accordance with best commercial practice.

Blueprints

A blueprint is a copy of an engineering drawing. Blueprints and drawings serve to indicate the information required and to fabricate or construct the assembly or part that the drawing illustrates. A picture or photograph only show the unit's general appearance, while the blueprint reveals its shape, dimensions, material, and detailed assembly requirements.

As with any form of communication, standards are necessary to provide a complete understanding of the information contained on a drawing. This section was written for this purpose and should be referred to when in doubt.

All units will be fabricated to the specification called out on the applicable drawing or worksheets, which may include the shop order, pull sheet, operation sheet, and manufacturing variance.

Engineering instructions, either written or verbal, will be followed on special jobs such as development or prototype units.

Applicable drawings and worksheets will be checked for title, part number, model number, and change letter. Read all notes prior to assembly. Released products may be built to controlled prints only.

Drawings will be kept with unit and referred to when necessary.

Changes or red lines will be made on blueprints by engineering department personnel only.

Hand Tools

Tools will only be used for the job or function intended.

Tools will be maintained in a clean, neat, and orderly fashion.

Tools that will no longer perform their function properly should be returned to the toolroom for repair or replacement.

Screw-driver tips will have sharp, *square* edges and will be the proper size for the screw being installed. Worn screw drivers burr the screw-head and thus must not be used.

The proper size and type wrench—open, box end, spin tight, or socket—will be used for the job at hand. Hardware shall not be marred nor shall unnecessary strain be placed on tool.

Adjustable wrenches will only be used in cases of emergency.

Typical Hand Tools Used in Electronics

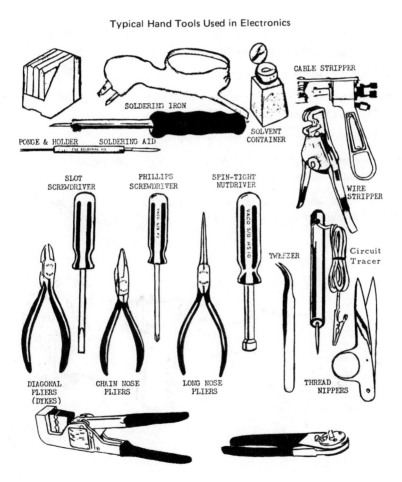

The use of holding or gripping pliers, especially those having serrated jaws, will be held to a minimum to avoid damage to parts and material.

Tools must not be placed on unprotected, finished surfaces.

Tools must not be altered in any way, without the expressed approval of the manufacturing engineering department.

Tools must be returned to the toolroom when there is no present use for them. Someone may be in need of those tools.

Files must have handles on the shank end, and broken files must not be used.

Power Tools

Do not operate any power tool you are not familiar with. Obtain instruction from your supervisor on the operation of new machinery.

Do not remove guards or guides from tools.

When working with power tools, use vices and holding fixtures, rather than holding parts with hands—particularly when drilling.

Never leave chuck key in drill presses or hand drills.

When changing bits in hand drills, always unplug the power cord to avoid injuries from accidently pulling the trigger.

When in doubt about the safety of a tool, such as a worn cord, faulty switch, arcing, or smoking motor, return it to the toolroom for repair.

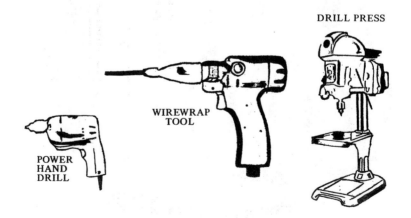

DRILL PRESS

WIREWRAP
TOOL

POWER
HAND
DRILL

Do not grind aluminum or copper on grinder.

Take all safety precautions when operating power tools. Wear your safety glasses.

Screws

Screws secured by nuts or other retaining devices that permit projection beyond the retaining device shall be long enough to permit a minimum projection of $1\frac{1}{2}$ threads and a maximum projection of $\frac{1}{8}$ inch plus $1\frac{1}{2}$ threads for screws up to 1 inch in length and $\frac{1}{4}$ inch plus $1\frac{1}{2}$ threads for screws over 1 inch in length. These requirements shall not be construed as precluding the use of screws assembled in blind tapped holes in castings, spacers, etc., or where the design restrictions require that the screw length be such as to require that the threaded portion of the screw be flush with the retaining device. (See Fig. 6-1.)

The ends of screws must not be clipped or deformed to secure the preceding conditions.

Note: If conflict exists between minimum and maximum length, the nearest standard screw on the maximum side shall be used.

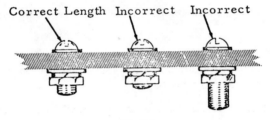

Fig. 6-1.

Exceptions to required screw-length projections are screws assembled in blind tapped holes in castings, spacers, etc., or where the design restrictions require that the screw length be such as to require that the threaded portion be flush with the retaining device.

Flathead screws used for flush finished surfaces must be the correct degree to properly fit the mating countersink they were designed to fit. (See Fig. 6-2.) All flathead screws shall be flush, or slightly below the surface, when properly set.

When fasteners that contain locking devices are used, the word *tight* means that the screws cannot be appreciably tightened further without damage or injury to the screw threads.

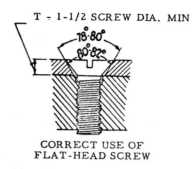

CORRECT USE OF
FLAT-HEAD SCREW

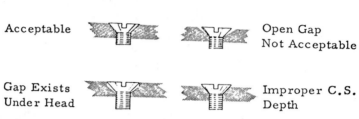

Fig. 6-2.

Screws installed into threaded (tapped) holes must have a minimum of three holding threads.

Care should be exercised during assembly to avoid damaging screw heads. Screws with damaged or burred slots are not acceptable. Take care in the selection of the proper size screwdriver or other tool.

Illustrations of acceptable and nonacceptable conditions of screw heads are shown in Fig. 6-3.

Nuts

Hexagonal nuts are preferred for general usage. Square nuts may be used only when they are captive or floating as part of a fastening device. (See Fig. 6-4.)

The use of floating basket nuts is preferred when a countersunk hole is to be matched with a tapped hole. (See Fig. 6-4.)

Nuts should be sufficiently thick and strong so as not to fail before the bolt or screw fails in tension.

Self-locking nuts shall not be used for ground connections. (See Fig. 6-5.)

ACCEPTABLE   NOT ACCEPTABLE

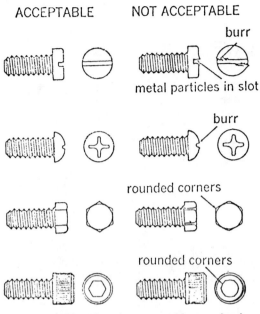

burr

metal particles in slot

burr

rounded corners

rounded corners

Fig. 6-3. Acceptable and nonacceptable screw heads.

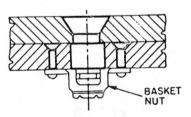

BASKET NUT

PREFERRED
Use floating basket nuts
whenever practical.

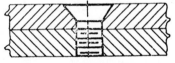

AVOID
Countersunk hole matching
tapped hole in mating part.
(Alignment problem)

ALIGNMENT OF COUNTERSUNK
HOLES

Fig. 6-4. Countersunk flathead screws.

Fig. 6-5. Locknuts.

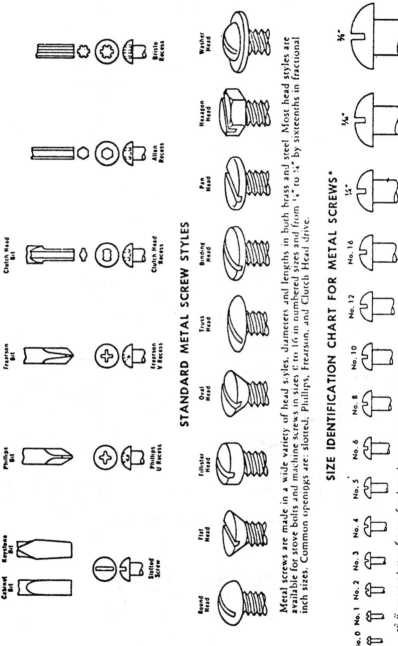

STANDARD METAL SCREW STYLES

Metal screws are made in a wide variety of head styles, diameters and lengths in both brass and steel. Most head styles are available for stove bolts and machine screws in sizes 0 to 16 in numbered sizes and from ¼ to ¾ by sixteenths in fractional inch sizes. Common openings are: slotted, Phillips, Frearson, and Clutch Head drive.

SIZE IDENTIFICATION CHART FOR METAL SCREWS*

*Full size reproduction for use of reader in determining screw sizes by comparison of actual screw with illustrations above.

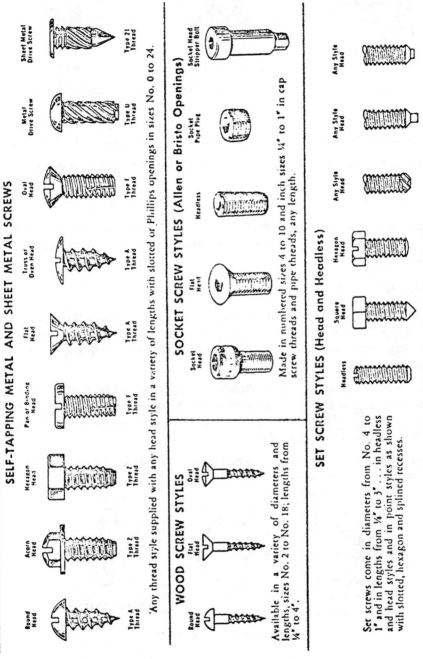

SELF-TAPPING METAL AND SHEET METAL SCREWS

Round Head — Type A Thread
Acorn Head — Type Z Thread
Hexagon Head — Type Z Thread
Pan or Binding Head — Type F Thread
Flat Head — Type A Thread
Truss or Oven Head — Type A Thread
Oval Head — Type F Thread
Metal Drive Screw — Type U Thread
Sheet Metal Drive Screw — Type 21 Thread

Any thread style supplied with any head style in a variety of lengths with slotted or Phillips openings in sizes No. 0 to 24.

WOOD SCREW STYLES

Round Head — Type A Thread
Flat Head
Oval Head

Available in a variety of diameters and lengths, sizes No. 2 to No. 18; lengths from ¼" to 4".

SOCKET SCREW STYLES (Allen or Bristo Openings)

Socket Head
Flat Head
Headless
Socket Pipe Plug
Socket Head Stripper Bolt

Made in numbered sizes 4 to 10 and inch sizes ¼" to 1" in cap screw threads and pipe threads, any length.

SET SCREW STYLES (Head and Headless)

Headless
Square Head
Hexagon Head
Any Style Head
Any Style Head
Any Style Head
Any Style Head

Set screws come in diameters from No. 4 to 1" and in lengths from ⅛" to 3" ... in headless and head styles and in point styles as shown with slotted, hexagon and splined recesses.

Types of Screwdriver Bits and Screw Openings

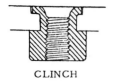
CLINCH

ANCHOR

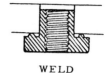
WELD

Fig. 6-6. Captive nuts.

Clinch, anchor, weld, and other types of secured nuts should not loosen when properly set. More intensive inspection of these types is required. (See Fig. 6-6.)

When self-locking nuts or nut plates are used, no other locking device is required.

The use of cross-threaded or otherwise damaged nuts will not be acceptable.

The stop-nuts on such parts as switches, pots, and lamp holders will be used to tighten parts in place, in order to avoid damage to the finished retaining nut or front panel.

WASHERS

A flat washer shall be used under the head of a screw, and ahead of lock washer to increase the load-bearing area, aid in tightening of nuts or screws, prevent loosening, and protect surfaces. Flat washers minimize buckling of thin materials.

A flat washer shall be used under the head of a screw/bolt when the head of the screw/bolt is normally the turning surface and the securing hardware includes a locking device. A flat washer and a lock washer shall be used when the securing hardware does not include a locking device. (See Fig. 6-7.)

A flat washer shall be used under the head of screws/bolts when it is evident that the mating surface should be preserved.

CORRECT USE OF LOCK WASHER

SPRING TYPE
LOCK WASHER

Fig. 6-7. Lock-washer installations.

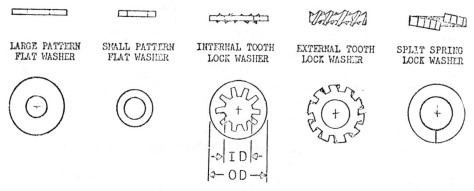

Fig. 6-8. Washer, standard types.

Lock washers shall be provided under all nuts, except those of the self-locking or castle type. (See Fig. 6-7.)

Lock washers shall be provided under the heads of all screws not secured by a locked-nut arrangement.

Lock washers with external teeth are preferred for making electrical bonds, noise suppression, and external grounding. (See Fig. 6-8.)

Lock washers will not be used on the exterior side of front panels unless so specified on applicable drawing.

PLATING AND FINISHES FOR FASTENERS

The information presented describes various fastener finishes and coatings used primarily for the purpose of increasing corrosion resistance. The terms "plating and finishes" and "coatings" are used interchangeably.

Applicable Documents

ASTM-A153	Zinc coating (hot dip galvanizing)
MIL-A-40147	Aluminum coating (hot dip) for ferrous parts
MIL-C-13924	Coating, oxide, black for ferrous metals
MIL-C-14328	Rubber sheet, synthetic, medium soft, general-purpose gasket material
MIL-C-14550	Copper plating (electrodeposited)
MIL-C-23422	Chromium plating, electrodeposited
MIL-C-26074	Coating, electroless nickel, requirements for
MIL-F-495	Finish, chemical, black, for copper alloys

MIL-G-45204	Gold Plating (electrodeposited)
MIL-L-13762	Lead alloy coating, hot dip (for iron and steel parts)
MIL-L-13808	Lead plating (electrodeposited)
MIL-P-14535	Plating, black nickel (electrodeposited)
MIL-P-14538	Plating, black chromium (electrodeposited)
MIL-P-18317	Plating, black nickel (electrodeposited) on brass, bronze, or steel
MIL-P-45209	Palladium plating (electrodeposited)
MIL-R-46085	Rhodium plating (electrodeposited)
MIL-S-5002	Surface treatments and inorganic coatings for metal surfaces of weapons systems
MIL-T-10727	Tin plating, electrodeposited or hot-dipped for ferrous and nonferrous metals
QQ-C-320	Chromium plating (electrodeposited)
QQ-N-290	Nickel plating (electrodeposited)
QQ-P-35	Passivation treatments for austenitic, ferritic, and martensitic corrosion-resisting steel (fastening devices)
QQ-P-416	Plating, chromium (electrodeposited)
QQ-S-365	Silver plating (electrodeposited)
QQ-Z-325	Zinc coating (electrodeposited)

Coating Selection

Protective coatings, such as those shown in Table 6-1 are recommended for fasteners subjected to corrosive conditions. To achieve a lower cost, a protective coating is used on ferrous fasteners instead of corrosion-resistant material.

Protective coatings should be used only where the fastener will be subject to mildly corrosive conditions. For extremely corrosive conditions, a fastener made of material that exhibits inherent corrosion resistance should be used.

The selection of finishes and coatings should be based on the following factors:

1. Whether coating or finish is for decoration or protection, or both.
2. Type of material, or plated surface of the mating parts.
3. Required color match, if any.
4. Physical aspects that may limit the type of fastener material used.
5. Type of corrosion expected during service.
6. Availability of the finishing process.
7. Cost.

Platings

The recommended plating for alloy and carbon steel fasteners is conversion-coated cadmium; for corrosion resistant steels, black oxide coating. On oc-

Table 6.1. Fastener Finishes and Coatings.

Coating or Finish for Fasteners	Used On	Corrosion Resistance	Characteristics
Anodizing	Aluminum	Excellent	Acid electrolytic treatment for aluminum, frosty-etched appearance. Hard oxide surface gives excellent protection.
Black, oxide	Steel	Indoor satisfactory, outdoor poor. Protection afforded mainly by wax or oil coatings.	Hot-alkali chemical process, black can be waxed or oiled. Produces a rust-inhibited surface.
Cadmium plate	Most metals	Very good	Bright silver-gray, dull gray or black electroplated finish. Used for both decoration and corrosion protection.
Clear chromate finish	Cadmium and zinc-plated parts	Very good to excellent	Clear bright or iridescent chemical conversion coating applied to zinc or cadmium-plated surfaces for added corrosion protection, coloring, and paint bonding. The colored coatings usually have greater corrosion resistance than the clear.
Dichromate	Cadmium and zinc-plated parts	Very good to excellent	Yellow, brown, green, or iridescent colored coating same as clear chromate.
Olive drab, gold, or bronze chromate	Cadmium and zinc-plated parts	Very good to excellent	Green, gold, or bronze tones same as clear chromate

Table 6.1. (Continued)

Coating or Finish for Fasteners	Used On	Corrosion Resistance	Characteristics
Chromium plate	Most metals	Good (improves with increased copper and nickel undercoats)	Bright blue-white, lustrous electroplated finish. Has relatively hard surface. Use for decorative purposes.
Copper plate	Most metals	Fair	Electroplated finish. Used for nickel and chromium-plate undercoat. Can be blackened and relieved to obtain Antique, Statuary, and Venetian finishes.
Copper, brass, bronze, miscellaneous finishes	Most metals	Indoor, very good	Decorative finishes. Applied to copper, brass, bronze-plated parts to match colors. Color and tones vary from black to almost the original color. Finish names are: Antique, Black Oxide, Statuary, Old English, Venetian, and Copper Oxidized.
Lacquering, clear or color-matched	All metals	Improves corrosion resistance. Some types designed for humid or other severe applications	Used for decorative finishes. Clear or colored to match mating color or luster.
Bright nickel	Most metals	Indoor excellent. Outdoor good thickness is at least 0.0005 in.	Electroplated silver finish. Used for appliances, hardware, etc.
Dull nickel	Most metals	Same as bright nickel	Whitish cast. Can be obtained by mechanical surface finishing or using a special plating bath.
Passivating	Stainless steel	Excellent	Chemical treatment. Removes iron particles and produces a passive surface.

Rust preventatives	All metals	Varies with function of oil	Vary in color and film thickness. Usually applied to phosphate and black oxide finishes. Used to protect parts in transit or prolonged storage.
Silver plate	All metals	Excellent	Decorative, expensive. Excellent electrical conductor.
Electroplated tin	All metals	Excellent	Silver-gray color. Excellent corrosion protection for parts in contact with food.
Hot-dip tin	All metals	Excellent	Silver-gray color. Same as electroplated, but thickness is harder to control especially on fine-thread parts.
Electroplated zinc	All metals	Very good	Bright-blue-white gray coating. For corrosion protection of steel parts.
Electrogalvanized zinc	All metals	Very good	Use where bright appearance is not wanted. Dull grayish color.
Hot dip zin	All metals	Very good	For maximum corrosion protection. Dull grayish color. Use where coating thickness is not important. Corrosion resistance is directly proportional to the coating thickness.
Hot-dip aluminum	Steel	Very good	For maximum corrosion protection. Dull grayish color. Use where coating thickness is not important. Corrosion resistance is directly proportional to the coating thickness.

casion, however, the engineer may be compelled to use platings other than those recommended.

Cadmium Plating

Cadmium plating offers excellent corrosion resistance for steel and copper alloy surfaces, and affords protection in dissimilar metal situations. Since cadmium plate is soft, it has lubricating properties which are effective in antiseize and antigall applications, but does not resist abrasion. Cadmium is toxic in both solid and gaseous states, and should never be used when contact with food is anticipated. Cadmium melts at 610°F and sublimation takes place at lower temperatures. As a result, cadmium plating is restricted to a maximum operating temperature of 450°F. If parts are to be painted subsequent to cadmium plating, a pretreatment coating in accordance with MIL-C-15328 should be applied before the primer and final cost films. Aluminum parts requiring cadmium plating should specify 0.0005-inch electroless nickel plating in accordance with MIL-N-26074, Class I prior to cadmium plating.

The recommended specification for cadmium plating is QQ-P-416 Type II and Type III are preferred.

Types of Cadmium Plating

Federal specification QQ-P-416 categories cadmium by three types (I, II, and III) as follows:

Type I: Without supplementary surtace treatment.

Type II: With a supplementary chromate finish to retard or prevent the formation of white corrosion products on exposed surfaces. A temperature restriction of 300°F is imposed on chromate surface-treated parts.

Type III: With supplementary phosphate treatment.

Classes of Cadmium Plate

Cadmium plating is divided into three classes, according to plating thickness:

Class 1:   0.0005 inches in thickness
Class 2:   0.0003 inches in thickness
Class 3:   0.0002 inches in thickness

Silver Plating

Silver plate should be specified for threaded fasteners when temperatures are expected to exceed 450°F. In this application, silver plating provides excellent resistance to seizing and galling. Silver reduces friction of sliding parts. The usefulness of silver for both corrosion resistance and lubrication purposes appears to terminate at 1300 to 1350°F. Silver plate offers excellent electrical conductivity along with good corrosion resistance and solderability. For threaded parts, a 0.002-inch plating thickness is recommended, while 0.0005-inch plating thickness is used for general applications. Ferrous metals should be copper plated to a depth of 0.0002 inches prior to application of silver plate. Aluminum requires a 0.0005-inch electroless nickel plate in accordance with MIL-N-26074, Class 1, before silver plating. For silver plated parts requiring a hard, gall-resisting surface (e.g., sliding parts), a rhodium plate 0.00002 inches in thickness should be applied after the silver plating. Since rhodium cannot be soldered, it should not be specified when soldering is intended.

Silver plating should be in accordance with QQ-S-365. This specification categorizes silver plate by surface finish into three types as follows:

Type I: Matte finish
Type II: Semi-bright finish
Type III: Bright finish

Aluminum Plating

Electrodeposited aluminum coating is applied to steel, iron, copper, magnesium, silver, gold, zinc, and nickel to enhance corrosion resistance. When diffused into the base metal, this coating exhibits good thermal properties. Aluminum plating, however, has poor abrasion resistance and, consequently, is not used frequently.

Chromium Plating

Electrodeposited chromium is applied to both ferrous and nonferrous base metals to impart resistance to wear, abrasion, and corrosion, a low coefficient of friction, good adhesion, and high reflectivity. (Corrosion resistance, however, is poor where no nickel underplating is used.) Chromium plate is frequently used to build up worn surfaces and in bearing applications. The specifications controlling chromium plating are QQ-C-320 and MIL-C-23422. Black Chromium plate is controlled by specification MIL-P-14538.

Cobalt Plating

Electrodeposited cobalt is used infrequently because its properties are similar to those of the less costly nickel plating. Cobalt plate is occasionally used where hardness in the range of Knoop 250 to 350 is required, and on reflectors. It is frequently used to upgrade alloy electroplate properties. Cobalt plating is gray or bluish-white in appearance and exhibits good abrasion resistance.

Copper Plating

Electrodeposited copper can be applied to most metals, both ferrous and nonferrous. It presents a bright or semi-bright appearance (red or pink in color) and offers some corrosion resistance along with high electrical and thermal conductivity. Electrical resistance varies for 3 to 8 microhms per centimeter. Copper plating is used frequently as an undercoating to improve adhesion and protective ability of subsequent electrodeposited nickel or chromium. It is also used to prevent hydrogen embrittlement. Copper plating has good adhesion and the cost is relatively low. The specification controlling copper plating is MIL-C-14550.

Gold Plating

Electrodeposited gold can be applied to copper, brass, nickel, and silver. This plating offers excellent adhesion, but abrasion resistance ranges from poor to good. Gold plating resists tarnishing, chemical attack, and high-temperature oxidation, and exhibits good ductility, thermal reflectivity, and electrical conductivity. Gold plating is relatively high in cost and is controlled by MIL-G-45204.

Indium Plating

Indium electroplating is applied to silver-plated steel and lead-bearing metals. It is soft and, unless diffused, has poor abrasion resistance. Indium plating is silver-white in color, tarnish resistant, malleable, and ductile. When diffused, it offers excellent adhesion. The most common use for indium plating is as an overlay diffusion coating on silver-plated steel bearings.

Lead Plating

An electroplated lead coating is most frequently used on copper and ferrous metals because of its resistance to many acids, hot corrosive gases, and corrosive atmospheres. It offers poor abrasion resistance, but good adhesion. Lead is more frequently applied by hot dipping. Its appearance is gray and the cost is low. The specification controlling lead plating is MIL-L-13808.

Nickel Plating

Electrodeposited nickel can be applied to most metals, both ferrous and nonferrous. Adhesion is very good, abrasion resistance good to very good, and hardness can be varied from 140 to 500 DPH. Nickel can be plated with either a dull or bright blue-white appearance and offers good corrosion resistance to many chemicals and corrosive atmospheres. Nickel plating is frequently used to build up worn or mismatched parts. The cost is moderate. The controlling specification for nickel plating is QQ-N-290. Black nickel is controlled by MIL-P-14535. The electrodeposition of black nickel on brass, bronze, or steel is controlled by MIL-P-18317.

Palladium Plating

Although costly to apply, palladium plating presents a good, white appearance. Careful application is required to prevent peeling of the plating from the base copper and copper alloys and underplating of silver, gold, and platinum. Although corrosion resistant, palladium plating tends to tarnish. It can be used alone or under rhodium on electronic equipment. Palladium electroplating is controlled by MIL-P-45209.

Platinum Plating

This very expensive plating offers a good, bright gray appearance and corrosion resistance to gold and copper alloys. Platinum plating tends to tarnish and abrasion resistance is poor. The most frequent use of platinum plating is protection of surfaces from unusual corrosive environments.

Rhodium Plating

Rhodium can be electrodeposited on most ferrous and nonferrous metals. Adhesion is good; abrasion resistance, high; and the cost, moderate. Rhodium plating offers corrosion resistance and good electrical conductivity. The electrical resistance of rhodium is 4.7 microhms per centimeter. Rhodium plating is used for electrical contacts, reflectors, mirrors, laboratory equipment, and optical goods. Rhodium plating is controlled by MIL-P-46085.

Tin Plating

Tin electroplating is usually applied to ferrous metals. Unalloyed, tin offers poor abrasion resistance, but this characteristic is greatly improved by alloying with copper, nickel, or zinc. Adhesion of tin plating is good; the appearance, bright white; and the cost, moderate. Tin plating is attractive in appearance, hygienic, solderable, and ductile. Tin plating is controlled by MIL-T-10727.

Zinc Plating

Zinc electroplating is usually applied to ferrous metals. It offers excellent adhesion and high corrosion resistance, but poor abrasion resistance. Application of zinc plate is rapid, easy, and inexpensive. Zinc plating is controlled by QQ-Z-325.

Surface Treatments

Surface treatment for fasteners is usually applied for the purpose of enhancing corrosion resistance when, for technical or economic reasons, electrodisposition plating is impractical. Surface treatments include passivation, black oxide coating, and hot dipping.

Passivation

Passivation is the application, to the surface of corrosion resistant steels, a nitric acid solution for the purpose of removing contaminants (such as free iron) from the surface of the metal. Passivation should be specified under any of the following conditions.

1. After machining operations.
2. When the most uniform and corrosion-resistant surface is required.
3. After any mechanical cleaning or surface operation where steel wool or wire brushes were employed.

When contaminants and residues have been removed and the clean, bare, corrosion-resistant steel is subjected to the oxygen of the air, an invisible protective oxide film forms a continuous, permanent covering. When in a condition of maximum corrosion resistance, the material is said to be passive. The specification controlling the passivation process is QQ-P-35, or MIL-S-5002.

Black Oxide Coating

Copper Alloys

Black oxide coating should be specified for copper alloy parts where it is necessary to match black surfaces, and to present a uniformly black color. The oxide coating may be used as a paint base, since it does not chip, flake, or peel, and will withstand severe deformation. Copper alloy parts joined by silver brazing should specify a 0.0002-inch copper flash before application of the black oxide coating. Under no circumstances should black oxide be used for items coming into contact with food or water supplies. The specification controlling black oxide surface treatment is MIL-F-495.

Ferrous Metals

Black oxide surface treatment is suitable for application on ferrous metal parts where dimensional build-up inherent in other treatments cannot be tolerated (such as for moving parts). The specification controlling black oxide surface treatment for ferrous metals is MIL-C-13924, Type 4.

Hot Dipping

Aluminum

The process of hot-dipping aluminum is known as "aluminizing," and consists of a 2-mil coating of pure aluminum over a 2-mil sublayer on steel or cast iron base metal. This coating offers protection from chemical attack at temperatures up to $1000°F$, and is instrumental in eliminating high-temperature oxidation in the base metal. Parts must be formed prior to aluminum hot dip coating. The aluminum hot dip coating process is controlled by MIL-A-40147.

Tin

Hot-dipped tin can be applied to steel, cast iron, copper, and copper alloys, and will withstand severe deformation. It offers very good resistance to tarnishing, but thin coatings may be porous. Solder coatings contain 10 to 60 percent tin, while terne coatings generally contain 10 to 15 percent tin, the balance lead. Hot-dipped tin coating is controlled by MIL-T-10727.

Lead

The hot-dipped lead process is applied to steel and copper base metals. Hot-dipped lead will withstand severe deformation and the protective oxide regenerates if damaged. It exhibits poor adhesion, wear resistance, and abrasion properties. The specification controlling the lead hot dipping process is MIL-L-13762.

Zinc

The process of zinc hot-dipping is known as "galvanizing," and the base material is steel. Zinc coatings offer high corrosion resistance and low cost, but are affected by environments containing sulfur and acid gasses. The process is controlled by MIL-Z-17871.

Other Coatings

In addition to electrodeposition, hot dipping, and surface treatments, design requirements may necessitate consideration of immersion, diffusion, vapor-deposited, or flame-applied coatings. The materials which can be applied by these processes are listed below:

1. Immersion coatings encompass electroless nickel, tin, copper, gold, silver, and platinum.

2. Diffusion coatings include aluminum, carbon, chromium, carbon and nitrogen, nickel-phosphorus, nitrogen, zinc, silicon, silicides, and iron.
3. Vapor-deposited coatings may consist of aluminum, cadmium, gold, silver, platinum, palladium, nickel, chromium, cobalt, silicon, zirconium, and refractory and precious metals.
4. Flame-applied coatings are applied ceramics and cermet coatings.

Corrosion

A protective coating cannot be selected without consideration of the different types of corrosion to which the fastener may be subjected. The types of corrosion commonly experienced by fasteners are uniform corrosion, galvanic corrosion, pitting, stress corrosion, and dezincification.

Fastener Corrosion Minimization

To minimize corrosion in fasteners, the applicability of the following considerations should be evaluated.

1. Select the fastener metal or alloy which is most likely to resist the corrosive environment to which it will be subjected.
2. Use metal combinations narrowly separated in the galvanic series.
3. Avoid fastener combinations where the area of the less noble material is relatively small.
4. Avoid irregular stresses in design.
5. Paint, coat, or insulate dissimilar metals.
6. When use cannot be avoided, keep dissimilar metals separated by a barrier material or insulator.
7. Use less noble metals in applications which are not functional to the item to be protected. These can then corrode sacrificially.

Plating Threaded Fasteners

Electroplating a threaded fastener can change the thread angle. (See Fig. 6-9). On external threads, the plating will be thickest at the top of the thread. On internal threads, the plating will reach maximum thickness at the root diameter and taper down to the minor diameter. This alters the thread angle slightly, but in the direction opposite to that of the plated external thread. Class 2A/2B unified threads have an allowance on the external thread which will accommodate electroplating and, therefore, these threads usually require no additional processing. Class 3A/3B unified threads have no allowance on the external thread and, consequently, the bolt threads must be undercut or preferably, the nut tapped oversize to assure proper parts.

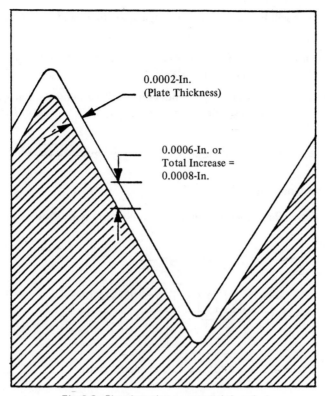

Fig. 6-9. Plated coating on external thread.

Installing Components and Hardware

Units will be fabricated in strict accordance to applicable drawings, in such a manner that no potential hazards exist to parts, components, or personnel under normal assembly or operating conditions.

All parts, components, and hardware will be stored, handled, and installed in such a manner as to avoid damage or loss during fabrication, inspection, test, and shipment.

Sharp edges and burrs shall be filed smooth unless said condition must exist to perform properly, in which case the hazard shall be padded or protected in an approved manner.

Due caution shall be exercised to avoid damage to switches, knobs, handles, shafts, and windows. Excessive weight or abrasive action may require a padded work station.

Finished surfaces will be free of scratches, dents, or stains, and no corrosion shall exist.

Fasteners, such as screws, bolts, and rivets, shall be as tight as possible without damage to fastener or adjacent materials.

All surfaces to be used as electrical conductors shall be bonded—that is, free of paint, oil, and corrosion in order to provide a suitable connection for coaxial jack or ground lugs, etc.

When possible, the screw or bolt will be held, and the nut shall be tightened.

Hardware Combinations

Acceptable nut-washer or nut-washer-screw combinations are shown in Figs. 6-10 through 6-17.

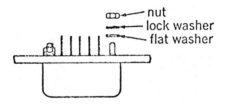

Fig. 6-10. Assembly of relay to metal.

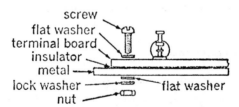

Fig. 6-11. Assembly of terminal board to metal.

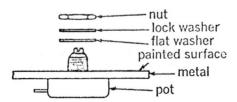

Fig. 6-12. Assembly of pot to front panel.

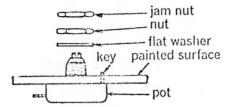

Fig. 6-13. Assembly of pot to front panel.

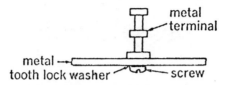

Fig. 6-14. Assembly of metal terminal to metal.

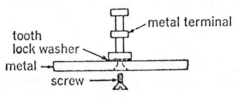

Fig. 6-15. Assembly of metal terminal with flathead screw.

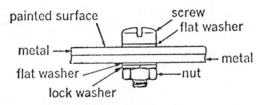

Fig. 6-16. Assembly of metal to front panel.

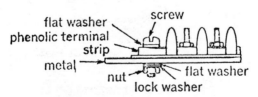

Fig. 6-17. Assembly of terminal strip to metal.

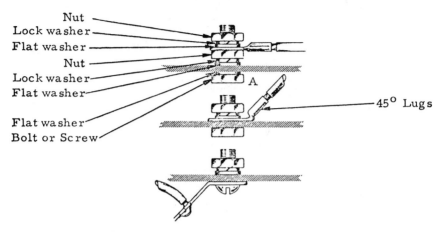

Fig. 6-18. Acceptable chassis ground connections.

Acceptable chassis ground connections are shown in Fig. 6-18.

Riveting

Heads of any type rivet must be properly formed and squarely seated, without damage to the parts, in order to provide a tight joint. (See Fig. 6-19.)

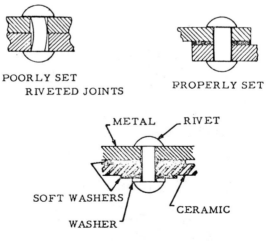

Fig. 6-19. Rivet setting.

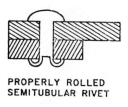

PROPERLY ROLLED
SEMITUBULAR RIVET

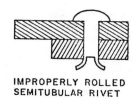

IMPROPERLY ROLLED
SEMITUBULAR RIVET

Fig. 6-20. Eyelet setting.

The material and finish of a rivet or eyelet must be compatible with the materials it contacts in an assembly. Normally, the rivet, as assembled, should be harder than the material it is fastening.

Rivets and eyelets shall be tight. The staking or rolling shall be uniform and reflect high quality of workmanship. Rivet heads that are not fully seated will be rejected.

Cracks or splits in the rolled or flared portion of eyelets shall be permissible with the following limits: (1) no single part may have more than one crack or split per end, and (2) no crack or split shall extend into the shank. Excessive splitting usually indicates defective tooling or material. (See Fig. 6-20.)

A flathead rivet should be countersunk flush or slightly below the surface. Slight protrusion above the surface is permissible if interference with another part does not occur or appearance is not degraded.

Terminals and Eyelets

The following conditions describe acceptable criteria for installing turret terminals in printed circuit boards (see Fig. 6-21):

1. The shoulder of the terminal shall be flush to the plated surface.
2. The formed head shall be swaged as flush as possible to the plated surface (no damage to circuitry or base material).

ACCEPTABLE

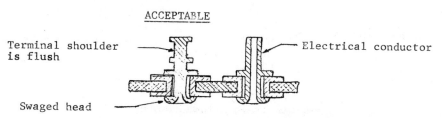

Terminal shoulder
is flush — Electrical conductor

Swaged head

Fig. 6-21. Acceptable conditions for installing terminals.

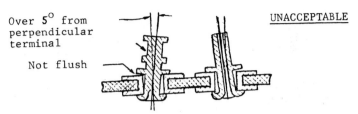

Over 5° from
perpendicular
terminal

UNACCEPTABLE

Not flush

Fig. 6-22. Unacceptable conditions for installing terminals.

3. The terminal shall be perpendicular (within 5°) to the board.
4. The terminal shall be spaced $\frac{1}{16}$ inch (minimum) away from the nearest electrical conductor.

Figure 6-22 illustrates unacceptable conditions for installing terminals:

1. The terminal is offset more than 5° from the perpendicular plane of the board.
2. The terminal is less than $\frac{1}{16}$ inch from the nearest electrical conductor.
3. The shoulder of the terminal is not flush to the plated surface.

In Fig. 6-23, excessive swaging pressure has damaged the circuitry and the board. This condition is unacceptable.

A maximum of two radial cracks in the swaged (formed) head shall be acceptable; however, a radial crack shall not extend into the shaft area. (See Fig. 6-24.)

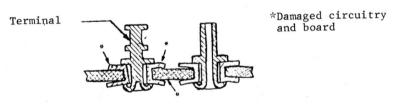

Terminal

*Damaged circuitry
and board

Fig. 6-23. Improper setting of swage-type terminals.

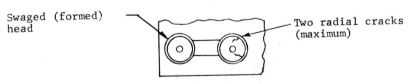

Swaged (formed)
head

Two radial cracks
(maximum)

Fig. 6-24. Swaged terminals and eyelets.

Excessive cracks ——————

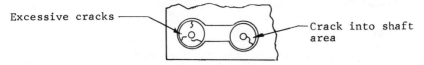

Crack into shaft
area

Fig. 6-25. Unacceptable swaging of terminals and eyelets.

More than two radial cracks in the swaged head and/or a radial crack that extends into the shaft area shall be unacceptable. (See Fig. 6-25.)

Eyelets and terminals used in printed circuit boards may show either a complete rollover or a flare. Eyelets may be of the funnel type. When used as part of a conducting circuit and when soldered to the printed conductors, the eyelet or terminal must be tight in the board before soldering. Prior to soldering, the rolled eyelet may show radial cracks that do not extend below the board surface.

Terminals and eyelets used to hold clips shall be completely rolled and must be tight but they may not have more than two minor splits, located no closer than 45° in 10 percent of the rivets or eyelets. A minor split is one that reaches only to the crown of the roll.

Locking Devices

A locking device shall be included in the mounting hardware of all screw/bolt/stud installations unless otherwise specified in the engineering drawing or assembly instructions.

A liquid thread-locking material shall be used, when it is impractical to use a locking device and it is specified on the engineering drawing or manufacturing instructions. Examples would be as follows:

1. Threaded inserts of transistor heatsinks.
2. Screws going into blind tapped holes where it is not desirable to use a lock washer under the screw head (for example, nameplates on front panels).
3. Adjustment screws on potentiometers (once the final testing and adjustments have been accomplished).

A drop or two of the thread-locking liquid should be applied to the male threads before engaging. In the case of adjustment screws, the thread-locking liquid should be applied at the intersection of the male and female threads.

COMPONENTS

The purpose of this section is threefold:

1. To promote a greater understanding of basic electronic components
2. To provide convenient reference material, illustrating their use, identification, and construction
3. To establish standards governing their handling and installation.

Note: This section may be used as a guide in the absence of clear or specific instructions on engineering drawings by all manufacturing department personnel.

As defined and used within this section, a *component* is a part that is generally supported and attached by its electrical leads or terminal wires, such as resistors, capacitors, diodes, transistors, small transformers, trimpots, and coils.

Resistors

General

Resistors are delicate units used to drop a given voltage to that required by a particular circuit and to limit current flow within a circuit.

The physical location of resistors in a unit is extremely important because the voltage dropped is dissipated in the form of heat. Any wiring routed to or near a resistor should be dressed in such a manner to prevent any thermal hazards.

Resistance is measured in ohms, usually with an ohm meter or the ohms scale of a multimeter. When a more accurate measurement is required, a Wheatstone bridge or similar network may be employed.

Substitution of one type of resistor for another is not permitted. Some characteristic differences between one type and another demand strict adherence to engineering drawings and specification.

Qualities pertinent to all resistors and generally standardized throughout the industry are:

1. Resistance: color-coded and written out.
2. Tolerance: 20, 10, 5, 2, and 1 percent sign as indicated.
3. Wattage: determined by physical size.

Carbon-Composition Type

Molded Form.   The most common carbon-composition resistor in use today is constructed of a molded resistive element made of carbon granules with one tinned lead imbedded in each end. It is then covered and protected by a hard resinous binder on which color-coded bands are painted. This type is quite suitable when high-resistance and low-power requirements up to 2 watts are needed.

Deposited Film. Deposited film resistors are very popular for close tolerance from above 2 megohms down to 0.1 ohm resistance. These resistors have a crystalline carbon film deposited on a ceramic shell, which in turn is covered with another ceramic shell. The air space between the two shells gives greater accuracy and better stability under changing temperatures and moisture conditions. They have tolerances of 1 and 0.01 percent and come with color code or value stamped on body.

Wire-Wound Type

Precision. Widely used by electronic industries for precise resistive values and high reliability for sustained periods of time. Constructed by wrapping fine lacquered wire around a bobbin or a core of various insulators, such as porcelain, phenolic, or enamel, it usually has an outer coating of the similar material. Some are dipped in sand while coating is wet, leaving a rough finish that aids in heat dissipation.

Power. Used in high-current applications, power supplies, resistor banks, and loading requirements. Usually constructed of heavy-resistance wire wrapped around a hollow core of porcelain or ceramic. This may be covered with a thick, brown layer of porcelain, leaving open ends for greater heat dissipation. Extreme care must be taken when working with power resistors. Being brittle, they break or crack quite easily, and excessive heat generated in use may present a hazard to associated parts or personnel.

Variable Type (Potentiometer)

Used to determine unknown resistance needs, or where a variance or manual adjustment is required. Manufactured in a wide range of resistance values and wattage ratings, constructed of both carbon composition and wire-wound material in two common shapes.

The Long-Bar Shape. With some area o   ie resistive material exposed, a movable contact can be adjusted to give the   esired resistance. Due to fine materials, the adjustable contact should be lc⌣sened before moving and then tightened afterwards.

The Round Shape. Having the resistive material applied in the shape of a disk (carbon) or donut (wire-wound). This type has a slidder actuated from the center in a circular motion and generally controlled with a knob. Both the long-bar and the round type may be used as a voltage divider by using the third lead as a second return. Most potentiometers are attached to the unit by a large nut and washer around the control shaft, which will be covered by the knob when installed. The leads are usually connected with solder or by mechanical means.

Trimpots. Extensively used in miniature design of circuit boards, where power requirements generally do not exceed 1 watt. Physical appearance will vary according to manufacturer and type, but usually can be identified by its three leads and the adjustment screw. As a rule, trimpots are mounted by their leads only, although some have mounting holes provided in case.

Capacitors

General

Capacitors were once referred to as "condensers," which is now considered obsolete. Capacitors are electrical devices used primarily to filter, bypass, or couple certain frequencies in electronic equipment. Generally, they block direct current (dc) and pass alternating current (ac) or pulsating current. Capacitors can store a charge or control the frequency of a given circuit.

Capacitance is measured in picofarads (pF) or microfarads (μF) on a capacitance meter or bridge.

Substitution of one type or value of capacitor for another is prohibited. Capacitors must be installed with regard to polarity.

Capacitors with welded leads or glass end seals are extremely delicate. They must be handled and soldered with great care.

Qualities pertinent to all capacitors and generally standardized throughout the industry are capacitance, working voltage, characteristic, and tolerance. All of these must be indicated by color code or written on body of component.

Due to the wide variety of type and configurations, only the most common will be covered herein.

Tubular (Foil Type)

The most common type of capacitor in use today is constructed of two conductive strips of foil with a dielectric or insulative material in between. The most common dielectrics used are paper and plastic film (Mylar). The laminated strip is then formed into a small roll. The leads are connected to the end of each foil strip. Next, a cover of wax, paper, plastic, ceramic, and sometimes metal will be placed over the roll and the ends sealed. A stripe or band, when used, indicates the outside foil lead.

Mica-Dielectric (Domino or Puff Type)

Very common for high-voltage and low-capacitance application. Consisting of mica wafers with the opposite ends dipped in a conductive alloy. The nonconductive mica between the two dipped ends serves as the dielectric. The leads are affixed to the silver alloy, either axially or radially. It is then molded or dipped with phenolic or plastic to seal and protect the capacitor against moisture and damage.

Ceramic Dielectric (Disk)

Very popular for circuit-board application. Constructed of two metal disks. Fused together with a thin film of ceramic-dielectric between. Leads may be axial or radial. The outer covering is a thin coating of phenolic (which is very brittle and must be handled with care). Values are generally written on the body although some smaller sizes are color-coded with dots.

Electrolytics (Aluminum and Tantalum)

Generally used only when high capacitive values are required. High critical values can now be obtained with relatively small space requirements. Construction may be of foil or an electrode. A thin film of gas is formed on the surface of the foil by the electrolyte and serves as the dielectric. One type, in which the electrolyte is a liquid, is called *wet*. The other, in which the electrolyte is in paste form, is called *dry*.

Electrolytic capacitors used for line filters often come with two or more capacitors in one case, with terminals marked for installation.

Polarized (Positive-Negative)

Electrolytics are polarized + (positive) and - (negative), and must be installed as called out on the drawing. Damage may result to component and unit if assembled improperly. Also, capacitors having glass end seals or welded leads must be handled with extreme care.

Variable

Most variable capacitors have air as their dielectric. As the air space between the conductors is varied, the capacity is also changed. The most popular design has one or more conductors called *rotor plates*. These turn with an insulated shaft, while their mating conductors, called *stator plates*, remain stationary. Some common uses for this type of capacitor are tuning-and-trimmer or padder capacitors for standard radios.

DIODES

A diode (germanium and silicon) is a semiconductor with a cathode and an anode lead shown in Fig. 6-26.

Diodes must be installed properly with regard to direction or polarity.

The most common case or shell used in diode construction is glass, which is adversely affected by excessive heat or vibration and must be handled with care.

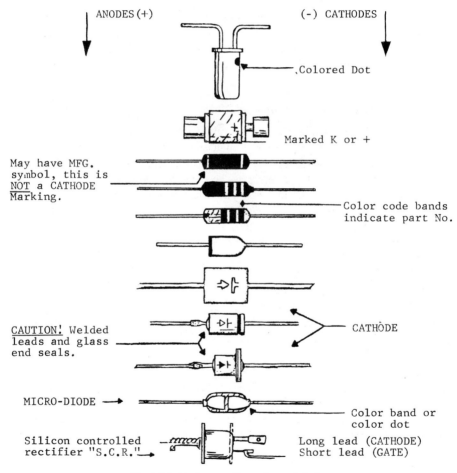

Fig. 6-26. Various configurations of diodes.

Under no condition shall diode leads be subjected to twisting, bending, or pulling at the body of the component. Cracked glass often causes a defective diode.

TRANSISTORS AND INTEGRATED CIRCUITS

Transistors and integrated circuits (IC's) are truly remarkable components. They fill the need for small, light, active, and reliable electronic controlled devices having low power requirements, low heat dissipation, and almost infinite life.

Both are adversely affected by heat and vibration. Due caution must be exercised to avoid excessive heat or physical shock.

Transistors and integrated circuits must be installed with close adherence to engineering drawings. Special care will be taken when working with components with nonstandard lead arrangement and pin location. Standard orientations are shown below and on p. 158. Nonstandard patterns should be on the drawing.

Leads shall not be subjected to pulling, twisting, nor bending at component body.

Power transistors that use the case as one lead shall be insulated with a mica washer when required or noted on the drawing.

When required, heat sinks shall be installed in such a manner as to provide maximum heat dissipation, but in no way damage component.

The following diagrams of integrated circuits show pin orientation as viewed from top or installed position.

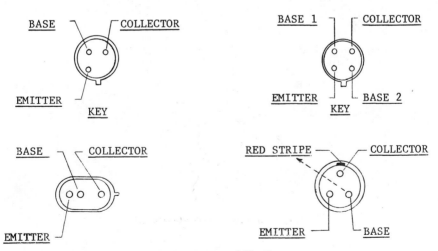

Typical Integrated Circuit

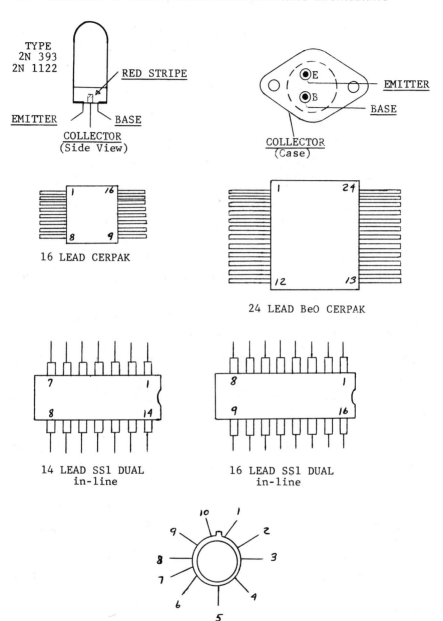

TYPE
2N 393
2N 1122

RED STRIPE

EMITTER   BASE
COLLECTOR
(Side View)

EMITTER

BASE

COLLECTOR
(Case)

16 LEAD CERPAK

24 LEAD BeO CERPAK

14 LEAD SS1 DUAL
in-line

16 LEAD SS1 DUAL
in-line

Typical Integrated Circuit (*Continued*)

Color Code

Resistor chart

RHYME	COLOR	VALUE	COLOR	VALUE	COLOR	MULTP'R	TOLERANCE	
Bad	Black	0	Black	0	Black	.0		
Boys	Brown	1	Brown	1	Brown	0.	Brown	±1%
Run	Red	2	Red	2	Red	00.	Red	±2%
On	Orange	3	Orange	3	Orange	000.	Gold	±5%
Your	Yellow	4	Yellow	4	Yellow	000,0.	Silver	±10%
Grass	Green	5	Green	5	Green	000,00.	No Band	±20%
But	Blue	6	Blue	6	Blue	000,000.		
Pretty	Purple	7	Purple	7	Gold	Multiply by .1		
Girls	Gray	8	Gray	8	Silver	Multiply by .01		
Won't	White	9	White	9				

EXAMPLES

1.

Yellow - Purple - Orange - Silver
 4   7   1,000   ±10%

$47,000 \pm 4,700$ Ohms resistance

A wide white band indicates solderable leads.

2.

Red - Gray - Gold - Gold
 2   8   .1   ±5%

$2.8 \pm .14$ Ohms resistance

Color-coding of wires facilitates wiring, testing, and localizing faults in the manufacture. The following chart shows the numbering system used to identify the basic colors.

Rhyme	Color	Value
Bad	Black	0
Boys	Brown	1
Run	Red	2
On	Orange	3
Your	Yellow	4
Grass	Green	5
But	Blue	6
Pretty	Purple	7
Girls	Gray	8
Won't	White	9

Component Mounting

Components shall be installed in accordance with applicable drawing, and mounted in a manner in which, under normal operating conditions, no hazards exist (such as open circuits, short circuits, broken wires, or damaged or lost components).

Components shall be mounted with regard to polarity, indicated by a plus (+) sign.

Components shall be mounted so that all values or part numbers may be read in the same direction. Where mounting of component is such that the part is not visible, a verification of the component part should be made by an inspector and noted on the in-process card during in-process inspection. (See Fig. 6-27.)

Component value or ID markings shall not be rubbed off during assembly.

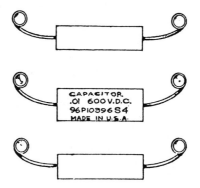

Fig. 6-27. Mounting axial lead components.

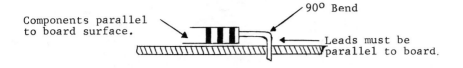

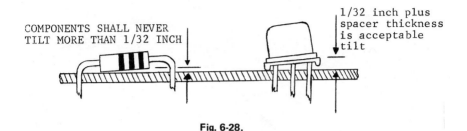

Fig. 6-28.

Components should be flush against board, but shall never exceed $\frac{1}{32}$-inch tilt (see Fig. 6-28). Exceptions:

1. The thickness of clamps, spacers, and mounts will be acceptable.
2. All power devices of greater than 2 watts shall be set up a minimum of 0.032 inches from the printed-circuit board.

No pressure points shall exist between components or adjacent hardware. (See Fig. 6-29.)

To prevent damage, component leads shall be bent with a round, smooth-finished tool. The radius of the bend shall be equal to or greater than twice the lead diameter. The minimum distance from component end seal to the start of the bend shall be $\frac{1}{16}$ inch. On components that have a welded lead, such as tantalum capacitors, the start of the bend shall be $\frac{1}{16}$ inch or more from the weld. (See Figs. 6-30 through 6-32.)

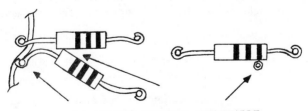

PRESSURE POINTS ARE NOT ACCEPTABLE

Fig. 6-29. Component interference.

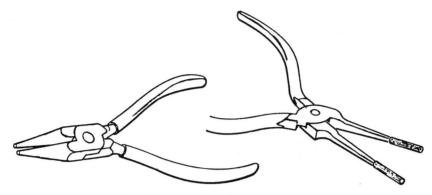

Fig. 6-30. Component-lead forming tools.

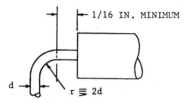

Fig. 6-31. Minimum lead bend.

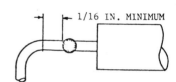

Fig. 6-32. Welded lead with proper bend.

Lead length of self-supporting components shall be as short as possible, consistent with quality workmanship. A component body shall be centered between bends of the axial component leads.

Component leads shall not be pulled tight, nor subjected to unnecessary strain, twisting, or bending, especially near component body or weld. (See Figs. 6-33 and 6-34.)

> *Note:* Extreme care shall be exercised when handling components having glass end seals and welded leads. Bend lead end *around* plier, not *with* plier.

Component leads passing within $\frac{1}{32}$ inch of any bare conductor shall be sleeved.

Component leads shall not be used as a jumper from one terminal to another. Buss wire will be used and sleeved when necessary.

Excessive or prolonged heat application must be avoided. To minimize component damage, heat shunts shall be used where required.

Component movement after soldering shall be avoided.

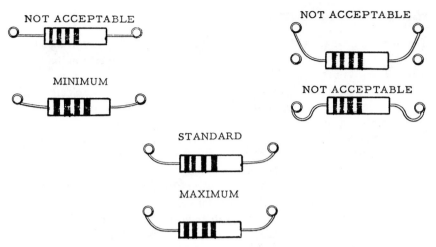

Fig. 6-33. Axial lead component mounting.

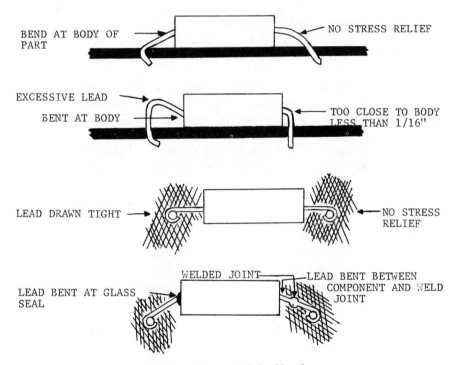

Fig. 6-34. Unacceptable lead bends.

SOLDER REQUIREMENTS

The increasing complexities of the electrical and electronic systems being used in industry today have made it necessary to exercise more care in joining the particular parts of these systems.

One of the most common means of joining metal parts is by soldering. This method is relatively simple, but requires that the solderer have considerable skill and ingenuity in order to consistently produce matings of maximum physical strength and electrical continuity. It is advisable to have a thorough knowledge of the fundamental properties of solder alloys and methods of applying them in order to solder efficiently at production speeds.

In soldering, a metallic substance of lower melting point than the metals to be joined is made to flow between the metal surfaces. The metals being joined remain unmelted, but are solidly mated when the joining metal solidifies. Soldering provides a completely metallic continuity and insures a permanent and constant electrical conductive medium.

The essential feature of a good solder joint is that each of the joined surfaces is wetted by a film of solder, and the films of solder are continuous with the solder filling between them. The films are formed by intermetallic combing and to some extent by metal solvent action, which occurs during soldering. These films will not form if the surfaces involved are dirty and/or coated with oxides.

The practical considerations involved in the development of soldering techniques are covered in this section. Attention is given to the selection of the correct solder composition for a specific application, surface preparation, and the proper iron for the job involved.

The technical aspects of soldering considered herein emphasize the advantage of proper soldering. Improper soldering effectively reduces the reliability of the equipment produced.

A typical instruction class should provide the trainees with the following information:

A. What is soldering, and how do we apply it?
  1. Soldering is the joining of two metals. It is usually a wire to a terminal on a piece of equipment such as a relay coil or capacitor, bonded together with an alloy called *solder*. The fusion of metals into one common mass is brought about by the melting of solder over and around the wire onto the terminal. If the solder is kept in a molten state and is in contact with these metals long enough, it will tend to unite with the outer layers of the metal being soldered.

  Flux is a compound that is used to clean and break down the oxides that have formed on the surfaces to be joined. The flux used in the electronics industry is noncorrosive, such as Kester 1544 or Superior No. 30. Extra flux should not be used unless absolutely necessary. Excessive residual flux is generally caused by using too much flux. Do not use additional flux unless absolutely necessary.

Clean material, the correct soldering iron, solder, flux, and a good operator should make a good soldered connection without damage to component or wire.

2. Reasons for soldering:
   (a) To provide a good electrical connection in the smallest space possible.
   (b) To produce electrical connections economically.
   (c) Protection of the connection by encapsulation in a cover of solder.
3. Procedure of soldering:
   (a) Have the correct iron, both in thermal capacity and size. Make sure the iron and point are large enough to do the job properly.

   With the iron joint cleaned and tinned, position the soldering iron point so that it is in contact with the wire and terminal to be joined. Apply the solder on the opposite side of the joint from the soldering iron point. The solder should flow quickly and easily, covering the wire and a section of the terminal with an even coat of solder. The soldering iron must be withdrawn as soon as the solder has encapsulated in the joint. During the time the solder is solidifying, the joint must not be disturbed. Disturbing the connection can cause it to become defective.

   The amount of solder used to make a good connection is best described as follows: The connection should be covered with sufficient solder to cover the entire joint, but still show the outline of the wire.

   The size of solder used depends upon the size of the parts to be joined—i.e., for soldering 12-gauge wire, .080-inch-diameter solder is preferred. For soldering 22-gauge wire, solder of .032- or even .020-inch diameter is preferred. The diameter of the solder makes it easier to apply sufficient solder to the heated connection in the least amount of time.

B. Types of Solder
   1. Roll
      (a) Roll solder has a rosin core, made in various diameters (i.e., .020, .032, .050, etc.). The alloy content is also variable, but the one generally used is 60/40 (60 percent tin, 40 percent lead). A 63/37 tin-lead eutectic is used on printed circuits and miniature assemblies.

2. Bar
   (a) Bar solder is used for solder pots or large soldering jobs, such as 2/0, 4/0 wire and connectors.

C. Soldering Iron and Points
   1. Size application. The correct-size iron and point must be determined from the work to be done. Generally, large, heavy-gauge wire requires a high-wattage soldering iron with a large point ($\frac{1}{4}$-inch chisel point) in order to achieve the proper thermal capacity.

   At times, a heavy-duty soldering iron and point may be necessary to solder a small wire and terminal (i.e., ground wire crimped to a ground stud). The reason here is that a large amount of heat is dissipated through the ground stud. Therefore, in order to overcome the heat loss and still solder the wire and stud, a heavy-duty iron and point are required.

   A small soldering iron is generally used for very small wire and small terminals. The recovery time is greater for the small iron, and therefore care must be taken in selection.

   2. Care of Iron Points
   (a) Point must be cleaned before using.
   (b) Point must be well tinned.
   (c) Copper points must be removed from iron and cleaned once daily. Care must be taken to insure proper metal-to-metal contact at tip and heating element.
   (d) Reshape point as required (copper only).
   (e) Other type points cannot be reshaped.

Soldering Methods

Soldering Iron

Heating of soldered joints with conventional soldering irons is the oldest soldering method. The most common electrical irons are heated by internal elements. Hand-held irons are generally used where joints are few in number or where they are so varied that another method is not as versatile. Component and wiring layout must provide for easy soldering without damage to adjacent parts by intimate contact. Heat sinks can be used to minimize damage. If wiring is complex or many closely spaced components are involved, the order in which joints are soldered can also become important. (See Fig. 6-35.)

Soldering irons of 60-watt rating are recommended for general use. Soldering irons of higher wattage may be used if necessary for some applications, but extreme care must be exercised to prevent overheating in all soldering applica-

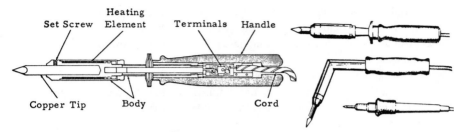

Fig. 6-35. Types of soldering irons.

tions. An iron of lesser wattage should be selected for miniature and subminiature work.

Wherever possible, the use of heat shunts shall be mandatory during soldering and tinning operations to minimize possible damage to components and insulation.

Soldering iron tips are made in various shapes, including pointed, chisel, bent, blunt, and concave, for general or specific use. The shape tip used shall be carefully selected for the item to be soldered. The use of a tip other than one that will fit the item to be soldered will result in an unsatisfactorily soldered connection. For general-purpose soldering, the pointed or chisel shapes are recommended; the flat narrow tip is for terminals of multipin connectors; etc. The bent tip is suited for connections difficult to reach. (See Fig. 6-36.)

To help achieve optimum soldering:

Do use a suitable stand, which will prevent the barrel and/or tip from touching metal parts, thereby limiting tool performance from heat sink. Heat-sinking a controlled tool results in shorter element and tip life.

Do keep tip tinned, and wipe only before using.

Do use rosin or activated rosin fluxes for electrical soldering.

Do remove tips, and clean tips and sockets regularly. Frequency of cleanings

POINTED   CHISEL   BENT   BLUNT   CONCAVE

Fig. 6-36. Types of iron tips.

should be determined by the type of work and usage. Tips in constant production should be cleaned at least once a week.

Do use a suitable cleaner for rosin-based fluxes, such as isopropyl alcohol or equivalent.

Don't clean tip with abrasive materials.

Don't use chloride or acid containing fluxes, as this reduces tip life (except those applications where special fluxes are necessary).

Don't remove excess solder before storing heated tool.

Don't file or attempt to reshape tip; this will destroy tip coating.

Don't use antiseize compounds on tip or socket; these parts are already protected from oxidation.

Dip and Wave Soldering

In dip soldering, the solder and heat are applied simultaneously to the work. All prefluxed metallic surfaces coming in contact with the molten solder are rapidly wetted. This allows high production volume at minimum cost of equipment and joint. Wave soldering was developed to overcome some of the drawbacks of dip soldering: to eliminate dross and skimming, and to shorten soldering time because of the dynamic movement of the solder over the surface (a 75 to 80 percent time reduction). This shorter time reduces warping, air, flux, vapor entrapment, etc.

In wave soldering, solder is lifted to the connections by one or more standing waves of molten solder. All joints are formed during the passage of an assembly over the waves. The operation, which is continuous, is limited in speed only by time-heat requirements. Because the wave is formed of continuously circulated, fresh, hot solder raised to the surface from the depths of the heated reservoir, drossing and local chilling are eliminated. At any given time, only a small band of the assembly is immersed in the wave; excess heat exposure is therefore eliminated. (See Fig. 6-37.)

• Prior to soldering, boards shall be free of oil, dirt, and foreign material.

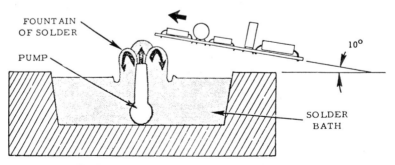

Fig. 6-37. Wave or fountain method of dip soldering.

- All circuitry, pad, and contact areas that are not to be soldered shall be covered with heat-resistant masking tape in a manner to prevent seepage of flux or solder.
- Circuit boards shall be dried and preheated with radiant heat just prior to soldering as directed on MPI sheet. *Caution:* Cold or moist objects may cause a violent reaction when placed in molten solder.
- Only approved bar-solder, thermostatically controlled at temperatures preset by manufacturing engineering department, shall be used for machine soldering.
- Dwell time or wave-solder rate of speed shall be set and controlled by the operator per MPI.
- Wave-soldered boards shall be allowed to solidify before removal from machine. Approximately 15 seconds will do.
- Remove masking promptly after solder solidification.
- After soldering, circuit boards shall be promptly cleaned of flux, oil, and all foreign matter, either by hand or in an approved cleaning system.
- Hand touch-up operations shall be employed when necessary to meet quality control requirements.
- Lead protrusion shall not exceed 0.062. Exposed copper on ends of cut leads after wave soldering need not be resoldered.
  Note: Due to board construction and the use of transistor mounts and pads, the need for heat shunts has been greatly minimized. However, some components, such as trimpots, crystals, lamps, and holders, must be soldered by hand, and heat shunts are required where necessary to protect heat-sensitive parts.
- Dip-solder pots for tinning shall have a temperature range sufficient to properly maintain the heat range for the type of solder used. (See Fig. 6-38.)

Fig. 6-38. Soldering pot.

- Solder shall be completely molten and stirred before use each day.
- When solder becomes contaminated with foreign material, remove solder and renew.

Resistance Soldering

An electrical current generates heat by flowing through a series circuit. A pair of electrodes can be applied across the joint, or one of the materials to be soldered can be connected as an electrode, requiring only a single probe to complete the circuit. Electrodes vary from 0.078-inch-diameter metal probes to large carbon blocks. Resistance soldering has the advantage of high joint production, no warm-up requirements, and instant, controlled heat. Joints can be made in from 0.5 to 2 seconds, depending on wire size and materials involved. Another advantage is the degree of miniaturization possible; tiny electrodes can be used to connect modular circuitry components or other closely spaced components. (See Fig. 6-39.)

Electrodes used on resistance-soldering tools shall be inspected for damaged or worn tips prior to use.

Selection of the proper size electrodes is important, in order to fit the size pin to be soldered.

No warm-up of a resistance-soldering tool is required.

Solder must be applied directly to the heated connection.

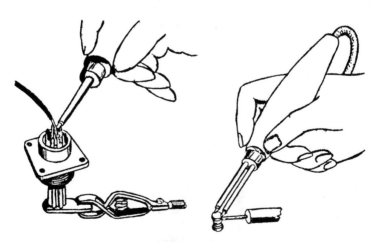

Fig. 6-39. Resistance hand soldering.

Induction Soldering

Electromagnetic-induction soldering uses the part to be soldered as the heating element. Current-induction coil heats the joint to a depth that depends on frequency. Because of skin effects, the higher the frequency, the more heat will be confined to material surfaces—an extremely useful phenomenon in soldering, because distortion and oxidation of the base metal are minimized.

Flux and solder are applied to the joints prior to joining, and preformed solder shapes can be used. (See Fig. 6-40.)

For induction soldering, as in other methods, surfaces must be relatively free of oxides.

Preformed solder rings, washers, and similar shapes that fit about the junction to be soldered shall be used.

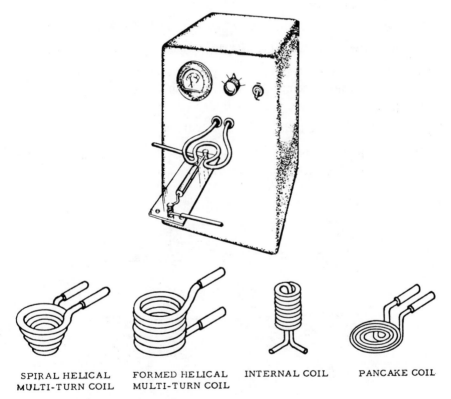

SPIRAL HELICAL   FORMED HELICAL   INTERNAL COIL   PANCAKE COIL
MULTI-TURN COIL   MULTI-TURN COIL

Fig. 6-40. Induction soldering unit and typical induction soldering coils.

Insufficient clearances, which may prevent the solder from penetrating properly, should be avoided.

Where practical, maintain a clearance of 0.003 inch for solder penetration.

Even temperatures should be maintained throughout the workpiece.

Tinning

All stranded conductors must be tinned prior to wrapping or soldering termination. (See Fig. 6-41.)

Tinning with a light coating of solder will be performed by dip or iron method, and should be done quickly to avoid insulation damage.

Surface contaminants (*dross*) shall be removed from solder pot, just prior to dipping or tinning, with an approved skimmer.

While tinning, solder shall flow no closer than $\frac{3}{32}$ inch to insulation.

After tinning, wire strands must remain visible and conform to original lay.

Only rosin core solder of an approved type and source shall be used for hand soldering and shall be compatible when used in conjunction with dip-solder operations.

Additional flux, if used, shall be compatible with the solder core type.

Solder cups (when tinned or filled) should contain sufficient solder to have a concaved appearance after wire is inserted.

ACCEPTABLE

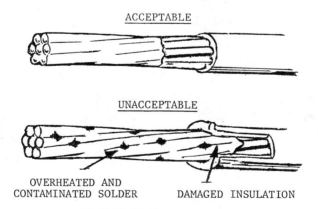

UNACCEPTABLE

OVERHEATED AND
CONTAMINATED SOLDER   DAMAGED INSULATION

Fig. 6-41. Tinning stranded wire.

General Soldering Requirements

While soldering, care must be exercised to avoid damage to components, insulation, and circuit-board material, especially heat-sensitive components. When necessary, heat shunts or other means of protection shall be used.

Only approved soldering irons of sufficient size and wattage needed for the particular job will be used. Irons will be kept in a clean, well-tinned condition.

Surfaces to be soldered shall be as close together as possible and be free of all oil, dirt, and corrosion in order to insure proper wetting action.

Soldering iron will contact both conductor and terminal on the side furthest from component or insulation. (See Fig. 6-42.)

Only SN-60 or SN-63 rosin core solder will be used unless otherwise specified on engineering drawings.

When used with flux-cored solder, liquid flux shall be chemically compatible with the solder core flux.

After cooling, solder joints shall be cleaned and have a smooth, uniform surface forming concaved fillets with no voids or solder peaks. A shiny appearance is preferred.

Solder shall not flow down terminal onto base material nor beyond or between printed-circuit or pad areas.

A minimum space of 0.0625 inch shall be maintained between conductors not intended to be connected. (Circuit boards are excepted. On circuit boards, the spacing between conductive paths must not be reduced by excess solder.)

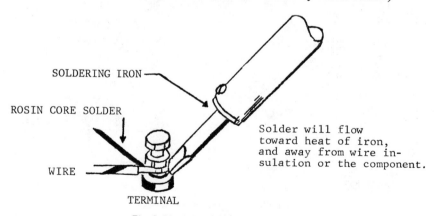

SOLDERING IRON

ROSIN CORE SOLDER

WIRE

TERMINAL

Solder will flow toward heat of iron, and away from wire insulation or the component.

Fig. 6-42. Soldering wire-to-terminal.

Connectors must be submitted for solder inspection before sleeving is pushed down over pins or back shell is closed or potted.

The soldering iron shall not dwell any longer than is absolutely necessary to flow the solder.

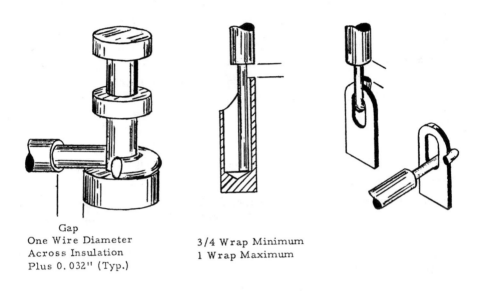

Gap
One Wire Diameter
Across Insulation
Plus 0.032" (Typ.)

3/4 Wrap Minimum
1 Wrap Maximum

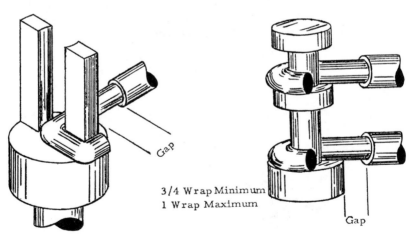

Gap

3/4 Wrap Minimum
1 Wrap Maximum

Gap

Fig. 6-43. Insulation gap and wrap.

The insulation shall not be damaged during soldering by contact with the soldering iron nor during the removal of spot ties.

Insulated magnet wire shall be carefully handled so that no damage occurs. Solder connections shall be made so that all of the insulation is burned off and does not remain in the connection.

All conductors terminating in a connector cup (solder pot) shall bottom in the cup.

When insulated wire is to be used, the insulation shall be stripped back a sufficient distance so that, when soldered, the insulation is not in the solder joint nor is the gap more than one wire diameter, across the insulation, plus 0.032 inch from the joint. (See Fig. 6-43.)

A solder connection shall not rely on solder alone. However, connections to connector-type pins are acceptable where a mechanical connection is not feasible. (See Fig. 6-44.)

Wicking (flow back) should be avoided during soldering wires to their termination point. Wicking is normally caused by allowing the soldering iron to dwell on the solder connection for a longer time than is required to properly form the solder joint. Minimum wicking shall be acceptable on wires that are adequately supported by clamps, lacing, or potting compound.

Thermal shunts (or heat clamps) shall be used when soldering precision resistors, diodes, transistors, and glass-bead capacitors in order not to damage the components. If excessive heat is applied, these parts will be damaged beyond usefulness. Thermal shunts shall not be removed for a minimum of 15 seconds after soldering. This 15-second delay is essential to permit maximum heat transfer to the thermal shunt. (See Fig. 6-45.)

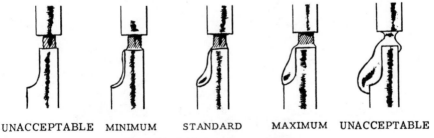

UNACCEPTABLE   MINIMUM   STANDARD   MAXIMUM   UNACCEPTABLE

Fig. 6-44. Soldering wire-to-solder cups.

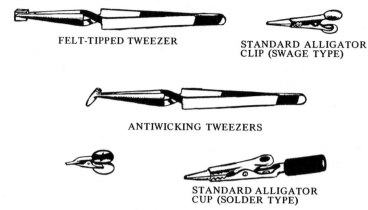

FELT-TIPPED TWEEZER

STANDARD ALLIGATOR
CLIP (SWAGE TYPE)

ANTIWICKING TWEEZERS

STANDARD ALLIGATOR
CUP (SOLDER TYPE)

Fig. 6-45. Thermal shunts.

Visual Requirements for Solder Joints

Several factors affect the geometrical configuration of the solder joint:

1. It is essential that the electrical connection has enough current-carrying capacity for the joint.
2. It should be strong enough from a mechanical standpoint.
3. It should have *inspectability,* a term that originates from the fact that we are using the external appearance of the solder joint to evaluate its quality. It is therefore essential that the criteria that determine the quality be plainly visible.

Thus, a fillet that is well feathered, continuous, and shiny can be defined as a good, soldered connection. However, it is also important that the contours of the two surfaces be visible through this fillet; otherwise, we are masking the properties that we are looking for. Figure 6-46(a) shows a wire inserted and

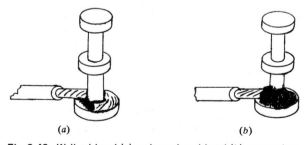

(a)   (b)

Fig. 6-46. Well-soldered (a) and poorly soldered (b) connections.

looped around a terminal. The fillet is well feathered, and the contours of both the terminal and the wires are plainly visible. This is by definition a well-soldered connection.

Figure 6-46(b) shows the same basic configuration with an excessively large solder fillet. Here, the contour of the wire is not visible, but in spite of this, the fillet is smooth, continuous, and feathered out on the terminal. This connection has no inspectability; therefore, it is a poorly soldered connection.

Good wetting.   In good wetting, solder is feathered out, indicating a small dihedral angle. The solder surface is bright and smooth, with few or no pinholes.

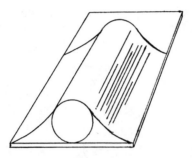

Poor wetting.   In poor wetting, solder makes a large contact angle $(75°$ to $90°)$. Solder surface is not continuous; irregularly round, nonwet areas are exposed. Normally, a small amount of this condition can be tolerated.

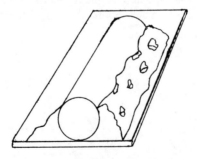

Dewetted.   Dewetted solder forms in balls, does not competely cover the surface, and has contact angle greater than $90°$. This condition should not be tolerated.

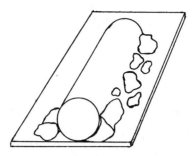

Insufficient heat. When heat is insufficient, solder solidifies in a nonadhering continuous layer because flux is not properly activated. Usually, a film of tarnish is still present on the work, and the solder can be pried loose. The contact angle is large and uneven.

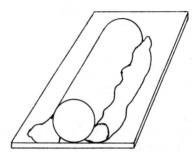

Rosin. Bond is achieved through a layer of solidified flux, usually a rosin type. In its worst form, this joint has no metallic or electrical continuity, and has little physical strength. Contact angle is poorly defined, although fillet is continuous.

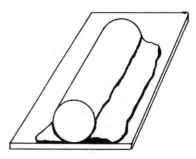

Cold. Joint appears frosty and granulated because of movement during solder solidification. In its worst form, it is the "fractured joint" shown in the diagram. The contact angle is usually even.

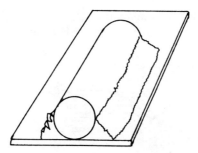

Overheated solder joint can be detected by a dull and crystallized joint caused by overheating the solder joint. *Not acceptable.*

Disturbed solder joints are reflected by a coarse crystalline appearance, fractured or separated at the junction of the elements soldered. *Not acceptable.*

Solder projections are caused from improper withdrawal of soldering iron. *Not acceptable.*

Insulation must be an adequate distance from connection prior to soldering. Insulation embedded in the solder connection is *not acceptable.*

Excessive solder can be detected by peaks, domes, or an overflow of solder. *Not acceptable.* When a sufficient amount of solder is used, the joint should be covered and the contour of the wire should be visible.

Insufficient solder is readily noticed when wire is exposed at the point of mechanical connection. *Not acceptable.*

Proper solder joint must wet the upper conductor and form a uniform concave fillet between the conductor and the lead wire. The conductor lines on the lower surface of the board must be covered with a thin, uniform film of solder.

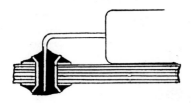

Eyeletted
P.C. Board

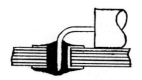

Plate-Thru
P.C. Board

Typical solder joints.

All "dip side" pads shall have a minimum of 90 percent fillet around all leads. All "component side" pads shall have greater than a 60 percent fillet.

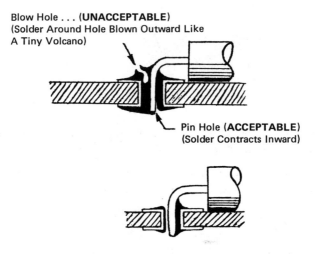

Blow Hole . . . (**UNACCEPTABLE**)
(Solder Around Hole Blown Outward Like A Tiny Volcano)

Pin Hole (**ACCEPTABLE**)
(Solder Contracts Inward)

Depleted Joint—(Chimney Void)—**UNACCEPTABLE**

Penetration. The solder must flow up through the plated-through interface connection, wetting the conductor surface and the component lead wire.

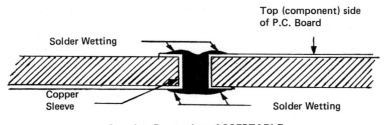

Solder Wetting

Top (component) side of P.C. Board

Copper Sleeve

Solder Wetting

Complete Penetration—**ACCEPTABLE**

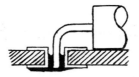

Incomplete Penetration—**UNACCEPTABLE**

Appearance.   The solder joint must be clean and free from any trace of flux residues or foreign materials.

Foreign Material - <u>UNACCEPTABLE</u>

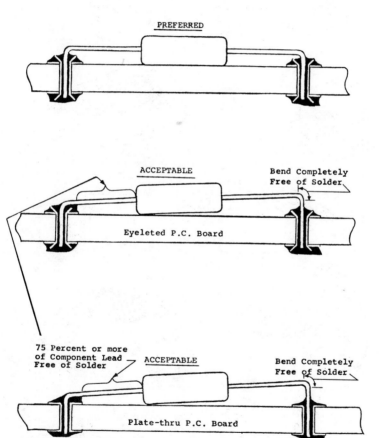

Installation position of axial lead components.

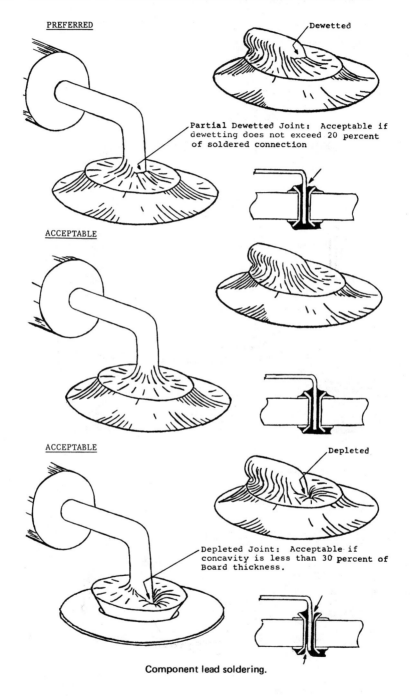

PREFERRED

Dewetted

Partial Dewetted Joint: Acceptable if dewetting does not exceed 20 percent of soldered connection

ACCEPTABLE

ACCEPTABLE

Depleted

Depleted Joint: Acceptable if concavity is less than 30 percent of Board thickness.

Component lead soldering.

UNACCEPTABLE

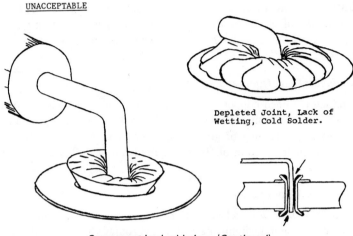

Depleted Joint, Lack of
Wetting, Cold Solder.

Component lead soldering. (*Continued*)

WIRING

This specification was written:

1. to provide standards covering wire routing and mechanical termination.
2. to promote higher product reliability and uniform construction.
3. to provide a convenient reference guide in the absence of clear or specific instructions on engineering drawings.

In the event of any conflict between the requirements of this specification and the engineering drawing, the drawing shall take precedence.

Note: The word *wire* as used herein refers to the conductor and the insulation when purchased as one item.

Wire Types

General

Wire is used as a principal conductor (carrier) of electrical current from one point in a circuit to another. Most electrical wire is made of copper because of its low resistance and economical cost. Modern manufacturing methods have improved the basic copper wire by adding a tin or silver coating that reduced copper oxidation and increased solderability. New insulative materials such as fluorinated ethylene propylene (Teflon), which will withstand temperatures above those required for soldering, have advanced quality standards and ease of soldering. However, nylon and vinyl insulations are still in use and melt at low temperatures, so care must be taken when soldering all wires. The current-carrying capacity of electrical wire is determined by its American Wire Gauge rating, such as #18, #22, and #24 gauge. The higher the gauge, the smaller the wire diameter.

Solid Conductor Wire, Noninsulated

Solid conductor, tinned copper wire without insulation, this wire is commonly called *buss wire*. It is primarily used in small sizes for short jumpers and in larger sizes as busbars for feeding or grounding of other circuits. This wire is very practical where little or no insulation is required and flexibility after fabrication is not necessary.

Solid Conductor Wire, Insulated

Solid conductor wire with white Teflon insulation is used mostly for Logic wiring where short and medium-length wires are routed through narrow channels to terminals. Here they are hand-wrapped and soldered. The prime advantage is its overall space requirement and solderability in confined areas. Used extensively for wire-wrap terminations.

Stranded Conductor Wire, Insulated

Stranded conductor with white Teflon insulation. Used for all general wiring and cable fabrication where flexibility is required.

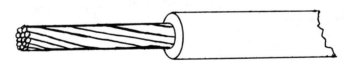

Wire, Twisted Pair

Two stranded conductor insulated wires twisted together. The number of twists per inch must be per engineering specification used in signal applications. The twist reduces noise, similar to the hum in a radio.

Wire, Twisted Trio

Same as twisted pair except with three conductors.

Magnet Wire

Copper conductor coated with a baked film of polyurethane-type resin with a superimposed film of baked nylon. Used in core-memory stacks.

Cable, Shielded

Stranded two-conductor cable with polyvinyl chloride (PVC) insulation. Two nylon inner jackets shielded with tinned copper braid and covered with a PVC outer jacket. Shielded wire is used to minimize induction between wires in a cable or chassis.

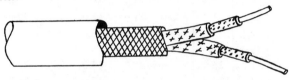

Cable, Coaxial

The dielectric or inner insulation is made of Teflon and is surrounded by silver-plated copper braid. The outer jacket is white taped Teflon. Used to carry high-frequency signals without distortion.

Cable, Power

Two or more color-coded, rubber-insulated conductors. Hemp cord is used as a filler to add strength. The outer covering or jacket is black neoprene rubber. Cable of this type is used to carry power from source to required areas within unit or system.

Table 6-2. Safe Current Ratings for Wire and Cable.

			Maximum Current in Amperes—Ambient Temperature 25°C (78°F)							
Number of Conductors			1				2-3			
			INSULATION							
Awg Size Solid Strand	Diam. in Inches	Resistance Ohms per 100 ft	PVC 80°C	PVC 105°C	Polyethylene	Teflon	PVC 80°C	PVC 105°C	Polyethylene	Teflon
24	.020	25.7	3.7	11.5	3.7	23.5	2.3	7.2	2.3	14.7
22	.025	16.2	4.6	14.9	4.6	31.2	2.9	9.3	2.9	19.5
20	.032	10.1	6.1	19.0	6.1	41.0	3.8	12.0	3.8	26.0
18	.040	6.39	8.3	25.0	8.3	56.0	5.2	16.0	5.2	35.0
16	.051	4.02	11.0	35.0	11.0	68.0	6.8	22.0	6.8	43.0
14	.054	2.52	14.0	46.0	14.0	92.0	9.1	29.0	9.1	58.0
12	.081	1.59	19.0	60.0	19.0	118.0	12.0	38.0	12.0	74.0
10	.102	.99	25.0	83.0	25.0	160.0	16.0	52.0	16.0	100.0
8	.129	.63	36.0	113.0	36.0	212.0	23.0	71.0	23.0	133.0
6	.162	.39	49.0	152.0	49.0	283.0	31.0	95.0	31.0	177.0

Values apply to individual wires.

Recommended Safe Current Ratings for Wire and Cable

Before a safe current rating can be applied to a wire or cable configuration, careful review must be given to the following factors. Because the characteristics of all wire insulating compounds are ultimately affected, it is important to select the proper one with the following in mind.

- The temperature of an electrical conductor will increase as the square of the applied current.
- Since conductor heat is usually dissipated from the insulation surface, ambient temperature and the amount of exposed surface available for heat dissipation greatly affect current carrying capacity.
- Wire bundles or multi-conductor cables will experience a greater temperature rise due to passage of current because the wire surface available to radiate heat is greatly reduced.

Table 6-2 of safe current values recognizes that there will be a rise in conductor temperature commensurate with the magnitude of the applied current and the

Table 6-2. (*Continued*).

Awg Size Solid Strand	Diam. in Inches	Resistance Ohms per 100 ft	Maximum Current in Amperes—Ambient Temperature 25°C (78°F)							
			Number of Conductors 4–5				6–15			
			INSULATION							
			PVC 80°C	PVC 105°C	Polyethylene	Teflon	PVC 80°C	PVC 105°C	Polyethylene	Teflon
24	.020	25.7	1.8	5.8	1.8	11.8	1.6	5.0	1.6	10.3
22	.025	16.2	2.3	7.4	2.3	15.6	2.0	6.5	2.0	13.7
20	.032	10.1	3.0	9.6	3.0	20.0	2.7	8.4	2.7	18.2
18	.040	6.39	4.2	12.0	4.2	28.0	3.6	11.0	3.6	24.0
16	.051	4.02	5.4	17.0	5.4	34.0	4.8	15.0	4.8	30.0
14	.054	2.52	7.3	23.0	7.3	46.0	6.4	20.0	6.4	40.0
12	.081	1.59	9.6	30.0	9.6	59.0	8.4	26.0	8.4	51.0
10	.102	.99	12.0	41.0	12.0	80.0	11.0	36.0	11.0	70.0
8	.129	.63	18.0	56.0	18.0	106.0	16.0	49.0	16.0	93.0
6	.162	.39	24.0	76.0	24.0	141.0	21.0	66.0	21.0	123.0

Values apply to individual wires.

ambient temperature. Current values are listed which, at the ambient indicated, will not result in temperatures sufficiently high enough to damage the insulation. However, many military and commercial codes and specifications detail the requirements for their particular applications and should be carefully consulted during circuitry design in instances where they govern.

Wire-Stripping

Wire-stripping is the removal of unwanted insulative material from wires and cables without damage to the conductor or remaining insulation. The use of either power- or hand-strippers may be employed. However, only hand-strippers that have been tested and approved by the manufacturing engineering and quality assurance departments will be covered here. These are:

1. *Thermal strippers* are approved for all types of wire except glass braid and shall be used extensively for stripping shielded wire and coaxial cable.
2. *Calibrated cutting strippers* (Ideal) are approved for use on all standard

Table 6-3. Conductor Data Chart.

Wire Size Awg	Conductor Stranding	Nom. Cdr. Area in Circ. Mils	Nom. Str. Dia. in Inches	Conductor Dia. in Inches	
				Nom.	Max.
32	Solid	63.2	0.008	0.0079	0.0083
	7 × 40	69.2	0.003	0.0093	0.0096
30	Solid	100.5	0.010	0.0099	0.0103
	7 × 38	112	0.004	0.0120	0.0124
28	Solid	160	0.013	0.0126	0.0130
	7 × 36	175	0.005	0.0150	0.0154
26	Solid	254	0.016	0.0159	0.0164
	7 × 34	278	0.006	0.0189	0.0195
	19 × 38	304	0.004	0.0188	0.0205
24	Solid	404	0.020	0.0201	0.0207
	7 × 32	441	0.008	0.0240	0.0247
	19 × 36	475	0.005	0.0248	0.0255
22	Solid	643	0.025	0.0254	0.0261
	7 × 30	707	0.010	0.0300	0.0309
	19 × 34	754	0.006	0.0312	0.0320
20	Solid	1022	0.032	0.0320	0.0329
	7 × 28	1120	0.013	0.0378	0.0389
	19 × 32	1216	0.008	0.0396	0.0405
18	Solid	1624	0.040	0.0403	0.0415
	7 × 26	1778	0.016	0.0477	0.0491
	19 × 30	1900	0.010	0.0495	0.0505
16	19 × 29	2426	0.011	0.0559	0.0570
14	19 × 27	3831	0.014	0.0703	0.0715
12	19 × 25	6088	0.018	0.0886	0.0905
10	37 × 26	9354	0.016	0.1102	0.1124
8	133 × 29	16983	0.011	0.1644	0.1693
6	133 × 27	26818	0.014	0.2066	0.2128

Reference: Alpha Wire Corp., Elizabeth, New Jersey
Storm Products Corp., Inglewood, California
Beldon Wire & Cable Corp., Richmond, Indiana
Mil-Spec Wire Inc., Van Nuys, Calif.

single or multistrand wire provided that there is no evidence of wire damage. To insure satisfactory results when stripping wires, the following should be observed:

a. Use the proper stripper hole.
b. Be sure stripper is adjusted correctly.
c. Check cutting edges of stripper for sharpness, nicks, and burrs.

Table 6-3. (Continued).

| Max. Resistance of Finished Wire (Ohms/m ft. 20°C) | | | | | | Max. Conductor Weight (Pounds/1000 ft.) | | |
| Soft or Annealed Copper | | | High Strength Cu Alloy | | | | | |
Tin	Silver	Nickel	Tin	Silver	Nickel	Tin	Silver	Nickel
176.1	168.1					0.19	0.18	0.19
177.1	169.1	182.0				0.23	0.21	0.20
110.8	107.0					0.30	0.28	0.26
107.7	100.3	110.2	129.5	120.5	133.9	0.40	0.38	0.36
68.9	67.0					0.48	0.46	0.44
68.2	63.6	67.6	82.0	76.4	83.0	0.61	0.58	0.56
43.3	41.8					0.78	0.75	0.73
40.9	40.9					0.86	0.82	0.79
40.1	37.3	41.0	47.7	44.4	49.3	1.08	1.00	0.99
27.3	26.8					1.22	1.14	1.13
25.1	24.8					1.37	1.29	1.28
25.4	23.6	25.1	30.2	28.1	30.6	1.64	1.54	1.53
16.8	17.1					1.95	1.85	1.83
16.2	15.2					2.18	2.10	2.08
15.9	14.8	15.7	18.9	17.6	18.8	2.55	2.42	2.40
10.6	10.5					3.09	2.96	2.94
10.1	9.46					3.45	3.32	3.30
9.76	9.09	9.67	11.6	10.8	11.3	4.03	3.87	3.85
6.64	6.68					4.92	4.76	4.74
6.32	6.26					5.49	5.33	5.31
6.22	5.80	6.03	7.40	6.89	7.18	6.22	6.01	5.99
4.82	4.54	4.73	5.72	5.38	5.58	7.89	7.66	7.63
3.05	2.87	2.99	3.61	3.40	3.50	12.5	12.0	12.0
1.92	1.81	1.88	2.36	2.22	2.23	19.3	18.5	18.4
1.26	1.19	1.24	1.50	1.41	1.44	32.4	29.7	29.6
0.702	0.661	0.688	0.842	0.792	0.822	56.8	55.2	55.0
0.444	0.418	0.436	0.531	0.500	0.515	90.1	86.8	86.5

3. *Cable strippers* (Taca) are approved and required for removing jackets and outer insulations from multiconductor cables, power cables, etc. *Caution:* Check blade depth adjustment prior to each use.

After stripping, insulation must show a clean termination with no frayed or drawn edges. Slight discoloration after thermal stripping is permissible.

On insulated solid conductor wire that has been stripped, no damage to the conductor is acceptable. On insulated stranded wire, the number of damaged or severed strands in a single lead shall not exceed one strand for 7-stranded wire and two strands for 19-stranded wire.

When stranded conductors are stripped, the strands have a tendency to unwind. If this should occur, the strands should be twisted back into their former position prior to tinning.

LUGGING OF WIRE AND CABLE ASSEMBLIES

Lugs are primarily used to terminate conductors where reliable connections and convenient repair or replacement are required. Many types, sizes and configurations have been developed for industrial use. The following are the most common types used in modern electronics assemblies.

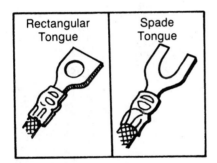
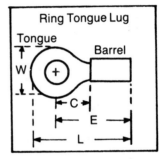

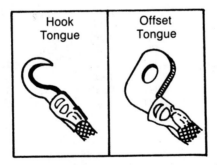

The stripped end of the wire should be inserted into the lug until the insulation rests against the inner shoulder. If no shoulder exists, make sure that the wire insulation is positioned so that it will be held securely when crimped. There should be no evidence of bare wire between the lug barrel and the wire insulation.

The conductor should extend through the lug inspection hole from flush to a maximum of $\frac{1}{18}$ inch, but must never interfere with the installation hardware.

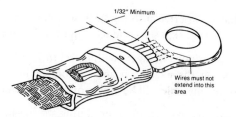

Lugs should be installed on the terminal post in the order of their current carrying capacity. The largest lug will always be on the bottom.

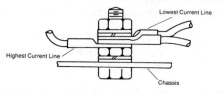

When two wires approach a terminal from the same side, the lugs should be installed back-to-back with the barrel of the bottom lug facing down. Single lug installations will always have the barrel up with the code marks showing.

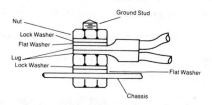

Ground legs should never be terminated to a component or part mounting hardware, nor should they be installed between a conductive and insulative material.

Where space does not allow for lugs, the wire may be stripped, tinned and formed around a binding screw in the direction that will tighten as the screw is installed. (Usually clockwise.)

Lug Termination

Lugs are primarily used to terminate single conductors when reliable connections and convenient repair or replacement are required. Many types, sizes, and configurations have been developed for industrial use.

Solderless terminals

Solderless terminals manufactured by Aircraft Marine Products (AMP) or equivalent are used wherever practicable. These lugs are easily attached and frequently permit a saving of time during assembly and disassembly of units.

Stripped wire shall be inserted into lug until wire insulation rests against inner shoulder. After crimping, exposed conductor must not be visible at lug entry.

Terminals shall be positioned in lugger with barrel toward handle so the crimp or indentation will be on the exact top and bottom of lug barrel. When lugged properly, crimp or indent code will be visible on top of barrel near front.

Lug must be held firmly against tool stop during the lugging operation to avoid movement of terminal or a mis-crimp.

The conductor shall extend through the lug inspection hole from flush to $\frac{1}{16}$ inch, but must never interfere with the installation of proper hardware.

Only approved luggers of the appropriate size shall be used. Lug shall grip both *conductor* and *insulation*. Periodic inspection and calibration shall be maintained to insure proper working action.

The swage or crimp should be entirely on the barrel or grip of the terminal and should be deep enough to hold the wire(s) securely in the terminal. (See Fig. 6-47.

A lug shall be the correct size required for the specific wire gauge and terminal post size. Lugs must not be modified or deformed.

Lugs shall be limited to the termination of a single wire unless otherwise noted.

Lugs shall be installed on a terminal post in the order of their current-carrying capacity, with the largest on the bottom.

When two wires approach a terminal post from the same side, the lugs shall be installed back to back with the barrel of the bottom lug facing down. Single lug terminations shall have barrel up (i.e., code marks shall be visible).

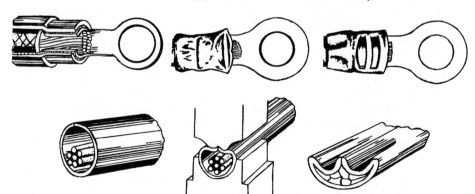

Fig. 6-47. Crimp terminal installation.

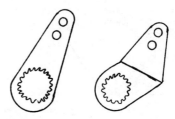

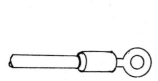

Fig. 6-48. Insulation support.   **Fig. 6-49.** Noninsulated support.

Lugs shall not ride terminal nodes or barrier posts. In no case will preloaded, cracked, or broken terminal strips or posts be acceptable.

Ground lugs should not be terminated at component- or part-mounting hardware nor between a conductive and insulative material.

Where space does not allow for lugs such as power plugs, etc., wire may be stripped, tinned, and formed around the binding screw in the direction that will tighten wrap while tightening the screw (generally clockwise).

Solder-Type Terminals

Solder-type terminals come in two types: insulation support, as illustrated in Fig. 6-48, and a noninsulated support, as illustrated in Fig. 6-49.

The noninsulation supporting terminals are usually flat and contain one or more punched holes for conductors.

Where practical, the wire(s) should come into the terminal from the back or bottom side, as illustrated in Fig. 6-50, so that the insulation is away from the soldering iron when soldering the terminal.

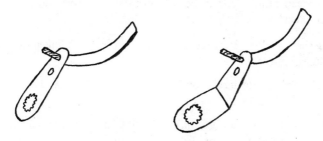

Fig. 6-50. Preferred noninsulated terminal assembly.

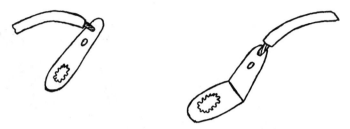

Fig. 6-51. Alternate assembly method for noninsulated terminals.

If the preceding method cannot be used, the wire(s) may then be brought into the terminal from the top or front side, as illustrated in Fig. 6-51. In this case, the wire insulation is no more than $\frac{1}{16}$ inch away from the edge of the terminal.

Crimp Termination

Terminals for removable contact connectors shall be crimped and installed using the appropriate tools. Substitutions are prohibited without prior approval from the manufacturing engineering department.

Crimping or installation tools must not be modified or adjusted in any manner except by authorized personnel.

Terminals or pins shall be positioned correctly in tool locator. Stripped conductor shall be bottomed in terminal and held firmly until crimping operation is completed.

Insulation shall be up to but not beyond edge of pin.

Crimped terminals shall not be bent or otherwise deformed before or after installation. Damaged terminals shall be removed and replaced, using care to avoid damaging connectors.

Connectors shall be free of cracks and all foreign material and must mate with corresponding part easily.

When no other means of insulation is provided, sleeving shall be installed over pins.

Connector must be handled and installed with care to avoid breaking brittle insulated case or connector body. *Do not force fit.*

Shield Termination

Shield termination is the connection or joining of the shields and the conductors therein without damage to either the conductor, insulation, or shield. This shield usually comes in the form of fine, tinned, copper wire braided around and over the primary conductor and the dielectric (the insulation between the wire and the shield).

There are several methods for preparing a metal shield of a wire for termination. The most common methods are shown in the following pages.

Shield terminations shall not cause damage nor be a potential cause of damage to conductor insulation.

The length of unshielded conductor shall be as short as possible, but shall not exceed 2 inches unless otherwise noted.

Thermal strippers shall be used to remove insulations from coaxial cables and shielded wire.

Strip outer insulation to expose required amount of shielding.

Push the shielding back so that it loosens about the cable. (See Fig. 6-52.)

Use a blunt-pointed tool and work a hole in the braid at the point where the pigtail will leave the wire. Do not break any of the strands. (See Fig. 6-53.)

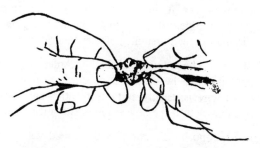

Fig. 6-52. Shield preparation—loosen shield.

Fig. 6-53. Shield preparation—open shield.

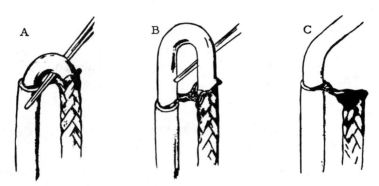

Fig. 6-54. Shield preparation—pigtailing shield.

Bend the wire so that the outside of the lead is opposite the hole. Work the shield back on both ends of the wire, then extract the wire through the hole with a blunt, round tool (awl or soldering aid). Do not damage insulation or shielding. (See Fig. 6-54.)

Straighten the wire, and smooth shield down on the wire. Stretch the shield to its entire length.

A hole is then opened in the shield from $\frac{1}{2}$-inch minimum to $\frac{3}{4}$-inch maximum from the point the shield leaves the wire.

The shield should not be cut at this point. The extra length of shield is used to hold the strands of the shield in place until the soldering operation has been completed, thus eliminating the possibility of the shield flaring.

Strip the ground $\frac{1}{4}$ inch. The ground lead shall be American Wire Gauge (AWG) stranded black wire and shall be of the same current-carrying capacity as the shield. AWG No. 20 wire is usually suitable for this application. Insert the stripped portion of the wire into the hole made in the shield. (See Fig. 6-55.)

Pull the shield in the opposite direction as the shield will fit snugly over the stripped portion of the ground lead. Bend the shield at a right angle, and even it

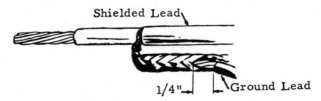

Shielded Lead

1/4" Ground Lead

Fig. 6-55. Ground lead to shield termination.

Fig. 6-56. Ground lead installation.

with the ground lead insulation level. Shape the shield, where the ground lead is inserted, to form a round form. (See Fig. 6-56.)

Flow solder to a point where the ground lead is inserted. Solder should be visible on all surfaces of the connection and must be sweated on rather than built-up, as this would constitute an excessive amount of solder. (See Fig. 6-57.)

The excess should be trimmed off following the contour of the connection and gently round any sharp points with needle. (See Fig. 6-58.)

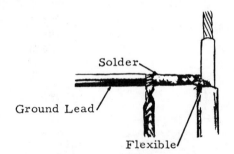

Fig. 6-57. Ground lead termination.

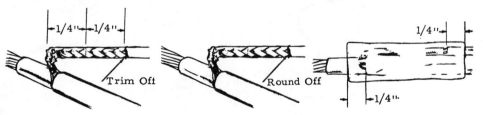

Fig. 6-58. Trimming excess shield.

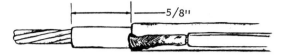

Fig. 6-59. Insulation stripping.

Unless otherwise specified, insulation should be stripped from flush to $\frac{5}{8}$-inch maximum from the conductor where the shield leaves the wire. It is advisable to adhere to the $\frac{5}{8}$-inch maximum whenever practical to facilitate ease of servicing. (See Fig. 6-59.)

When it becomes impractical to use conventional methods, shields can be terminated as shown in Fig. 6-60. Shields must be insulated with shrink sleeving from one another when shields are at a different potential or when the method shown in the figure is used to decrease bulkiness of wires in connectors and/or cables.

When the engineering drawing indicates several pigtails and/or unnumbered pigtail wires connected to a common point, the shields may be connected together in an approved manner. (See Fig. 6-61.)

Shielded wire ferrules may be used for grounding in electronic equipment only when specified in the engineering drawing. Two sleeves are used for each assembly. The inner (smaller) sleeve is slipped under the braided shield, while the outer (larger) sleeve is slipped under the shield and ground lead. The inner sleeve acts to prevent damage to the insulation during installation. AWG stranded

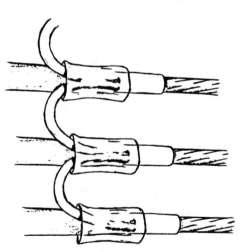

Fig. 6-60. Multiple shield terminations.

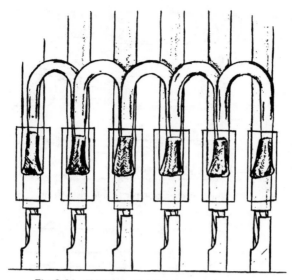

Fig. 6-61. Shield terminations to a common point.

black wire is usually suitable for this application. The ground shall be of the same current-carrying capacity as the shield. The sleeves also anchor the shield and prevent fraying and slipping.

To prepare the leads for ferrules, cut the braid off approximately 1 inch from the end of the conductor insulation making sure that the conductor insulation is not damaged in the process. Place the outer sleeve over the braid and rotate the cable in a circular motion to flare the braid and enable the inner sleeve to slip under the braid, extending a maximum of $\frac{1}{16}$ inch from the end of shield. Push both the outer sleeve and the groundwire over the entire shield and inner sleeve, and install with proper hand or bench tool. The insulation on the ground wire shall be even with the ferrule or may have a maximum exposed wire length of $\frac{1}{16}$ inch. (See Fig. 6-62.)

In order to determine the proper size of ferrules to use, the following method is suggested.

Measure the outer diameter of the conductor insulation. For two conductor cables, measure the widest point across the insulation of the two conductors.

Add 0.025 inch, and select an inner sleeve with a total diameter equal to the total of these two figures. If this does not give a standard size, select the next largest size.

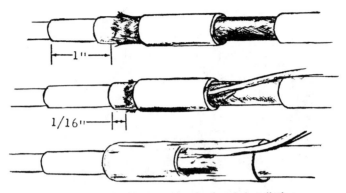

Fig. 6-62. Shield preparation for ferrule installation.

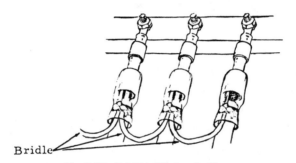

Fig. 6-63. Shield bridle termination.

To determine the outer sleeve, measure the outer diameter of the inner sleeve, and add 0.025 inch to this figure for a single braided shield.

If a ground wire is to be used, add 0.04 inch for a No. 18 wire size or smaller. The total of these figures gives the inner diameter of the outer sleeve. If the outer sleeve is not stranded, select the next largest size.

Ferrules shall be installed a minimum of $\frac{1}{2}$ inch and not more than $\frac{3}{4}$ inch from the terminal.

Pigtails may be used to form a bridle to reduce termination difficulties. (See Fig. 6-63.)

Ribbon-Cable Termination

In ribbon-cable termination, there are two different types of connectors. The methods for assembling these connectors are given in the following pages.

Connectors must be assembled either with an assembly press or with an arbor press or similar parallel-action device fitted with suitable fixtures.

Printed-Circuit–Board Type Connectors

General

The connector body must be soldered to the printed-circuit board *before* assembly. The heat and pressure applied to an assembled connector during soldering can cause internal misalignment and shorting of conductors.

Standard hand-, wave-, or flow-soldering techniques can be used on the unassembled connector. The connector body must be weighted during wave soldering to prevent lifting.

Assembly

1. If an assembly press is used, it should be checked for adjustment. Refer to the instruction sheet included with the press for adjustment procedure and to the chart for correct settings.
2. Cut cable or individual wires to length if connector is being used to terminate. Cable should always be cut with a vertical shear device. The sheared end should be 90° to cable edge. (See Fig. 6-64.)

Note: Do not cut cable with a scissors-type device nor trim cable or wires flush with connector after assembly. These actions can dislodge or smear the ends of the conductors, resulting in poor connections.

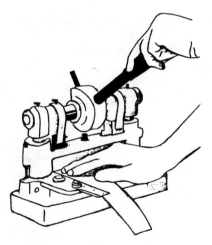

Fig. 6-64. Cut cable with vertical shear.

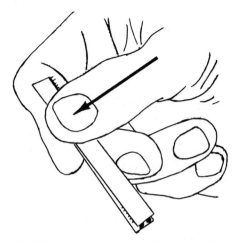

Fig. 6-65. Remove protective liner from cable.

3. Remove protective liner from adhesive on cover. Lateral thumb pressure on linear will displace it enough to permit it to be grasped and peeled off. (See Fig. 6-65.)

Note: Connector covers for use with individual wires have deeper grooves to hold wires in position during assembly and do not have adhesive. Individual wire covers should be used only with individual wires, and flat cable covers should be used only with flat cable.

4. Press deeply ribbed side of cable or individual wires into alignment grooves of cover. Use a hand tool or finger pressure to seat cable firmly into cover. Check visually for correct alignment of cable ribs in the grooves of the cover. (See Fig. 6-66.)

5. Place printed-circuit board into position on press so that solder pins of connector mate with relief slots in locator plate. (See Fig. 6-67.)

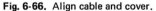

Fig. 6-66. Align cable and cover.   **Fig. 6-67.** Arbor press.

Fig. 6-68. Align cover and cable over connector.   **Fig. 6-69.** Press cable and connector.

6. Place cover and cable on connector body. Align the holes in the cover with the posts on the body, and press into place with light finger pressure. (See Fig. 6-68.)

7. Lower handle of press to complete connection. The handle of the assembly press should be lowered until the arm contacts casting. (See Fig. 6-69.)

8. Remove assembly and inspect to insure that:

   a. Cover is fully seated and is parallel with base.

   b. Cable ribs or wires align with grooves in cover.

   c. Cable or wires are even with the edge of the connector if it is used to terminate.

Wrap-Post Connectors

Assembly

1. See p. 201, for steps 1 through 3, and follow the information given there.

2. Press deeply ribbed side of cable or individual wires into alignment grooves of cover.

3. Place cover and wire or cable into position on locator (base of press), with wiring up.

4. If a flat cable is being used, press it firmly into position with a hand tool or finger. Check visually for correct alignment of cable ribs in the grooves of the cover. (See Fig. 6-70.)

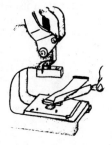

Fig. 6-70. Check cable/connector alignment.

Fig. 6-71. Remove assembly from press.

5. Place connector body into press with "U" contacts down. Light finger pressure will insure proper alignment of body with locating pins on cover. (See Fig. 6-71.)
6. See p. 203, and follow the information given there.
7. Raise handle, and remove assembled unit by lifting on connector. Do not remove by pulling on cable.
8. Inspect visually to insure that:
   a. Cover is fully seated and is parallel with base.
   b. Cable or wires align with grooves in cover.
   c. Cable or wires are even with the edge of the connector if it is being used to terminate.

Coaxial Cable Termination

Coaxial Cables

Coaxial cables are used for many specialized purposes. They are commonly used where radio-frequency power, pulse, synchronizing, video- or audio-signal energy must be channeled to another circuit or distant location. Coaxial cable is also used to carry television programs between stations and offers minimum losses in a circuit. It reduces radiation and prevents noise-energy pickup along its length, when properly assembled, it presents a completely enclosed and shielded circuit whose losses can be closely calculated from available data.

The insulation (polyethlyene dielectric) may become soft at approximately the temperature of boiling water. Particular care should be taken during the soldering operation to handle the cable in such a manner that the conductor cannot change its location while the dielectric is in a plastic state. (See Fig. 6-72.)

Coaxial cable must be protected against small radius bends. Bending radii less than 10 times the maximum cable diameter can cause trouble.

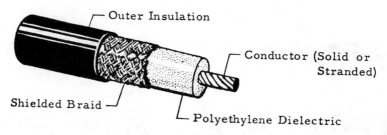

Fig. 6-72. Typical coaxial cable.

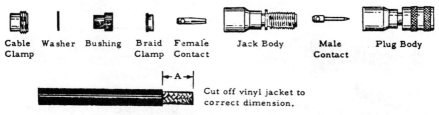

Fig. 6-73. Typical coaxial cable termination hardware.

Figures 6-73 and 6-74 show the assembly of a typical series BN coaxial connector. Table 6-4 should be used as a reference to determine the correct length of A, B, and C.

Special care should be taken to insure that the shield braid is carefully folded back on the braid clamp to make contact for its entire circumference. The braid must continue an unbroken circuit to the connector body with no wire separation at any point. The return circuit and shielding could be rendered inadequate and the circuit operation may become faulty due to radiation losses or noise pickup or poor electrical connection.

Table 6-4. Cable Prep for Coaxial Connector.

Dimension	RG-/U Cable	Plug	Jack
A	58, 58B	$\frac{7}{16}$	$\frac{15}{32}$
	59A, 62A, 71	$\frac{23}{32}$	$\frac{3}{4}$
B	58, 58D	$\frac{5}{16}$	$\frac{11}{32}$
	59A, 62A, 71	$\frac{19}{32}$	$\frac{5}{8}$
C	58, 58B, 59A	$\frac{1}{4}$	$\frac{9}{32}$
	62A, 71		

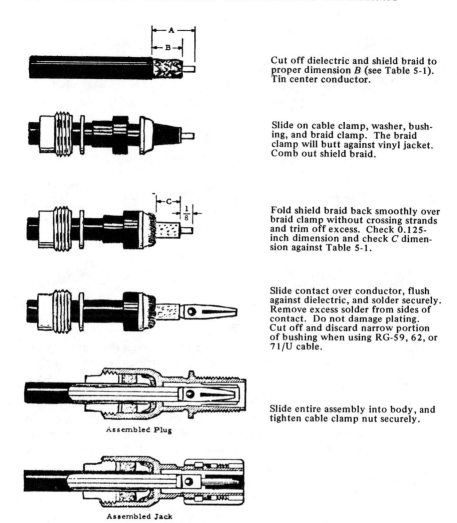

Cut off dielectric and shield braid to proper dimension *B* (see Table 5-1). Tin center conductor.

Slide on cable clamp, washer, bushing, and braid clamp. The braid clamp will butt against vinyl jacket. Comb out shield braid.

Fold shield braid back smoothly over braid clamp without crossing strands and trim off excess. Check 0.125-inch dimension and check *C* dimension against Table 5-1.

Slide contact over conductor, flush against dielectric, and solder securely. Remove excess solder from sides of contact. Do not damage plating. Cut off and discard narrow portion of bushing when using RG-59, 62, or 71/U cable.

Slide entire assembly into body, and tighten cable clamp nut securely.

Assembled Plug

Assembled Jack

Fig. 6-74. Coaxial cable connector assembly.

When handling coaxial cable, take care to avoid denting, distortion, and kinking of cable. When ties are absolutely necessary, the tie shall fit in such a manner that the outer conductor is not in any way deformed.

Keeping the cable at elevated temperatures longer than required can cause shift of the conductor within the dielectric, resulting in troublesome circuit operations. When soldering coax cables, keep in straight line until solder joint has cooled; then form.

Shielded Cable Terminations

Pigtail Preparation

The insulation is stripped exposing from $\frac{3}{16}$" to $\frac{1}{4}$" of bare wire as shown in Fig. 6-75.

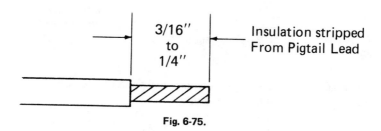

3/16"
to
1/4"

Insulation stripped
From Pigtail Lead

Fig. 6-75.

Strip the cable jacket $\frac{11}{16}$" to $\frac{3}{4}$" to expose the braided shielding. Trim the braided shield back exposing a minimum of $\frac{1}{2}$" of the inner insulation. From $\frac{3}{16}$" to $\frac{1}{4}$" of braided shield is now exposed as shown in Fig. 6.76.

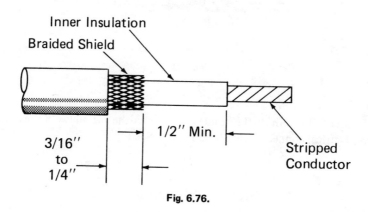

Inner Insulation

Braided Shield

1/2" Min.

3/16"
to
1/4"

Stripped
Conductor

Fig. 6.76.

Fan the end of the braided shield and loosen it enough to allow an insulation sleeve to be inserted under the shield. The sleeve should be a maximum of 0.020" greater than the wire O.D. Slide the sleeve back until it butts firmly against the inner shoulder of the braided shield. See Fig. 6-77. (The sleeve should extend approximately 0.06 beyond the end of the braided shield.)

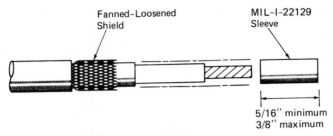

Fig. 6-77.

Install a pigtail lead between the braided shield and the sleeve as shown in Fig. 6-78. Trim and straighten the braided shield.

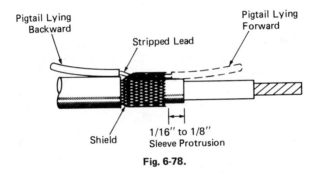

Fig. 6-78.

Solder the full length of the braided shield and the pigtail lead to insure that the best electrical connection has been made. (See Fig. 6-79.)

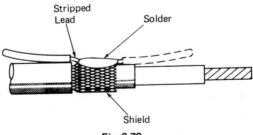

Fig. 6-79.

Cover the pigtail lead termination with heat-shrinkable insulation tubing as shown in Fig. 6-80.

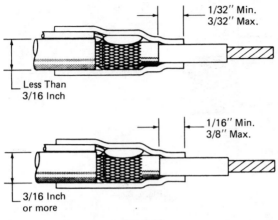

Fig. 6-80.

Dead-Ended Floating Shield Terminations

Shield Preparation

The shield should be loosened and pushed toward the cable jacket so that it forms a ridge approximately $\frac{3}{8}''$ from the end of the jacket. (See Fig. 6-81.) The shield is than cut leaving approximately $\frac{3}{8}''$ exposed beyond the cable jacket.

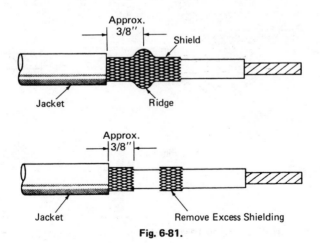

Fig. 6-81.

Fold the braided shield over the cable jacket as shown in Fig. 6-82 and trim the shield to $\frac{1}{8}''$ to $\frac{3}{16}''$.

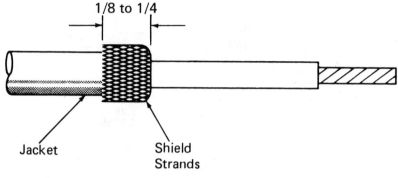

1/8 to 1/4

Jacket

Shield Strands

Fig. 6-82.

The dead-ended (floating) shield is than covered with heat-shrinkable sleeving. The heat-shrinkable insulation sleeving is centered over the braided shield with the front end extending approximately $\frac{1}{8}''$ to $\frac{1}{4}''$ beyond the shield. (See Fig. 6-83.)

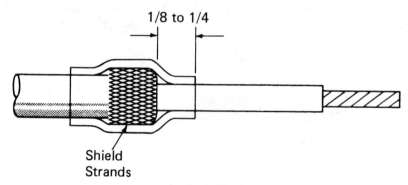

1/8 to 1/4

Shield Strands

Fig. 6-83.

Wire Routing

General

Wiring routing, as discussed in this section, pertains to wiring done directly to a chassis or assembly and not to a prefabricated cable or harness assembly.

Wire shall be routed to facilitate efficient operation and ease of fabrication. It shall have a neat appearance. The following will serve as a guide:

1. Drawing requirements
2. Assembly notes
3. Wire lists
4. Manufacturing operation sheets
5. Previously fabricated units.

Point-to-point (PTP) wiring shall take the most practical direct path from one terminal to another. No service loop is required but no strain shall exist on wire or terminals. (See Fig. 6-84.)

Routed wiring, unless otherwise noted, shall pass through the nearest, convenient bundle or channel to its destination. Service loops are required on all but wire-wrap terminations. (See Fig. 6-84.)

Service loops shall be sufficient to allow wire to be cut from terminal, restripped, and soldered or lugged once without pulling wire tight.

Service loops shall follow the natural curve of the wire and shall not crowd adjacent terminals or form an S shape.

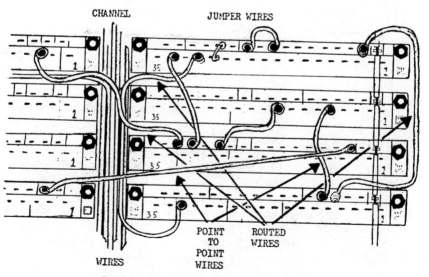

Fig. 6-84. Typical point-to-point wire routing.

Where possible, wires shall be routed in bundles in such a manner as to provide neatness and minimize individual spot ties.

Wiring shall be dressed paralleled and as straight as possible, with a minimum of crossover. All breakouts, bends, and turns shall be in the form of a smooth curve. *They are not to have sharp bends!*

Cable and Harness Assemblies

Cable Assemblies

Cable assemblies are normally used to transmit current, or signal from one assembly to another within a system, or from the system to an external source.

Cables generally consist of two or more conductors in a common covering, terminated at least on one end with a movable connector.

Usually, the connector or paddle board on one end is prewired with predetermined length wires, laid out on a bench or cable board to determine the exact overall length, and the second end terminated to a connector, paddle boards, terminal lugs, etc. (See Fig. 6-85.)

Cables covered with vinyl or zipper tubing shall be spot-tied outside the tubing and only at the cable ends.

Fig. 6-85. Typical paddleboard/connector cable.

I/O Cable Breakout Procedure

At the point of the I/O cable construction, where the wires have been laid out and strung to their proper lengths, terminate the breakouts as shown in the following steps.

1. Apply shrink tubing over each branch of the breakout. It must extend at least 1.5 inches on each side of location of the finished breakout. (See Fig. 6-86.)
2. Apply a 2-inch length of shrink tube on the main trunk of cable, with one end about 0.75 inch away from the breakout. (See Fig. 6-87.)
3. Apply braided shield as indicated on reference drawing. Allow the cut end of the shield to stop exactly at the breakout. (See Fig. 6-88A.)

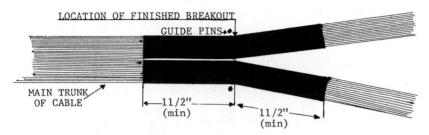

LOCATION OF FINISHED BREAKOUT
GUIDE PINS

MAIN TRUNK
OF CABLE

←—11/2"—→
(min)

←—11/2"—→
(min)

Fig. 6-86. Cable breakout.

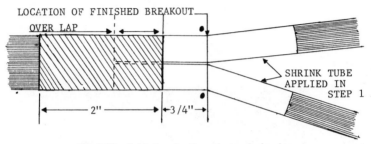

LOCATION OF FINISHED BREAKOUT
OVER LAP

SHRINK TUBE
APPLIED IN
STEP 1

←————— 2" —————→ ←3/4"→

Fig. 6-87. Cable breakout—main trunk sleeving.

Spot-tie shield tightly (to hold in place) 1 inch from breakout. (See Fig. 6-88B.)

Fold the end of the shield back over the spot tie smoothly and tight. (See Fig. 6-88C.)

Caution: Do not allow the wires in the shield to become unwoven or ragged.

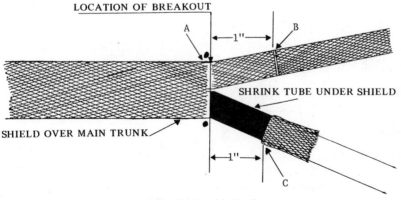

LOCATION OF BREAKOUT

A

←—1"—→

B

SHRINK TUBE UNDER SHIELD

SHIELD OVER MAIN TRUNK

←—1"—→

C

Fig. 6-88. Shield cable breakout.

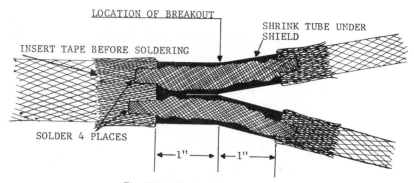

Fig. 6-89. Sleeving braid preparation.

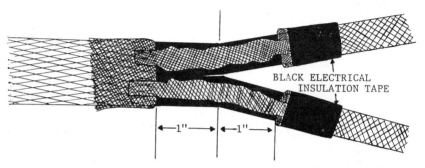

Fig. 6-90. Tape shield fold back.

4. Solder $\frac{1}{4}$-inch ground braid across unshielded gaps to create full continuity from all shields to ground.

Solder joint *must* be as flat and neat as possible and must be to folded back part of shield *only*.

Place folded piece of tape under fold back of shield to restrict solder flow. Widen ground braid before soldering to provide flexibility. (See Fig. 6-89.)

5. After inspection of solder joints, wrap each one with 0.75-inch black electrical insulation tape. (See Fig. 6-90.)

Harness-Making and Installation

The harness is a number of individual wires assembled into a cable, or series of branching cables, that will fit into a specific instrument. It is carefully made so that when the harness is installed, the free ends of its wires are neatly adjacent to their respective terminals and long enough to provide a service loop. When

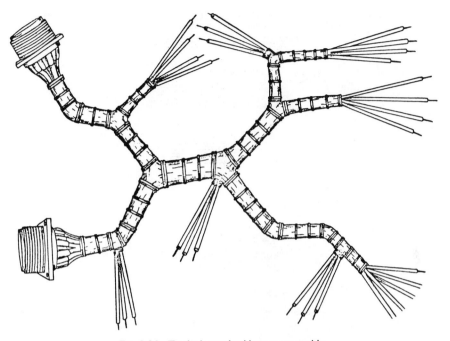

Fig. 6-91. Typical prewired harness assembly.

specified, the ends of the wires are fitted with terminals, lugs, taper pins, etc. The harness, as opposed to custom wiring, saves time in manufacture, reduces the chances of error, and assures uniform, good-quality instrument wiring. (See Fig. 6-91.)

Harness assemblies are usually fabricated on a harness board that has the exact configuration of the harness laid out showing the main trunks, branching trunks, breakouts, and terminating points of the wires. Harness posts or nails are driven into the board to form the harness paths. Following assembly instructions, the assembler lays in, routes, and spot-ties the wires on the harness board.

When two or more conductors are to be grouped together, they shall lie parallel to one another unless otherwise specified by engineering drawings or assembly instructions.

Conductors shall lie parallel to one another and shall not entwine other conductors. (See Fig. 6-92A.)

Crossovers must not cause excessive strain on insulation and should be avoided at breakouts where possible. (See Fig. 6-92B.)

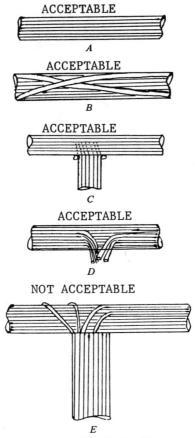

Fig. 6-92. Wire lay in cables.

Conductors shall break out at the bottom of the cable bundle wherever possible. (See Fig. 6-92C.)

Conductors may break out from top and center of bundle. This condition is not desirable but is acceptable if it cannot be avoided. (See Fig. 6-92D.)

Excessive and unnecessary crossover of wires to form a breakout is shown in Fig. 6-92E. This condition is not acceptable.

The drawing in Fig. 6-93 reflects examples of poor workmanship.

Wires shall be run to a common supported terminating point and shall not be spliced *unless otherwise specified.*

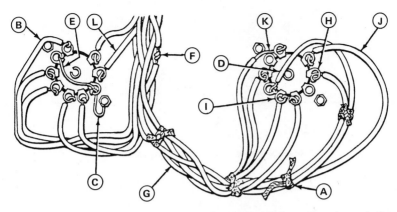

Fig. 6-93. Examples of poor workmanship. A, Poor tying. B, Screw threads are chafing the insulation. C, Wire end is grounding on mounting screw. D, Wire insulation being cut by sharp edge. E, Jumper wire is not protected by sleeving. F, Splice. G, Poor bundling. H, Wire is not wrapped around terminal. I, Insulation is dressed into terminals. J, Excessive wire length. K, Excessive insulation stripping is causing short. L, No slack in wire.

An attempt to connect wires before they are dressed in their proper position will result in an inferior product. Clamps and/or string ties shall be used to keep the harness and its wires properly dressed. Without clamps, the "bouncing" characteristically taken by an electronic unit in normal service could be detrimental to the functioning of the harness wires.

Leads shall be run by the shortest practical route to avoid excessive lengths of wire. The ends of all flexible wires shall have sufficient slack to avoid breaking wires or solder connections due to strain or vibration. (See Fig. 6-94.)

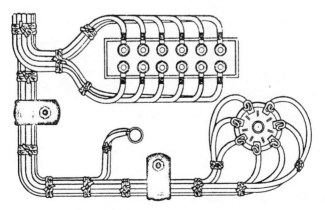

Fig. 6-94. Typical harness installation.

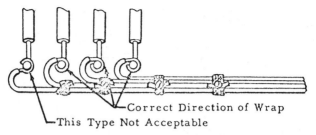

Correct Direction of Wrap
This Type Not Acceptable

Fig. 6-95. Cable lead breakouts with strain relief.

Lead breakouts and spot-ties shall be as pictured in Fig. 6-95, and care shall be taken so that correct direction of wrap is attained.

Wires in a cable of harness shall lay essentially parallel and not entwine other conductors with exception to twisted pairs used in filament leads, etc. Leads shall be branched in such a manner as to eliminate bulkiness in breakouts. (See Fig. 6-96 and 6-97.)

Simple Wire Bundle Calculations

Wire harnesses and cables are often left to the Manufacturing Engineer. For the initial planning phase in production the wire bundle diameter is required to establish the harness or cable routing. The following method is a simple analytical method for determining the wire bundle diameters.

For wires of the same diameter the outside diameter of the wire bundle which will pass through a circular hole of a specified size the following formula can be used:

Where:

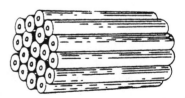

D = Diameter of wire bundle
d = Diameter of a single wire
N = Total number of wires

a. The expression for the outside diameter of a bundle of wires of the same diameter is:

$$D = 1.155d\sqrt{N}$$

b. The formula for the total number of wires of the same diameter which will pass through a given circular opening is:

$$N = \pi/4 \frac{D^2}{d^2} - 1$$

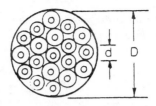

When a wire bundle consists of wires of different sizes (gauges) the outside diameter of the total wire bundle can be calculated by the following equations:

Let:

D = Outside diameter of wire bundle in inches
d_1 = Diameter of wire of first size
d_2 = Diameter of wire of second size
d_3 = Diameter of third size, etc., etc.
N_1 = Number of wires of first size
N_2 = Number of wires of second size
N_3 = Number of wires of third size, etc., etc.

$$D = \sqrt{21.3 N_1 (d_1)^2 + 21.3 N_2 (d_2)^2 + 21.3 N_3 (d_3)^2 \cdots 0.0625}$$

Example: Given 15 wires of 0.080 diameter, 8 wires of 0.120 diameter, and 20 wires of 0.150 diameter, find the outside diameter of the resulting wire bundle.

$21.3 \times 15 \times (0.08)^2 = 2.055$
$21.3 \times 8 \times (0.12)^2 = 2.454$ $D = \sqrt{(14.094)(0.0625)} = 0.938$ inches
$21.3 \times 20 \times (0.15)^2 = \underline{9.585}$
$ 14.094$

Allowable Bend Radius for Cables and Harnesses

The bend radius for cables and harnesses should be calculated and allowed for during the design phase to ensure that there are proper clearances and access for the wiring operations, and all cables and harnesses should not interfere with other parts.

The three controlling factors in cables and harnesses are:

- Unsheathed harnesses without coaxial cables: the minimum bend radius should be 10 times the outside diameter of the trunk (or branch), except that termination points may be 3 times the outside diameter when they are properly supported.
- When coaxial cable are used: the bend radius may be 5 times the outside diameter of the harness or cable.
- Harnesses and cables that cannot meet the minimum bend radius defined

above should be sheathed in their entirety in an insulation tubing without lacing, cord, or tape.

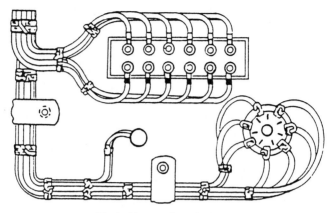

Typical harness installation.

Spot-Ties and Adjustable Cable Straps

The spot-tie or string tie is used almost exclusively for holding wire bundles together and in a desired position. The clove hitch is the most acceptable knot due to its self-holding ability. The double topknot is to allow for shrinking, which is prevalent with nylon cord. Tie each spot-tie next to the previous tie, then slip it forward to the desired location and tighten with a square knot. This procedure will aid greatly in maintaining a neat harness with straight parallel wires.

The use of individual spot-ties in electronic equipment is preferred to the running-stitch method because field-servicing is expedited by the removal and replacement of spot-ties only in the affected areas.

Black nylon lacing cord (ribbon type) shall be used in preference to round cord. This will reduce the possibility of cutting into the wire insulation or the possibility of spot-ties becoming loose in the handling of laced cables. Best suited for this purpose is 0.175-inch nylon or similar ribbon.

Nylon-ribbon lacing cord is available in several widths. Care shall be taken to use a cord width best suitable for the size cable to be tied. For example:

Approximate Cord Width (in.)	*Approximate Cable Diameter* (in.)
$\frac{1}{32}$	Up to 0.5
$\frac{1}{16}$	0.5 to 1.0
$\frac{1}{8}$	1.0 to 2.0

NOT APPROVED

APPROVED

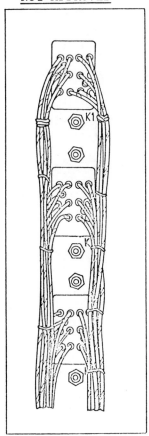

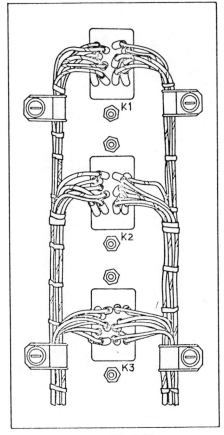

Fig. 6-96.

Fig. 6-97.

Fig. 6-96. Improper spot-ties: cable clamps omitted, sleeving not used, wires entwined in cable, no service loops, and terminals not accessible.

Fig. 6-97. Proper spot-ties: cable clamps used, sleeving used on closely spaced terminals, proper service loops, and terminals left accessible.

To prevent excessive waste, spot-tie cord should not be cut from spool until after completion of each tie.

Spot-ties shall not be used or placed inside conduit or sleeving.

The preferred method of forming a spot-tie is illustrated in Fig. 6-98. A clove hitch is formed around the cable, and a surgeon's knot is added to secure the spot-tie.

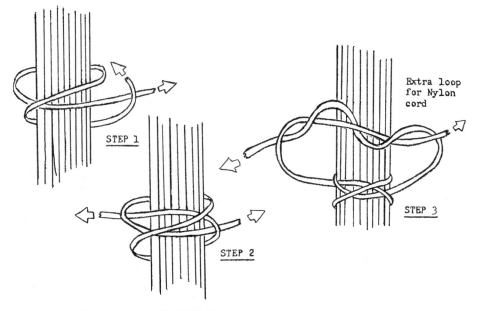

Extra loop
for Nylon
cord

STEP 1

STEP 2

STEP 3

Fig. 6-98. Preferred spot-tie knot.

The alternate method of forming a spot-tie is shown in Fig. 6-99. A clove hitch is formed around the cable, and a square knot is added to secure the spot-tie.

A spot-tie or adjustable cable strap shall be tight enough to hold conductors in place without deforming the insulation. (See Fig. 6-100.)

A spot-tie or adjustable cable-strap installation shall be unacceptable when the insulation is deformed as shown in Fig. 6-101.

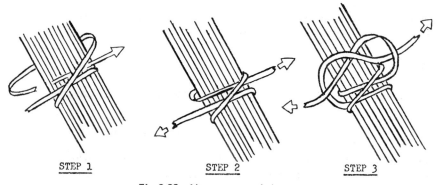

STEP 1

STEP 2

STEP 3

Fig. 6-99. Alternate spot-tie knot.

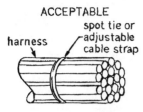

ACCEPTABLE

Fig. 6-100. Spot-tie.

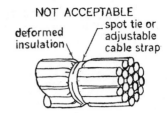

NOT ACCEPTABLE

Fig. 6-101. Unacceptable spot-tie.

The distance between spot-ties should be approximately twice the diameter of the bundle, but shall be no closer than necessary to hold harness neatly in place.

General requirements for ties are shown in Fig. 6-102.

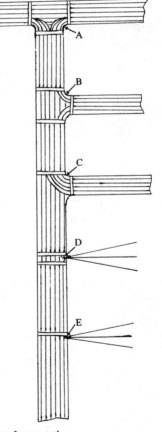

A. *Typical major junction:* Shall be spot-tied on each side of the junction and as close as practical to the junction.

B. *Typical minor junction:* Wires from two directions shall be spot-tied on each side of the junction and as close as practical to the junction.

C. *Typical minor junction:* Wires from one direction shall be spot-tied on one side and as close as practical to the junction.

D. *Typical breakout:* Wires from two directions shall be spot-tied on two sides and as close as practical to the junction.

E. *Typical breakout:* Wires from one direction shall be spot-tied on one side as close as practical to the junction.

Fig. 6-102. General requirements for spot-ties.

A. *Harness Trunk:* The main path or paths from which wires or groups of wires will branch off at junctions or breakouts.

B. *Junction:* (1) Major junction: when 30 to 50% of the wires branch off the trunk. (2) Minor junction: when 10 to 30% of the wires branch off the trunk.

C. *Breakout:* When one or more wires branch off the trunk and fan out to termination.

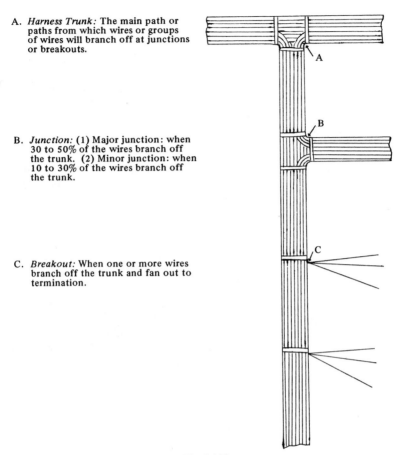

Fig. 6-103.

Definitions to supplement the previous are shown in Fig. 6-103.

Spot-tie ends shall be clipped approximately $\frac{1}{16}$ to $\frac{1}{8}$ inch long to allow for shrinkage while maintaining a neat appearance.

Location of spot-ties or adjustable cable straps on a harness bend is illustrated in Fig. 6-104.

Spot-ties or adjustable cable straps shall not be spaced so far apart that conductors will protrude from the harness bend as shown in Fig. 6-105.

When a breakout leg is formed off the main harness, spot-ties or adjustable cable straps shall be located as close to the breakout points as possible. (See Fig. 6-106.)

ACCEPTABLE

harness

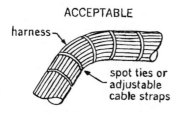

spot ties or
adjustable
cable straps

Fig. 6-104. Acceptable spot-
ties for cable bend.

NOT ACCEPTABLE

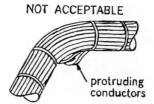

protruding
conductors

Fig. 6-105. Unacceptable spot-
ties for cable bend.

ACCEPTABLE

spot ties or
adjustable
cable straps

main
harness

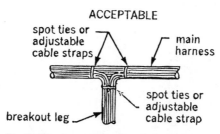

breakout leg

spot ties or
adjustable
cable strap

Fig. 6-106. Acceptable spot-ties for breakout.

Figure 6-107 illustrates spot-ties or adjustable cable straps located too far from
the harness breakout points and shall be unacceptable.

A spot-tie or adjustable cable strap shall be located a sufficient distance away
from a connector to prevent strain on the conductors. (See Fig. 6-108.)

Figure 6-109 illustrates a spot-tie or adjustable cable strap located too close
to the connector and causing strain on the conductors. This condition shall
be unacceptable.

When a service loop is formed to a connector, the spot-tie or adjustable cable
strap shall be located so that the connector may be coupled and uncoupled without strain on the conductors. (See Fig. 6-110.)

NOT ACCEPTABLE

main harness

spot ties or
adjustable
cable straps

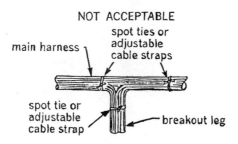

spot tie or
adjustable
cable strap

breakout leg

Fig. 6-107. Spot-ties too far from breakout.

ACCEPTABLE

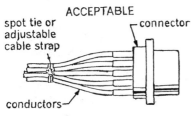

Fig. 6-108. Spot-tie for connector termination.

NOT ACCEPTABLE

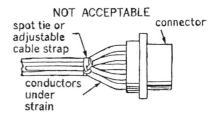

Fig. 6-109. Spot-tie too close to connector termination.

ACCEPTABLE

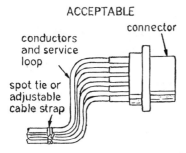

Fig. 6-110. Spot-tie with strain relief.

NOT ACCEPTABLE

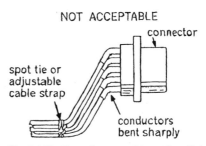

Fig. 6-111. Spot-tie to provide strain relief.

Figure 6-111 illustrates the routing of conductors without sufficient service loops, causing the conductors to bend sharply. This condition shall be unacceptable.

Continuous spot-ties (broom stitches) shall be used wherever it is necessary to reduce a harness diameter. The knot(s) shall be located between the harness bundles, except for the final knot, which may be on top of the bundle. (See Fig. 6-112.)

ACCEPTABLE

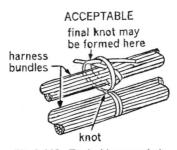

Fig. 6-112. Typical broom stitch.

ACCEPTABLE

NOT ACCEPTABLE

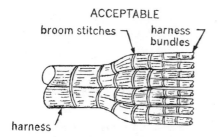

Fig. 6-113. Multiple broom stitch.

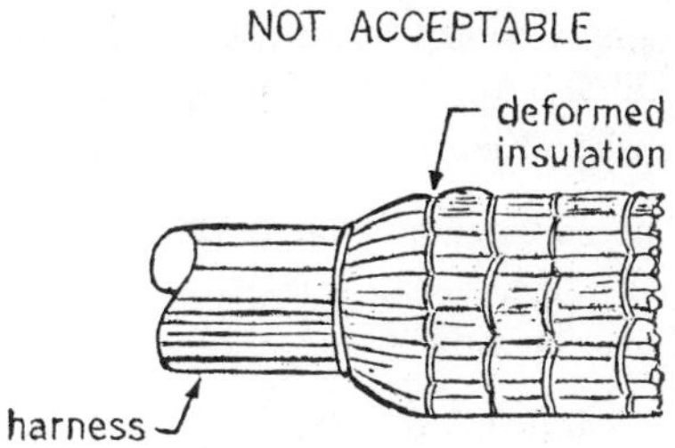

Fig. 6-114. Broom stitch too tight, causing harness deformation.

A large-diameter harness shall be separated into successively smaller bundles using continuous spot-ties (broom stitches) as shown in Fig. 6-113. The distance between the spot-ties should be approximately twice the diameter of the bundle, but shall be no closer than necessary to hold harness neatly in place.

Figure 6-114 illustrates (1) an unacceptable method of separating a harness into successively smaller bundles and (2) deformed insulation caused by overtight spot-ties.

Adjustable cable straps are also used for holding wire bundles together. Cable straps are available in varous sizes. The proper size to use is in relation to the diameter of the wire bundle to be secured.

The adjustable cable strap shall be positioned on the cable assembly as shown in Fig. 6-115A, with the rib side inside. The tail end of the self-locking cable strap shall be pulled through the eye of the head as shown in Fig. 6-115B.

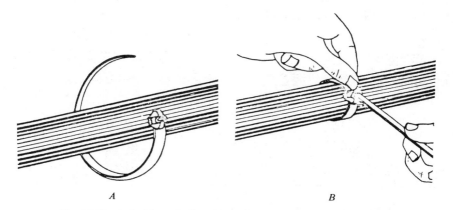

A

B

Fig. 6-115. *A*, Position of adjustable cable strap. *B*, Securing cable strap.

Using a hand tool, pull the tail end of the adjustable cable strap tight. The hand tool will automatically apply the proper amount of tension and cut off the tail flush to the head.

The adjustable cable strap shall not be capable of lateral movement under normal finger pressure, after installation. Rotational movement of the strap (using finger pressure) is acceptable.

The adjustable cable strap shall be tight, but not so tight that it damages the insulation of the wires.

Cable Clamps

Cable clamps are used to secure bundled wires in a desired position in a chassis or cabinet. Cable clamp sizes and locations are defined on the engineering drawings.

The diameter of the wire bundle will determine the size of the clamp to use to secure the cable without deforming the insulation. Cable can be slipped laterally through strap with finger pressure or rotated around wire bundle.

Correct size clamp shall be used, such that the clamp does not distort, deform, or compress cable. No wires shall be pinched. (See Figs. 6-116 and 6-117.)

Panel Routing Holes

A. If panel is less than $\frac{1}{8}$ inch thick, a grommet shall be installed to prevent chafing. (See Fig. 6-118.)

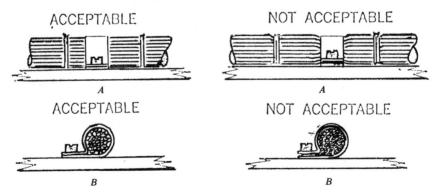

Fig. 6-116. Cable clamp, correct size.   Fig. 6.117. Cable clamp, under size.

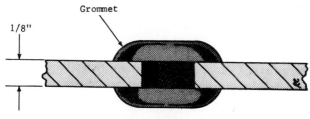

Fig. 6-118.

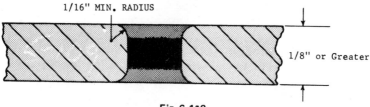

Fig. 6-119.

B. If panel is $\frac{1}{8}$ inch or more, the edges shall be chamfered with a minimum $\frac{1}{16}$-inch radius. A rubber grommet is an equally acceptable method on thicker panels. (See Fig. 6-119.)

C. Wiring shall also be insulated when passing over or around sharp corners. Acceptable methods of protection include:

1. Chamfering sharp edges with at least a $\frac{1}{16}$-inch radius.
2. Clamping wire at the sharp edge.
3. Placement of caterpillar insulation over sharp edge.

Note: Electrical tape is not a suitable insulation material to protect wiring from sharp edges of the chassis.

Insulation

Unless otherwise specified, all stranded wire used in cables or cable harnesses shall be Teflon- or Raychem-insulated, with a minimum breakdown voltage rating of 600 volts or greater.

Shrink or zipper tubing will be cut to the desired length for the particular application. There shall be no exposed wires visible at either breakout or connector termination. (See Figs. 6-120 through 6-125.)

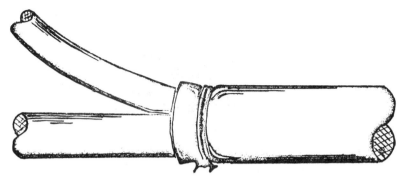

Fig. 6-120. Insulation at breakouts. Acceptable (preferred): Insulation at breakouts overlap to cover and give additional support to the wires.

Shrink tubing shall be the correct size for use on a particular wire bundle, and shall not be installed over spot-ties or other forms of clamps. (See Figs. 6-126 through 6-128.)

Zipper tubing shall be the correct diameter for the wire bundle. After installation, the seam or track shall be sealed with a cement specified for this purpose. The sealer will be applied to the track only. Excess sealer shall be removed.

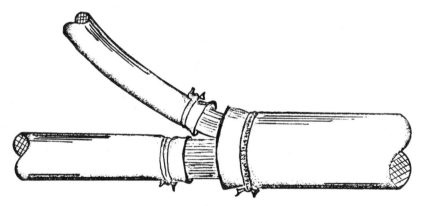

Fig. 6-121. Insulation cut too short. Not acceptable: Insulation cut too short for application, resulting in exposed wires at the breakout.

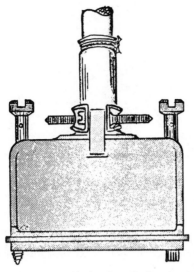

Fig. 6-122. Sleeving secured.   Acceptable (preferred): Sleeving is securely clamped or tied under the connector boot.

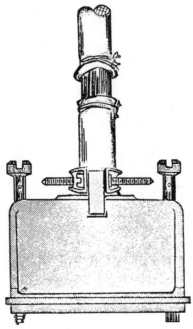

Fig. 6-123. Sleeving cut too short.   Not acceptable: Sleeving is cut too short for the desired application, resulting in wires exposed at the connector termination.

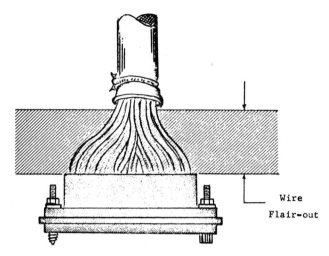

Fig. 6-124. Acceptable shrink sleeving termination. Acceptable (preferred): Shrink sleeving extends as close to the connector as possible.

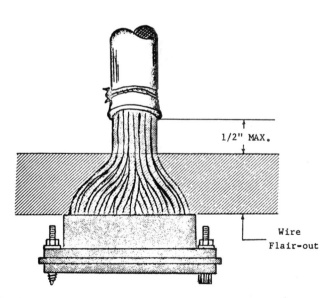

Fig. 6-125. Maximum clearance from wire flair-out. Acceptable: Sleeving shall be no farther than $\frac{1}{2}$ inch from the wire flair-out.

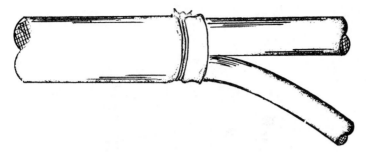

Fig. 6-126. Acceptable shrink tubing. Acceptable (preferred): Shrink tubing is snug and has the smooth contour of the wire bundle.

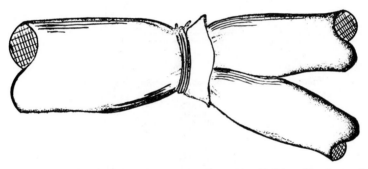

Fig. 6-127. Unacceptable shrink tubing.   Not acceptable: Shrink tubing is too large for application, resulting in a loose fit.

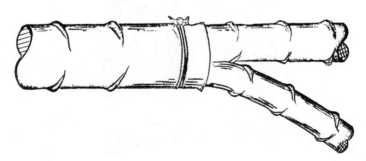

Fig. 6-128. Shrink tubing over cable ties.   Not acceptable: Shrink tubing is installed over cable ties resulting in a lumpy surface and possible damage to the tubing.

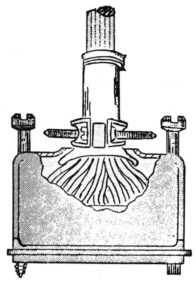

Fig. 6-129. Acceptable (preferred): Insulation boot is installed under the edge of the connector hood to protect the wires from the hood and strain-relief clamp.

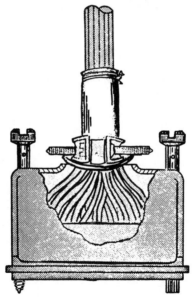

Fig. 6-130. Not acceptable: Boot is not installed under the edge of the connector hood. Wire chafing will result from this condition.

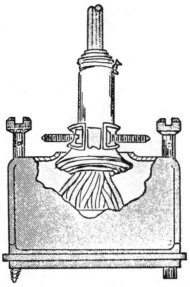

Fig. 6-131. Acceptable (reinforcement). If the wire bundle is too small for source clamping, additional boots shall be installed to build up the clamping surface.

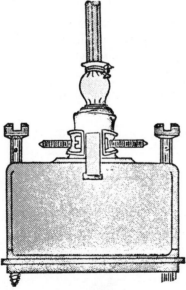

Fig. 6-132. Acceptable (reinforcement). If additional boots are not available, the clamping surface may be built up by wrapping shrink sleeving or other insulating material around the cable and existing boot.

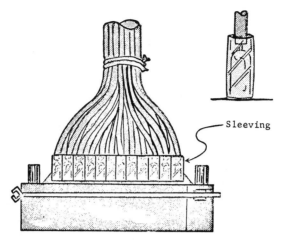

Sleeving

Fig. 6-133. Acceptable (preferred): Sleeving is cut to the correct size and is snug around the solder joint. (Note: Sleeving shall be no shorter than that required to cover the lug and strip-gap, and no higher than 0.25 inch above the lug.)

Shrink or zipper tubing shall be installed over all movable cables or harnesses. Zipper tubing shall be clamped or securely tied at each end.

Electrical tape shall not be used for insulation purposes except by agreement of the manufacturing engineering and quality control departments. Insulation boots shall be installed on all hooded connectors. (See Figs. 6-129 through 6-130.)

The wire bundle shall be securely clamped under the strain relief clamp (Figs. 6-131 and 6-132).

When engineering specifications require solder lugs or pins to be sleeved, the sleeving shall be selected and cut to the correct size before installation. (See Figs. 6-133 and 6-134.)

Insulation Damage

Flat spots or splits in wire insulation, caused by pliers or clamping devices, shall be kept to a minimum. Flat spots are not acceptable if the wire strands are visible through the insulation. (See Fig. 6-135.)

Split or torn insulation at solder lug terminations is acceptable (see Fig. 6-136) only under the following conditions:

A. Split or tear does not extend through the insulation to the wire strands.

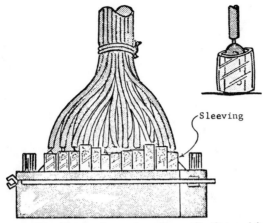

Fig. 6-134. Not acceptable: Sleeving is too large for application and is cut too short to cover the entire conductive area.

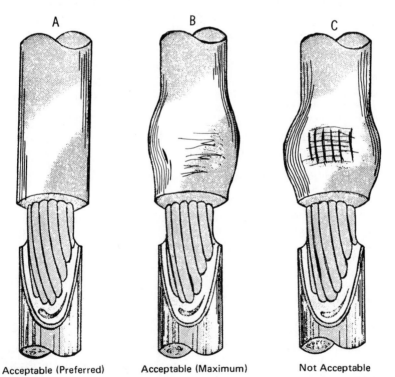

Acceptable (Preferred)	Acceptable (Maximum)	Not Acceptable

Fig. 6-135. *A,* Insulation is not damaged in any way be clamping devices. *B,* Flat spot in insulation does not reveal the wire strands. *C,* Flat spot in insulation has exposed the wire strands.

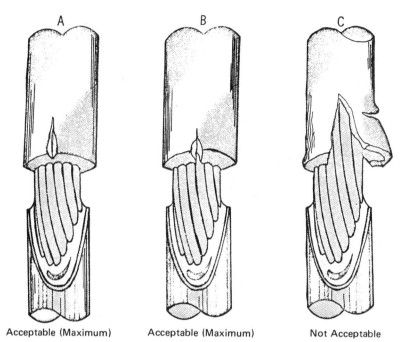

Acceptable (Maximum)   Acceptable (Maximum)   Not Acceptable

Fig. 6-136. *A*, Split insulation is surface damage only. *B*, Split insulation does not extend far enough to exceed the maximum strip-gap limit. *C*, Split insulation is in excess of the maximum strip-gap limit and is pulled away from the wire strands.

B. Split or tear does not extend far enough into the insulation to increase the strip gas past the 0.25-inch maximum limit.

C. Insulation is not separated from the strands.

An acceptable repair of flattened or split insulation is illustrated in Fig. 6-137. Zipper or shrink tubing shall not be melted, torn, punctured or otherwise damaged. (See Fig. 6-138.)

Melted, torn, or otherwise defective insulation boots are not acceptable. (See Fig. 6-139.)

Mass Termination Connectors

The flat cable connectors in use today have been specifically designed for the mass termination of flat cables. There is however, another factor that must be taken into consideration when the designer selects flat cables for use in a system, the human factor.

Service and test personnel have a tendency to use the flat cable as a handle for extracting the connectors. This has been a major problem area and the designer must provide a means of extraction and the extractor device must be very obvious.

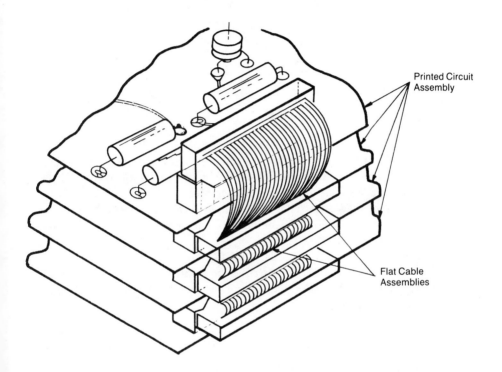

Printed Circuit Assembly

Flat Cable Assemblies

The problem to watch for is when a force is applied to disengage the connector, the resulting force against the connector header results in loosening the wires from the contacts and often causes cable damage as shown in the following figures. When a force is applied to disengage the connector using the cable as the extractor the resulting pressure against the connector header results in loosening the wires from the contacts. This is to be expected since the header was never intended to retain the connections under these conditions.

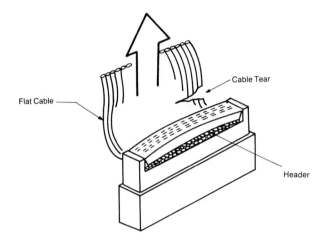

As tension to the cable is applied the leading edge of the connector header places a bearing surface against the cable which in most cases will result in cable tears where the cable enters the connector assembly.

The use of a bonded cable rather than a molded cable provides a stronger insulation which will reduce the forces applied to the conductors when they are used to extract the connector.

A typical digital assembly using flat cable interconnections in a limited space having no extraction capabilities must either be redesigned to employ new connectors with extractors or some type of corrective action must be taken to eliminate potential cable damage.

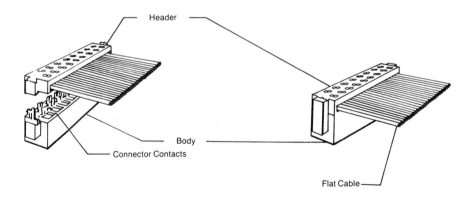

Applying a small amount of adhesive to the connector body between the rows of contacts to bond the flat cable insulation to the connector body will reduce the potential to disengage the contact to wire interface.

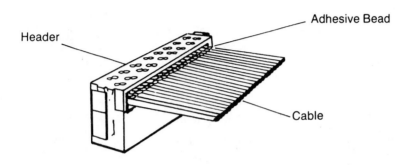

Connector assemblies that are already assembled can be strengthened by applying a bead of adhesive to lock the cable/header/connector interfaces.

All of the major manufacturers have qualified field engineers on staff with design backgrounds. In the opinion of Linda Stern, Field Engineer for the Burndy Corporation, if the design engineers would call on the field engineers early in the design phase many of the problems could be avoided. There is no cost to the customer for this consultation and the field engineers exposure to all the applications of their product is a valuable asset to the design engineer.

REQUIREMENTS FOR ELECTRICAL CONNECTIONS

This section defines the requirements for mechanically and electrically stable, solderless, wrapped, electrical connections made with solid,* round wire and appropriately designed square and rectangular terminals.

*Stranded wire may be used but only in special applications where flexing is prevalent and the use of solid wire would not be suitable. The type of wire to be used will always be specified by the engineering department.

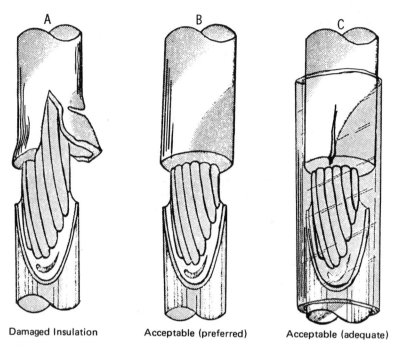

Damaged Insulation   Acceptable (preferred)   Acceptable (adequate)

Fig. 6-137. Damaged insulation (A) has been restripped and soldered in B. If the wire is too short to be restripped, the damaged portion may be covered with a piece of suitable sleeving (C).

Classes

1. *Class A Connection* (modified solderless wrapped connection). A Class A connection consists of a helix of continuous, solid, uninsulated wire tightly wrapped around a solderless terminal to produce a mechanically and electrically stable connection. The number of turns required will depend on the gauge of wire used (see p. 247). In addition to the length of uninsulated wire wrapped around the terminal, an additional minimum half-turn of insulated wire shall be wrapped around the terminal to help insure better vibration characteristics. To accomplish a half turn, the wire must be in contact with at least three corners of the terminal (Fig. 6-140).

2. *Class B Connection* (conventional solderless wrapped connection). A Class B connection is the same as a Class A connection except that the additional half-turn of insulated wire is not required.

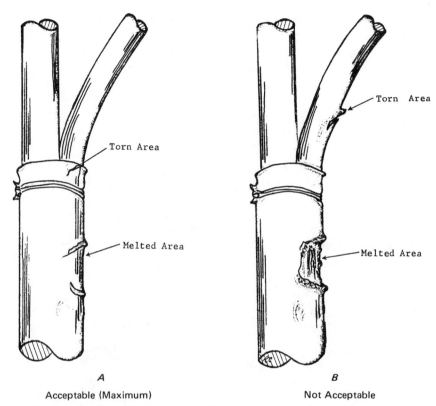

Torn Area

Melted Area

Torn Area

Melted Area

A

Acceptable (Maximum)

B

Not Acceptable

Fig. 6-138. *A*, Melted tubing (surface only). Damaged area does not expose individual wires. Tear is minor with no possibility of getting larger. *B*, Melted area is excessive enough to reveal and possibly damage individual wires. Tear reveals the wires and could become larger.

Terminals

The corners on the terminals must not have excessive burrs. The terminal edge radius shall not exceed 0.003 inch. The terminal shall be straight and parallel within 0.005 inch per inch.

Wire

Copper wire conforming to the minimum elongation listed in the following table will facilitate passing the strip-force test (p. 244) and the unwrap test (p. 247).

AWG No.	Minimum Elongation (%) in 10 In.
26	15
24	15
22	20
20	20
18	20
16	20

Connections (Fig. 6-141A-G)

The space between turns of uninsulated wire of a connection, except for the first and last half-turns of uninsulated wire, shall not exceed one-half the nominal uninsulated wire diameter.

There shall be no overlapping within the minimum specified number of turns of uninsulated wire except that the first turn of insulated wire in Class A con-

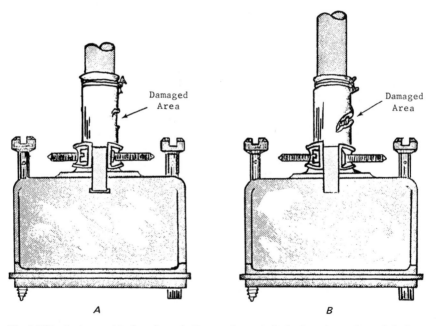

A   *B*

Fig. 6-139. *A*, Acceptable (maximum): Damaged area is limited to the surface of the boot and the individual wires are not visible. *B*, Not acceptable: Damaged area is completely through the boot, and the individual wires are visible. The only acceptable repair of a damaged boot is by replacement.

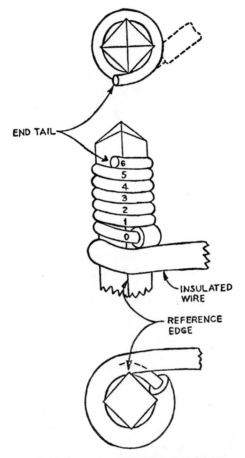

Fig. 6-140. Example of Class A connection (modified solderless wrapped connection): six turns of uninsulated wire plus four terminal corners.

nection may overlap the last turn of uninsulated wire in a connection below it on the same terminal.

The bottom surface of the insulation of the first wrapped connection shall be within $\frac{1}{16}$ inch of the top of the terminal insulator surface (Fig. 6-141B).

The termination of the last turn of a wrapped wire connection (end tail) shall not exceed away from the outside diameter of the stripped wire on the terminal

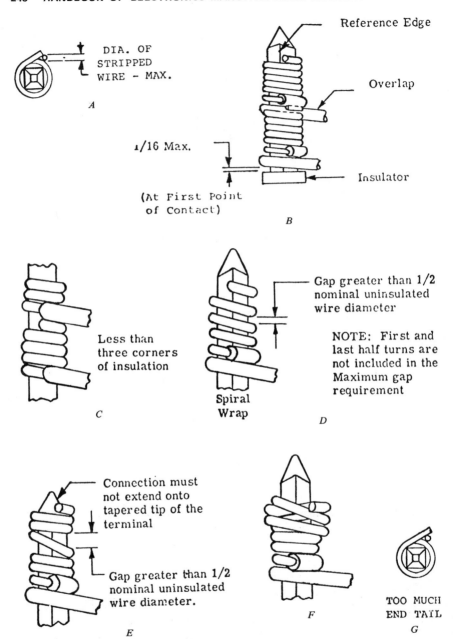

DIA. OF STRIPPED WIRE – MAX.

A

Reference Edge

Overlap

1/16 Max.

Insulator

(At First Point of Contact)

B

Less than three corners of insulation

C

Gap greater than 1/2 nominal uninsulated wire diameter

NOTE: First and last half turns are not included in the Maximum gap requirement

Spiral Wrap

D

Connection must not extend onto tapered tip of the terminal

Gap greater than 1/2 nominal uninsulated wire diameter.

E

F

TOO MUCH END TAIL

G

Fig. 6-141. Examples of Class A connections.

by more than the diameter of the stripped wire (Fig. 6-141). Dressing and clipping of wires must be accomplished so as not to loosen the helix connection.

The minimum number of turns of uninsulated wire per connection shall be as shown in the following table:

AWG No.	Minimum No. Uninsulated Turns
26	6
24	5
22	5
20	4
18	4

The required minimum number of uninsulated turns in a wrapped connection shall not extend beyond the wrapping length of a terminal.

The minimum number of turns of insulated wire shall be $\frac{1}{2}$, and the maximum shall be $1\frac{1}{2}$, on a Class A (modified) wire-wrapped connection.

There shall be no visible terminal fracture in a wire-wrapped connection.

Reduction in wire diameter due to nicks, scrapes, and deformation caused by the wrapping tool must not be more than:

AWG No.	Percent
30	10
26	10
24	20
22	25
20	25
18	25

Rewraps

It is not permissible to rewrap the portion of wire that has been previously wrapped on a terminal.

Rewraps can be made on the same portion of a terminal from which a previously wrapped connection has been unwrapped, provided the terminal meets the requirements previously mentioned. Prior to rewrapping, the terminal shall be checked for plating loss, corrosion, or other damage.

No attempt shall be made to move a wrap on a terminal after the wrap is made. The bond between wire and terminal is destroyed if any adjustments are made after the wrap is completed.

A combination of solder joints and wrapped-wire connections on a terminal is not acceptable.

Assemblies shall be clean and free from wire clippings and other foreign matter.

Wire must not be nicked when stripped *prior* to wire-wrap.

Wire insulation shall not be cut, torn, frayed, or otherwise damaged away from the post.

Insulation cut by a tool during the wrapping operation is acceptable if this cut is included as a portion of the wrap around the terminal.

Making the Wire-Wrapped Connection (See Fig. 6-142)

Strip the insulation from the wire to a length determined by the gauge and number of turns needed for the connection.

Insert the bare wire into the wire slot of the wrapping bit, butting the insulation of the wire against the bit face. The funnel shape of the sleeve permits easy wire insertion. The wire is then positioned in the notch of the wrapping sleeve.

Position the tool with the terminal hole of the wrapping bit over the terminal to be wrapped.

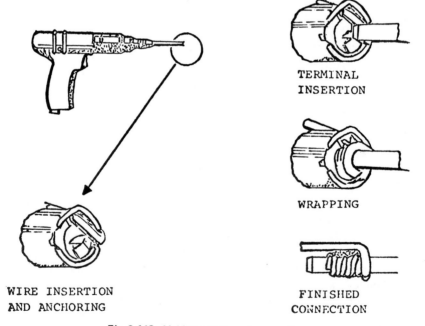

TERMINAL
INSERTION

WRAPPING

WIRE INSERTION
AND ANCHORING

FINISHED
CONNECTION

Fig. 6-142. Making a wire-wrap connection.

Squeeze the trigger, and apply a slight amount of back force (i.e., forward pressure applied by the operator, approximately 1.5 lbs).

The wrapping operation shall not twist the terminal (perpendicular to its axis more than 15°).

The last connection must not extend on the tapered tip of the terminal (Fig. 6-141E).

Wire Dress

The wire shall not be routed in any manner that will tend to unwrap the connection.

Wiring shall be routed to preclude strain on wire or terminal.

A wire routing between terminals of a panel assembly shall be such that the wire does not interfere with the normal routing of other wires.

A wire shall not be dressed tightly around the sharp corners of terminals or other hardware throughout its routing where penetration or damage to the insulation may result.

Test

Strip Force.   A completed wire-wrapped connection shall be capable of meeting the following minimum strip-force limits:

(Rate of Pull: 1 In. to 10 In. per Minimum)

Wire AWG No.	Minimum Strip Force (Lbs)
30	4.0
26	6.5
24	7.0
22	9.0
20	9.0
18	15.0

Jaws of strip-force tool shall have freedom of movement over terminal and as shown below.

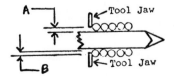

The maximum total clearance between jaw and terminal (A + B) shall not exceed 70 percent of the diameter of the wire.

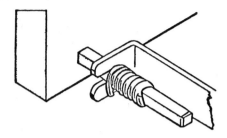

Note: This test shall be performed on samples only, not on items being prepared for shipment to the customer.

Unwrap

The wire of a connection shall be capable of being sufficiently unwrapped to free the wire from the terminal without breaking. The unwrapping shall be accomplished with a standard unwrapping tool to ensure the life of the wire-wrap terminal.

Gas Tight

The total gas-tight contact area between the terminal and the uninsulated wire in a wire-wrapped connection, excluding the first and last half-turns of uninsulated wire, shall be at least equal to the cross-sectional area of the uninsulated wrapped wire.*

Electrical

The voltage drop across the wrapped connections shall not exceed 4 millivolts when measured to MIL-STD-1130.

ELECTRONIC ASSEMBLER TEST

The following test was developed to allow manufacturing engineers to establish the capabilities of electronics assembly personnel, thus enabling them to efficiently assign tasks. In addition, it provides information regarding the type of training required.

*Ref.: Test Procedures, Paragraph 5.6.2 of MIL-STD-1130.

Indicate whether each statement is true or false by marking T or F in the parentheses.

1. ( ) Mechanical wire strippers must be recalibrated periodically.

2. ( ) The process of putting solder on the tip of a soldering iron is called *tinning.*

3. ( ) The strip length of a wire should be the length needed to make a three-fourth wrap and have an insulation gap.

4. ( ) A lock washer must be installed under a locknut.

5. ( ) Wire that is bent over a metal edge must have additional protection in the form of a grommet or electrical tape to insure that it would not short to the metal.

6. When reading wire colors, the _____ color is the first one named.

7. In contour soldering, the fillet should extend to_____ the diameter of the wire.

8. The maximum allowable insulation gap from a terminal is _____.

9. There must be _____ threads protruding through a nut on surface-based equipment.

10. The written instruction sheet telling you how to do your job is called a

_____.

11. Solder flux is used to
    a. Aid the flow of solder.
    b. Clean oxidation off of the metal being soldered.
    c. Improve the wetting action.
    d. All of the above.

12. Components such as resistors and capacitors are to be installed
    a. With a lazy bend in their leads.
    b. To cover the stencil identification.
    c. So the electrical value of the part is readable.
    d. All of the above.

13. When installing wires and components on a terminal
    a. The wire or wires are installed closest to the base.
    b. The components are installed closest to the base.
    c. Either may be installed closest to the base.

14. A flat washer is used
    a. To protect a nonmetallic surface.
    b. To cover an oversized hole.
    c. To protect a finished metallic surface.
    d. All of the above.

15. A heat sink is used to
    a. Wash printed-circuit boards after dip solder.
    b. Prevent capillary action in a wire.
    c. Help dissipate heat from a component lead.
    d. None of the above.

16. When is it necessary to file a permanent-type soldering iron tip?

17. What is meant by the term *wicking*?

18. What is meant by the expression "enough wire to allow two reconnections"?

19. What advantage is there in using a thermal wire stripper?

20. After you have made a spot-tie, what do you do to prevent it from coming open?

7
SOLDERING

DEFINITIONS

Soldering. Soldering is the joining of two metals, such as a wire or component and a terminal, by bonding them with an alloy called *solder.* Since a soldered connection should be a true alloying of the solder with the metal, a good joint should present an appearance indicating the "wetting" action of the molten solder on the metals joined. The degree of "wetting" action occurring may be determined from the curvature of the boundary between the solder and the terminal. This may be illustrated by comparing the junction of water and mercury with glass. Whereas water "wets" glass, mercury does not. The boundary between water and glass is concave; that between mercury and glass is convex as shown in Fig. 7-1.

Soldered Connection. A soldered connection is an electrical connection that employs solder for bonding two or more metals with a metal as shown in Fig. 7-2.

Lead Extension. A lead extension is that part of a soldered lead or wire that extends beyond the connection.

Flux. Flux is a substance used in soldering for the purpose of removing oxides and impurities from the components to be joined, thus promoting their union.

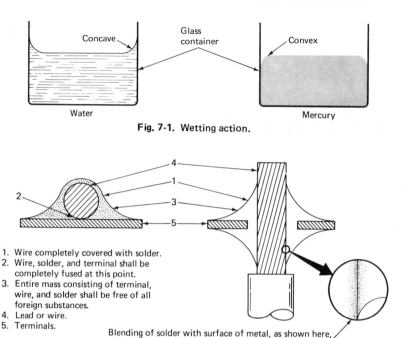

Fig. 7-1. Wetting action.

1. Wire completely covered with solder.
2. Wire, solder, and terminal shall be completely fused at this point.
3. Entire mass consisting of terminal, wire, and solder shall be free of all foreign substances.
4. Lead or wire.
5. Terminals.

Blending of solder with surface of metal, as shown here, is implied in all sectional views of soldered joints.

Fig. 7-2. Soldered joint (basic).

It promotes or accelerates the wetting of metals by molten solder. Flux is of two types, corrosive and noncorrosive. (*Only noncorrosive* flux is used on soldered connections.)

Corrosion. Corrosion is a gradual chemical or electrochemical attack on a metal by atmosphere or other agents.

Solder. Solder is a fusible metallic alloy used when melted to join metallic surfaces.

Pitting. Pitting is the process of becoming marked with scars, holes, or depressions.

UNACCEPTABLE SOLDERED CONNECTIONS

Excessive Solder Joint. An excessive solder joint (shown in Fig. 7-3) is one on which there is more solder than necessary. The solder should not exceed 50 percent of the thickness of the wire attached to the terminal with the outline of

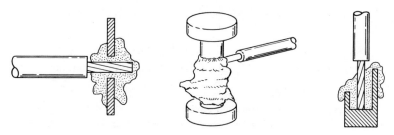

Fig. 7-3. Excessive solder joint.

the wire still visible. Joints with solder exceeding this amount would fall into this classification.

Cold Soldered Joint.   A cold soldered joint (shown in Fig. 7-4) is a soldered joint where the solder has flowed all around a wire or lead but has not bonded the wire and terminal together. It can be caused by insufficient heating of the part to be soldered, dirty surfaces, insufficient flux, poorly tinned iron, or a cold soldering iron. It has a milky appearance, lacks a metallic luster, and generally presents a rough, piled-up appearance.

Disturbed Joint   (sometimes referred to as *fractured joint*). A disturbed joint is one where there is little or no bonding between the terminal and lead wire because of movement of leads while the solder is at the critical solidifying stage. It will have an irregular or crystallized appearance and is not a good electrical connection. It can loosen later in service and introduce high resistance into a circuit or cause noise with the slightest vibration.

Rosin Joint.   A rosin joint is a joint where a thin film of flux separates the terminal from the lead, causing a loose nonconducting joint even though sufficient solder has been applied. This type joint is similar in appearance to a cold soldered joint and may also exhibit excessive flux residue. It is generally

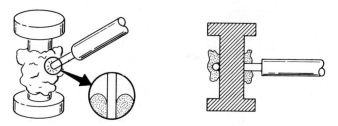

Fig. 7-4. Cold soldered joint.

caused by insufficient heat or dirty parts, which will not allow the flux to react properly.

Solder Short. A solder short is a soldered connection from which excessive solder has been allowed to flow to the extent that it forms a conductive path to some other circuit.

Solder Tit. A solder tit is a soldered connection that contains a sharp point. In high-voltage circuits, this type of connection can cause arcing or corona. See Fig. 7-5.

Wicking. Wicking is the drawing of molten solder into stranded wire due to capillary action (as water is drawn into a blotter). A soldered joint is unacceptable when wicking causes the solder to flow under the insulation of the wire, thus burning it and or preventing proper flexing.

Overheated Joint. An overheated joint is a joint that has been burned by excessive heating with the soldering instrument. The flux usually appears black, and the solder is dull and crystallized. (It is usually accompanied by wicking and by damaged insulation.)

Insufficient Solder Joint. Insufficient solder joint (Fig. 7-6) is defined as the lack of coverage by solder on the required wrap. It is a desired standard

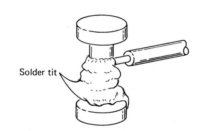

Solder tit

Fig. 7-5. Solder tit joint.

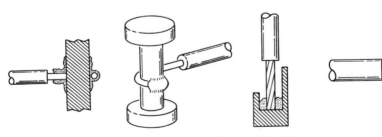

Fig. 7-6. Insufficient solder joint.

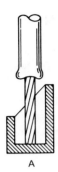  

A   B   C

Fig. 7-7. Damaged insulation.

of workmanship that the entire length of wire used in the wrap be properly
soldered.

Unsoldered Joint.   An unsoldered joint is one in which solder is required but
to which none has been applied.

Improper Wrap.   A joint not having the proper turn(s) around terminal. Detec-
tion of this defect is practical only where the amount of solder used has left the
wire contour visible.

Damaged Insulation.   Damaged insulation (Fig. 7-7) may be in the form of
melting caused by excessive heat or separation from the conductor due to excessive solder flow underneath it (Fig. 7-7A), cuts or severe crimping from pliers
used to hold the wire during soldering (Fig. 7-7B), or placement of the insulation
into the solder connection (Fig. 7-7C).

HAND SOLDERING

As the quality and reliability of an end item become dependent on the characteristics of soldered connections, operators and inspectors become responsible
for the workmanship involved. Formal training is needed. With a comprehensive program for the qualification and certification of personnel, quality can be
built into the product. The assembly operator must know what a reliable soldered connection is, how it is made, and the proper tools, methods and techniques
required.

The major cause of solder joint failure is generally attributed to poor soldering techniques such as cold joints, excessive heat, improper fluxing, etc. Yet
actual causes are more complicated. In Printed Circuit Board failures the cause
may be poor choice of board materials, improper solder-mask coatings, etc. The

soldering iron and the techniques used will determine the reliability of the solder joint.

Unfortunately, operators are too likely to select one iron and use it for all purposes. This compares to using a sledge hammer to drive tacks. No one tool is suitable for all uses.

There is no one best training and certification program. Organizations are different; each must decide what is suitable for its own product and customers. To build the most reliable product at the lowest cost with maximum customer satisfaction, the following points should be considered:

- Establish soldering, wiring and assembly standards.
- Train, qualify and certify employees for adherence to these standards.
- Monitor the program and take positive corrective actions as indicated by the changing technology.
- Ensure that only qualified personnel are performing specialized skills and processes.
- Provide employees with the required knowledge for competent job performance and evaluation of products.

The objectives from a trainee viewpoint are to:

- Develop an understanding of their responsibilities in meeting the company standards and the customer's requirements.
- Become familiar with the company methods and standards for wiring, soldering, and assembly.
- Demonstrate ability to consistently meet the expected reliability and performance standards.

Tools

Good work may occasionally be turned out with poor tools, but this is an exception. Using the wrong tool invariably requires more time, and the attempt increases the potential of poor quality. The correct tool, properly used and maintained, results in quality workmanship and contributes to the skill and the pride of the workman.

The following (Fig. 7-8) are basic individual tools and should be assigned to or owned by the operator:

- Round-nose bending pliers
- Long-noise pliers
- Diagonal cutting pliers
- Tweezers (including self-locking type)

- Soldering aid
- File (for dressing soldering irons)
- Scissors (small, thin blade)

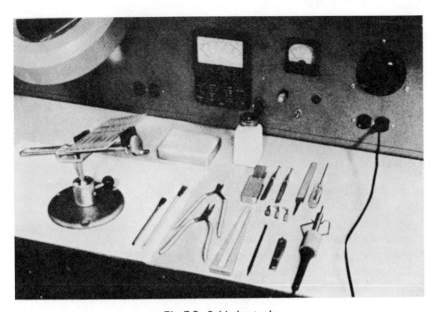

Fig. 7-8. Soldering tools.

The following tools may be necessary and should be issued as the job requires:

- Calibrated precision cutting-type wire strippers
- Thermal wire-stripping apparatus
- Variable voltage control transformer
- Soldering iron and spare tips
- Heat dissipator clamps (thermal shunts)
- Printed wiring board holding fixtures
- Tensioning spring
- Magnifying bench lamp (for miniature work)
- Resistance soldering apparatus
- Sponge (for cleaning soldering iron tip)

The following are expendable supplies, issued for specific job requirements:

- Lacing cord or plastic ties
- Core solder (Sn60 or Sn63)

- Liquid flux
- Alcohol or other approved cleaning fluid
- Bristle brushes
- Clear flexible tubing

The following special equipment, when available, is used in preparation and soldering:

- Automatic wire strippers and cutters
- Lead bending and cutting machines
- Automatic part insertion machines
- Automatic soldering machines
- Ultrasonic solder pots and ultrasonic cleaning equipment.

Work Area

Properly directed and sufficient light is a prime requirement for quality work. This may be provided by a combination of glare-free area lighting and adjustable supplementary light sources at the work surface. A light level of 100 footcandles at the work place is considered necessary for adequate visibility to assemble miniature parts and connectors.

A neat and orderly work area contributes to quality work by reducing confusion. Tools should be arranged within easy reach, and each tool should be kept in a specified place. Plastic trays or shallow pans lined with industrial towels are good for holding tools. Only those tools required for the current series of operations should be at the operator's position. Misplaced or inaccessible tools cause delays and tempt the workman to use the wrong tool. This can lead to rejections, reworks, delays, and degradation of the end product.

The safety of a work area depends on the development and maintenance of safety practices by all personnel in the area. A general list of these practices appears below; other practices, for special areas, should also be developed and maintained.

1. Place the soldering iron in a location that will not require reaching across or around it. Place the soldering iron in the upper corner of the work area when not in use.
2. When unsoldering a wire, make certain that there is no tension or spring to the wire. Safety glasses should be worn during this operation to shield eyes from hot solder.
3. Plug soldering iron into proper electrical socket and keep plug free from strands of wire or metal.
4. When cutting wires, keep the open side of the cutter away from the body,

keep the wire pieces in the work area, and keep wire pieces from other work areas from entering your area.

5. Disconnect all electrical power before working on a chassis.
6. Insure that blade-type screwdriver bits are square and sharp, and use screwdrivers only for the work for which they were designed.
7. Connect the airhose to the air supply nearest to the work area; keep the hose out of the aisle, and keep excess hose reeled in.
8. When cleaning with an air hose, use a pressure-controlled nozzle with minimum pressure, insure that the air stream is not directed toward personnel, and wear safety glasses. Do not clean clothes with air pressure.
9. When working with solvents, keep a minimum amount of solvent in a safety can. DO NOT SMOKE, and do not spill solvent over the work area.
10. Safety glasses should be worn when work such as cutting wires, soldering, or using solvent is performed overhead.
11. Women working in the solder area should wear low-heeled shoes with closed toes and heels. Other recommended apparel includes hair nets and slacks or coveralls. Loose clothing is dangerous around machinery.

Soldering Irons

Research has shown that the best solder joints are produced within a narrow range of temperature and application time. Joint temperatures of $500°$ to $550°F.$, applied for 1 to 2 seconds, result in the strongest connections. Size, wattage, and shape of the iron should be selected to approach these conditions as closely as possible. When the tip is applied, it should rapidly heat the joint to soldering temperature. The amount of heat capacity in the tip implies that the mass of the tip is large with respect to the mass of the metal being heated for solder application.

The tip must not be so large as to obscure visibility of the workpiece, or to cause damage to adjacent parts or wire insulation. The tip shape, such as spade, chisel, or pyramid, should be appropriate for the workpiece. It is good practice to keep an assortment of spare tips and to change the tip as necessary to meet the requirements of various types of work.

A variable voltage supply is recommended for controlling the soldering iron temperature when soldering printed circuits; it is also advantageous in many other soldering applications. By proper selection of tips and correct voltage adjustment, a single 50-watt pencil-type iron can be used for soldering miniature printed circuits or relatively large terminals. Thermostatically controlled irons require less skill in the judgment of heat but are somewhat less versatile.

The unplated copper tip will produce the best results, and therefore is recommended. This does not exclude the use of plated tips for production work, provided the quality of the solder connections can be maintained.

Copper tips in an unheated condition should be dressed smooth with a suit-able file. After the tip has been filed and the minimum temperature required to melt solder has been reached, apply solder to the dressed face of the tip. Clean the tip by wiping on a wet sponge or other suitable material before each connection is made.

NOTE: Do not change from unplated tips to plated tips during a single operation without allowing for the extra time required for heat transfer from the plated tip.

Heat Application with Soldering Iron

The soldering iron tip should be applied to the metal part having the greatest mass (Fig. 7-9). Ordinarily, in a wire-to-terminal connection, the greatest mass is the terminal. Fresh tinning on the tip of the iron provides quick transfer of heat to the workpiece.

Solder should be applied to the junction of the joint members (not to the soldering iron) as soon as they will readily melt the solder. The iron should be tilted sufficiently to allow the solder to be applied to the junction. The iron

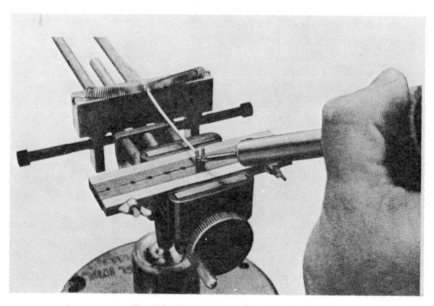

Fig. 7-9. Proper application of heat.

should be withdrawn as soon as the joint is complete to avoid overheating the molten solder.

The surface temperature of both metals being soldered must be above the solder melting point to expedite efficient wetting. Solder should not be permitted to flow onto a surface which is cooler than the solder temperature; this will cause cold or "rosin" joints.

Solder applied to a properly cleaned, fluxed, or heated surface will melt and flow without direct contact with the heat source. It will have a smooth, even surface, feathering out to a thin edge. A built-up, irregular appearance is an indication of improper solder application.

A good solder joint is characterized by a smooth surface, even distribution of solder to a feather edge at the base metal, no porosity, good fillet between conductors, and good adherence to both parts. Charred or carbonized flux residue indicates excessive application of heat. Solder obscuring the contour of the wire, and making visual inspection difficult or impossible, is classified as excess solder.

Resistance Soldering

Resistance soldering (Fig. 7-10) is an effective method for supplying heat to the metal to be soldered. This processes passes current through the metal; the heat

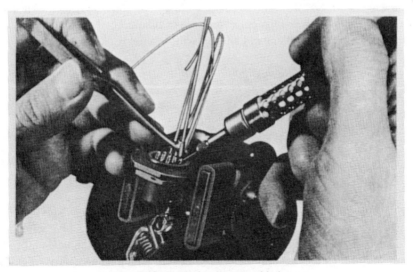

Fig. 7-10. Typical resistance soldering.

generated at the interface of the metal and the electrode is used to melt the solder. The proper mating pin or socket should be used to avert damaging the connector.

Since resistance soldering generates heat directly in the metal area to be soldered, it offers the advantage of localizing or confining heat to a selected area. This method is well adapted to the soldering of connectors.

One application of this method uses tweezer-type electrodes. When the metal is gripped between the electrodes, an electric circuit is completed through a low-voltage transformer and the metal between the electrodes is heated.

Another application uses a carbon pencil; the return circuit is a metallic connection. When contact is made by the carbon pencil, the metal of the terminal is heated at the point of contact (where the resistance is maximum).

Preparing Conductors and Part Leads

Insulation Stripping

To strip insulation, nonadjustable, factory set, cutting-type strippers should be used. Fig. 7-11 shows a typical hand stripping tool.

When using tools with multiple stripping holes, the correct hole for the gage of wire being stripped must be used. Tools equipped with single dies are recommended. For long production runs on single wire sizes, the unused holes in multiple wire strippers may be masked off to prevent accidental use of an undersized hole which could result in nicks, cuts, and scrapes to the wire strands. The calibration of all precision-type cutting strippers should be checked periodically. Out-of-tolerance strippers should be removed from the work area.

Fig. 7-11. Proper cutting-type stripper.

Thermal-type strippers (Fig. 7-12) may be used on certain types of wire insulation which can be effectively stripped by this method. Care should be taken to eliminate burned insulation residue which would impair solderability. Ventilation should be provided where thermal strippers are used. Inhalation of the

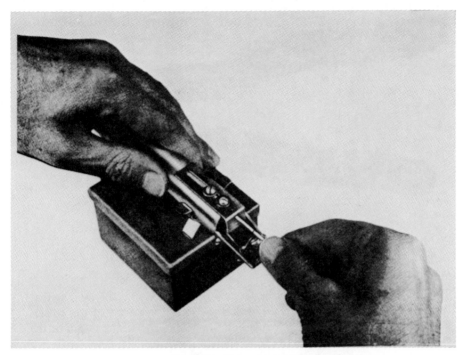

Fig. 7-12. Thermal stripper.

vapors from the breakdown of polymerized organic insulation can cause polymer fume fever.

If the stripping operation has caused the strands to become separated or disarranged, the strands should be restored to the original lay before the ends are tinned.

Stripped wire with nicked or cut strands is not acceptable because the stress concentration will cause failure during vibration. For this reason, cutting-type strippers (Fig. 7-13) must not be used under any circumstances. They will invariably cut, nick, or scrape strands of wire.

Insulation Damage

Excessive pressure by the gripping jaws of hand-operated strippers will crush the insulation at the wire end. Incorrect gripping blocks in the machine-type stripper will also damage the insulation. Wire with damaged insulation should not be used. Slight discoloration as a result of thermal stripping is acceptable.

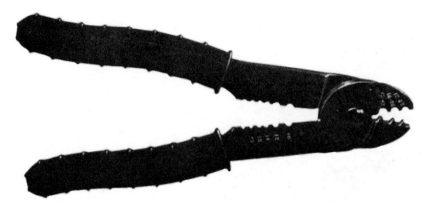

Fig. 7-13. Improper cutting-type stripper.

Insulation Gap

The end of the insulation should be far enough from the soldered joint so that the insulation cannot become embedded in the solder, yet not so far as to permit a short circuit between two adjacent wires. In general, the length of the gap should be a distance equal to approximately a wire diameter (Fig. 7-14).

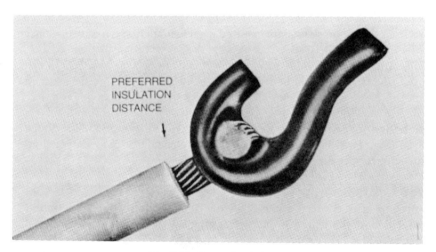

Fig. 7-14. Proper insulation gap.

Cleaning Part Lead

To assure a good wetting action between parts to be soldered, all impurities such as dirt, grease, or oxide film, must be removed. Surface contamination or corrosion formed on the part lead during processing, storage, or handling should be removed. A typical cleaning tool is shown in Fig. 7-15. The lead should be retinned with an alloy of the same composition as will be used in the soldering operation.

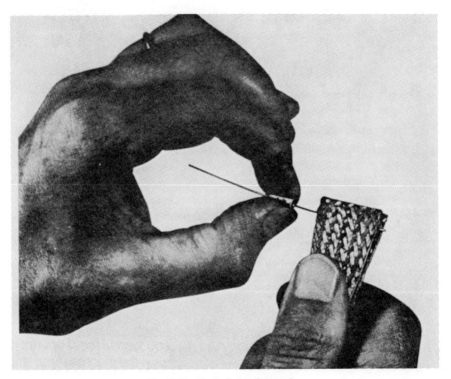

Fig. 7-15. Typical cleaning tool.

Stress Relief

An allowance for expansion and contraction during temperature cycling should be made in all part leads. This applies not only to parts with axial leads, but also to solid jumper wires which could transmit tensile or compressive forces.

Vibration Bend

Wires connected to a terminal board or to a part having fixed terminals should have slack in the form of a gradual bend to allow flexing during vibration (Fig. 7-16). When multiple wires are routed from a cable trunk to equally spaced terminals, a uniform amount of slack should be provided to prevent concentration of stress in the shortest wire. The cable trunk should be clamped or supported to avoid stresses on the electrical connection.

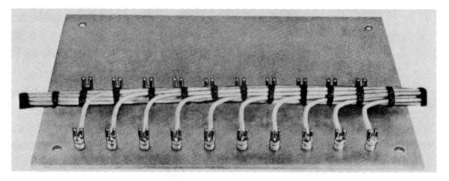

Fig. 7-16. Typical vibration bends.

Soldering

Tin melts at 450°F.; lead melts at 621°F. Eutectic tin-lead solder melts at 361°F. Solders other than eutectic melt at higher temperatures. The chart in appendix A provides information on the melting points of the various tin-lead compositions.

When heat is applied to solder other than eutectic, it becomes plastic and then liquid. Upon removal of heat, the order is reversed, the solder changing from liquid to plastic and then to a solid state. If either member of a joint is moved in relation to the other while the solder in the plastic state, the solder will become coarse grained, and the resulting connection will be physically weak and unreliable. Such a joint is called a fractured joint (Fig. 7-17) and should be reworked.

The length of time that solder is kept molten, and the temperature at which it is maintained while a liquid, are critical since molten solder absorbs gases. If excessive temperature has been used or the solder has been molten too long, the molten alloy will oxidize, and the solder will appear granular and gray when cool. The solder connection will be physically weak and unreliable.

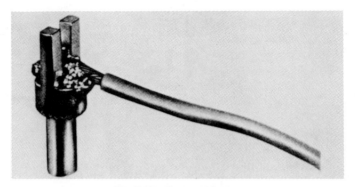

Fig. 7-17. Fractured joint.

Solder and Flux

The core of wire solder should contain only rosin or activated rosin fluxes. For general use, solder composition Sn60 or Sn63 (60 or 63 percent tin and 40 or 37 percent lead) should be used. On miniature printed circuits and in other applications where the heat should be minimized, eutectic solder (Sn63) should be used. Other compositions should be used only when specified. Soldering paste, acid-type fluxes, or other corrosive or conducting-type fluxes should not be used.

In some applications, such as removing excess solder by wicking into a stranded wire or into a piece of shielding braid, liquid-rosin flux may be used. To prevent undesirable chemical reaction, the activating material in the liquid-rosin flux must be compatible with the core flux of the wire solder used in the assembly.

Flux Application

Flux should be applied to a surface before the solder melting temperature is reached. The rosin in core solder melts before the solder and the application occurs automatically.

If excessive temperatures are used, rosin flux will carbonize and will hinder soldering rather than aid it.

Tinning Stranded Wire

Tinning of the wire should extend only far enough to take full advantage of the depth of the terminal or solder cup. The ends of the stranded wires should be stripped and dipped in liquid flux to the depth that tinning is desired. Tinning is then accomplished by either dipping the wire into a solder pot with a control temperature of $500° \pm 20°$F. (Fig. 7-18), or by using a soldering iron and cored solder (Fig. 7-18).

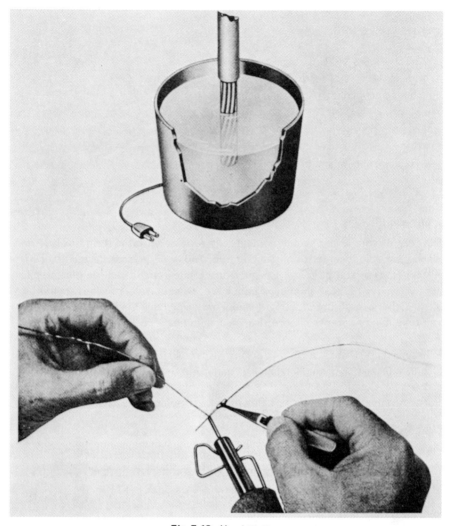

Fig. 7-18. Hand tinning.

If it becomes necessary to strip, tin and solder wires within an assembly, care should be taken to avoid dripping or spraying solder or dropping insulation residue and other impurities.

Holding

Materials being soldered should be held motionless with respect to each other. Depending on the type of termination, the tinned end of the wire should be formed into a hook or loop, and held firmly against the joining member during soldering, cooling, and solidification. Devices such as those shown in Fig. 7-20 are satisfactory to hold a wire motionless. Other holding devices will perform equally well.

Cleaning Soldered Joints

Flux residue is often tacky and tends to collect dust and other foreign matter which could cause electrical leakage paths across insulator surfaces. To prevent this the flux should be removed. After the solder has solidified, the flux residue should be dissolved by applying an approved solvent with a bristle brush or spray. The dissolved residues should be removed with industrial wiping tissues or other absorbent materials. A clean part of the tissue should be used each time to avoid the possibility of transferring the dissolved flux back to the work. (*Caution:* When switches, variable resistors, nonhermetically sealed relays, or any other nonsealed parts containing movable contacts are part of the assembly, do not use large quantities of solvent *or* dip assembly in solvent. The rosin dissolved in the cleaning solvent can enter spaces from which it cannot be removed and cause malfunction of the device).

Inspection of Completed Soldered Connections

The quality of a soldered connection can be determined by visual inspection. On miniaturized assemblies it is necessary to use magnification. Wires should not be pulled or bent, nor should force be exerted on the connection with a soldering aid or other tool to test the mechanical soundness of the connection. It may be necessary to inspect a solder joint at various stages of production when later assembly will make it impossible to inspect. If a joint must be reworked, it should be disassembled, the excess solder removed, and the area cleaned before reassembling in the same manner as a new joint.

Unsoldering

To remove wire from cup terminals, as in plugs or receptacles, the preferred method is the use of a resistance-type tool. A conduction-type soldering iron can also be used, with the tinned tip placed againt the lower side of the cup. In either method, heat is transferred to the cup until the solder has melted. A light, steady pull should remove the wire. The excess solder remaining in the cup may then be removed by either wicking into a stranded wire or by using a vacuum type solder remover. Avoid prolonged heating of the terminal.

The soldering iron, with a tip well-wetted with solder, should be used for unsoldering turret, split, pierced, or hook terminals. In close quarters, it may be desirable to wick or vacuum most of the solder from the joint before attempting to disengage the wire from the terminal. Where splashing of solder is not critical, the connections may be melted, the wire lifted off gently, and the excess solder then removed by wicking or vacuuming.

Terminal Connections

Design is responsible for insuring that the size and number of terminals are sufficient to accommodate the conductors of an assembly. Terminal slots and solder cups should not be modified to accept oversize conductors, and conductors should not be modified to accommodate the termination.

Hook Terminal Connections

The stripped and tinned wire should be bent a minimum of $90°$ and held against the terminal by tension during soldering. The end of the wire should extend no farther than a distance equal to one conductor diameter beyond the hook. A bend of approximately $180°$ (U shape) is also permissible (Figs. 7-14 and 7-19). No attempt should be made to squeeze the tinned wire end against the hook with pliers (because the components to be soldered might be damaged).

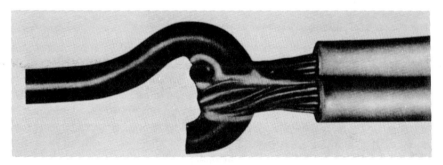

Fig. 7-19. Hook terminal connection.

Pierced Terminal Connections

A wire to be soldered to a pierced terminal should have a $90°$ minimum bend and should extend no farther than a distance equal to one conductor diameter beyond the terminal. The $180°$ bend is preferred. Both hook and pierced terminations should be protected by transparent flexible tubing (Fig. 7-20). The tubing should be cut to length, pushed up the wire before soldering, and slipped into position over soldered connection after the joint is cleaned and inspected. When in the final installed position, the tubing should overlap the wire insulation by a distance equal to at least a diameter.

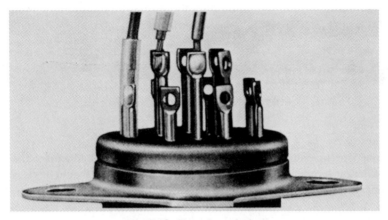

Fig. 7-20. Pierced terminal.

Solder Cups (Connector Type)

Solder Cup Termination

The connection should be firmly mounted with the open end of the cups facing the operator. The wire should be soldered in rows, progressing from the bottom to the top. The cups may be prefilled before any of the wires are inserted. The cups should be heated, either with a resistance type tool or by holding the flat side of a soldering iron against the lower side of the cup, until the solder is completely melted. Keep the heat on the terminal until all trapped flux comes to the surface. The tinned wire should be slowly inserted into the molten solder until the wire bottoms in the cup. The conductor should be in contact with the back wall of the cup.

A smooth fillet should be formed between the conductor and the inner wall of the cup (Fig. 7-21). The solder should follow the contour of the cup entry slot and should not spill over (exceed diameter of the cup as in Fig. 7-22) or adhere to the outside of the cup (except for the slight tinned effect where the

Fig. 7-21. Proper solder cup termination.

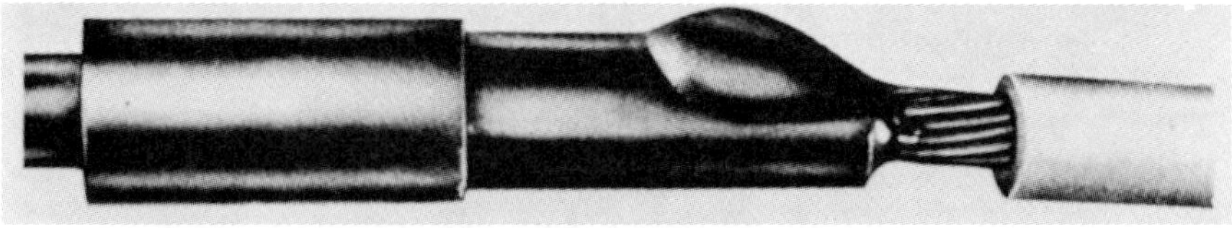

Fig. 7-22. Improper solder cup termination.

sodering iron tip contacts the side of the cup). Wicking of solder up to the point of insulation termination is permitted. All outside strands should be clearly discernible adjacent to the insulation.

Sleeving Over Cup Terminals

When the connector is to be plotted, a protective sleeve over the wire and terminal is not required. If the connector is not potted, the connections should be protected by clear, flexible vinyl tubing or sleeving. The sleeving should not be slid down over the terminal until after cleaning and inspection.

Protection of Connectors

Since connectors are an important delicate part of electronic equipment, probes of any kind should never be used for testing. A mating connector or a test box should be used. Plastic or metal dust covers should be installed and maintained on all connectors at all times and should be removed only when required by inspection or connecting to the mating part. They should be properly reinstalled after inspection.

Swage Cup Terminals

Bottom Entry

When the tinned conductor is inserted into the terminal from the bottom, it should be stripped far enough that the insulation does not extend into the barrel ($\frac{1}{32}$-inch minimum clearance is recommended). It should be bent through the side slot, extending a maximum distance of 0.03 inch, and soldered.

Top Entry

The insertion of conductors into the top of the swage cup is the same as insertion into a solder cup terminal except that the conductor should bottom on either the bottom route conductor or the cup bottom. See Fig. 7-23.

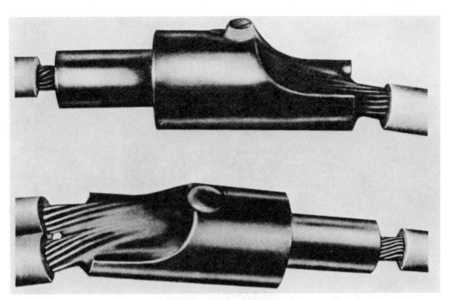

Fig. 7-23. Swage cup terminals.

Bifurcated Terminal Connections

The common methods of inserting conductors into a bifurcated terminal are the bottom route and the side route. If top entry must be used, provisions should be made to obtain good solder fillets to both prongs of the terminal.

Bottom Entry

When the tinned conductor is inserted into the terminal from the bottom (Fig. 7-24), it should be stripped far enough that the insulation does not extend into the barrel ($\frac{1}{32}$-inch minimum clearance recommended). The wire should be bent to lie flush against the should or one of the posts.

Side Entry

For the side entry, the tinned wire end should be brought through the gap and bent flush against the post (Fig. 7-25). A second wire should be bent in the other direction to lie flush against the other post (Fig. 7-26). Additional wires should alternate similarly. The gap may be completely filled, but under no circumstances should the last wire extend beyond the top of the posts. Wires smaller

Fig. 7-24. Bifurcated terminal—bottom entry.

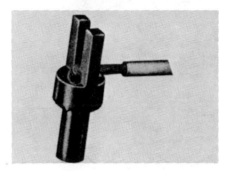

Fig. 7-25. Bifurcated terminal—side entry.

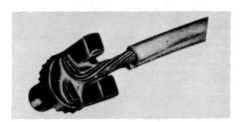

Fig. 7-26. Bifurcated terminal—side entry, two wires.

than AWG26 may be wrapped one-half turn around the posts if sufficient side clearance exists.

Top Entry

When conductors must be inserted from the top, the wire should fill the space between the posts so that a fillet is formed to each post. A small diameter wire conductor may be bent into the "U" shape, provided the combined diameter is sufficient to fill the tap. A filler wire (solid or stranded) may also be used to fill the space (Fig. 7-27).

Fig. 7-27. Bifurcated terminal—top entry.

Turret Terminal Connections

The end of the conductor should be stripped, tinned, and bent into a loop of approximately one-half turn and slipped over the guide slot, snug against the shoulder (Fig. 7-28). Wires smaller than AWH26 may be wrapped to a full turn but not to overlap. Wires larger than AWG26 should not be wrapped more than three-fourths turn.

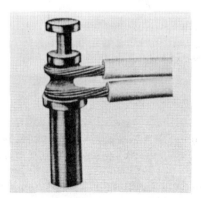

Fig. 7-28. Turret terminal.

When hollow post turret terminals are used, the wire should be stripped to a length that will allow at least $\frac{1}{32}$-inch insulation clearance from the swaged end of the terminal. The tinned end of the conductor is brought out the side slot at the top and wrapped as required for the side route.

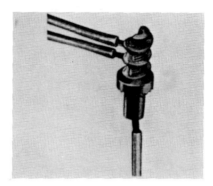

Fig. 7-29. Hollow post turret terminal.

Storage of Terminals

Terminals should be stored in sealed plastic containers with a small bag of desic-cant. Oxidized terminals should not be used.

Printed Wiring Boards

Board Holders

To prevent damage during assembly and inspection, the wiring board should be held by a fixture to prevent bending, warping, or deformation. A typical fixture (Fig. 7-30) will permit the wiring board to be held at each side and fixed in any desired position.

Printed Wiring Board Care and Storage

Boards should be kept in plastic bags between fabrication processes. After assembly and soldering operations are complete, the board should be cleaned and inspected. If required, a conformal coating should be applied to both sides of the board. Conformal coating techniques are governed by the type of material used. The work should not be performed in the soldering area.

Fig. 7-30. Typical board holder.

Boards with Wiring on Both Sides

When parts with conductive cases (such as metal-cased capacitors) must be mounted over conductor patterns, the parts should be insulated with transparent tubing (Fig. 7-31). When a part lead or wire is used to connect pads on opposite sides of the board, the lead or wire must be soldered on both sides.

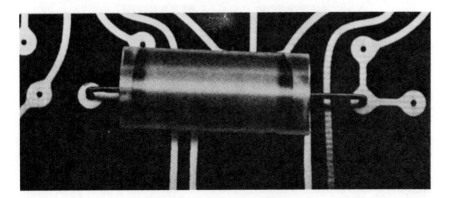

Fig. 7-31. Crossing conductor lines.

Bending Part Leads

Round-nose pliers, long-nose pliers with plastic protected jaws, or other suitable tools are recommended for bending the solid wire and part leads (Fig. 7-32). The bending tool should not flatten, nick, or damage the leads. The printed or lettered part identification should be visible after installation.

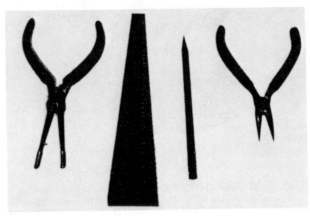

Fig. 7-32. Typical bending tools.

To prevent cracking of the end seals of a part body, the leads should be bent at a distance equal to twice the lead diameter away from the part body. See Figure 7-33, for proper and improper bends.

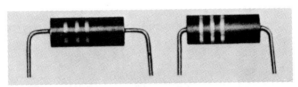

Fig. 7-33. Proper lead bends (left), and improper lead bends (right).

The minimum inside radius of the bend should be equal to one lead diameter.

After insertion, the part leads should be cut to the proper length and clinched. When the lead is to be soldered to a circular termination area, the clinch lead length should be equal to one to three times the radius of the pad. When the termination area is irregularly shaped, such as for shield and ground plane connections, the clinch lead length should be equal to two to four times the hole diameter. The lead should be clinched in such a direction that it does not overhang the edge of the pattern; on round pads this is parallel to and along the circuit pattern (Fig. 7-34). The end of the lead should be clinched with a nonmetallic clinching tool so it lies approximately parallel to the board surface, the natural spring back of the lead material being acceptable.

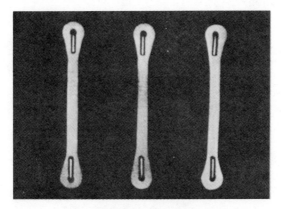

Fig. 7-34. Properly clinched leads.

Part leads which cannot be bent or clinched, should be cleaned and tinned and then cut to a length that will allow the end of the lead to protrude from $\frac{1}{32}$ to $\frac{3}{32}$-inch beyond the solder pad. The parts should be securely mounted to the board with clamps, or embedded in epoxy resin. The solder should form a fillet, but should not obscure the contour of the lead (Fig. 7-35).

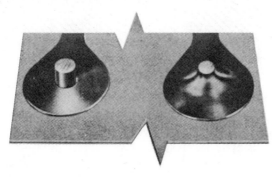

Fig. 7-35. Unbendable lead.

Removing Gold Plating

Immediately prior to soldering, gold plating must be removed from the areas to be soldered. An approved method of gold removal from pads is to use a white typewriter eraser (pencil shaped), applying light pressure to the pad until the base metal is exposed. Gold particles and eraser crumbs should be removed to avoid contamination.

Handling

Cleaned wiring boards, terminals, part leads, etc., should not be touched with bare hands. When handling is unavoidable, clean finger cots or white gloves should be used.

Soldering Printed Wiring Boards

Soldering to printed wiring boards should be accomplished only on base metal, hot tin-lead coating, or on tin-lead electroplated reflow surfaces.

It is essential that the soldering iron be of the correct wattage and tip shape, and temperature controlled.

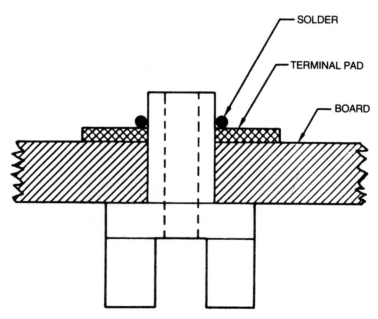

Fig. 7-36. Terminal insertion.

Satisfactory joints can be produced on printed wiring boards by holding the soldering iron at approximately 45°, with the tip of the iron making contact with both the lead and the pad. A small amount of fresh solder should be applied to the tip of the iron at the intersection of the lead and tip to promote heat transfer. The solder should then be moved quickly to the opposite side of the lead and continued in a circular motion from the tip of the lead toward the back of the pad until, a proper amount of solder is deposited.

Mounting and Soldering Terminals

When mounting and soldering terminals to printed circuit wiring boards, the following procedure should be followed:

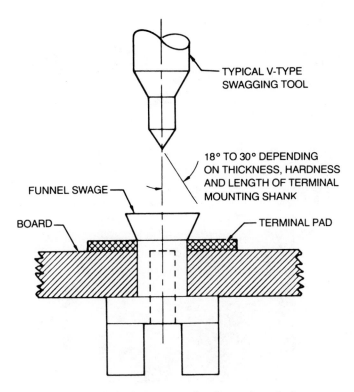

TYPICAL V-TYPE
SWAGGING TOOL

18° TO 30° DEPENDING
ON THICKNESS, HARDNESS
AND LENGTH OF TERMINAL
MOUNTING SHANK

FUNNEL SWAGE

BOARD

TERMINAL PAD

Fig. 7-37. Typical swage tool and swaged terminal.

1. Drill the pad hole to a diameter that will permit the terminal shank to be pressed through the board by hand. A press fit is not necessary, but the terminal should fit snugly enough to prevent it from falling out. (Fig. 7-36).
2. Remove gold plating from the terminal area; do not remove tin-lead coating.
3. Solder rings, if used, should be placed over the shank prior to swaging.
4. Swage the terminal with a funnel "V" swage (Fig. 7-37).

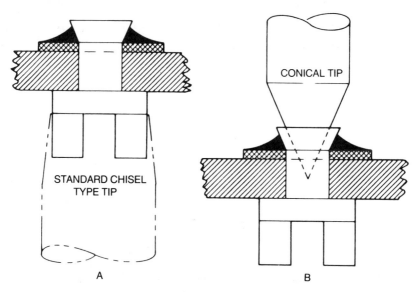

Fig. 7-38. Soldering a swaged terminal. An alternate method is shown on the right.

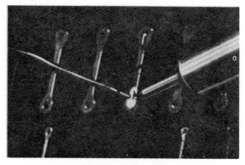

Fig. 7-39. Acceptable solder joints.

The finished connection should have a smooth even fillet on both sides of the clinched lead. See Fig. 7-39. Figure 7-40 shows that appearance and cross sections of good solder joints.

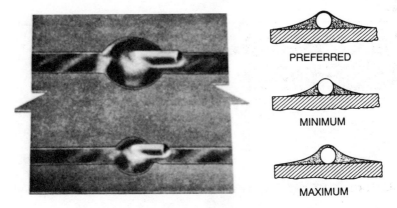

PREFERRED

MINIMUM

MAXIMUM

Fig. 7-40. Soldering method for printed wiring board.

Part Mounting

Tubular parts or parts with a flat surface should be mounted parallel to and in contact with the surface of the wiring boards.

Transistors mounted vertically should be positioned on special pads to allow room for a slight "C" bend which is necessary for stress relief. Odd-shaped parts should be mounted in clamps or embedded in potting compound (Fig. 7-41).

Heavy Parts

Parts that weight more than $\frac{1}{4}$ ounce per lead should not be supported by leads only, but should have a suitable clamp or bracket to be embedded in epoxy.

Fig. 7-41. Part mounted in clamp.

8
MECHANICAL ASSEMBLY

SECURING PARTS WITH THREADED FASTENERS

The most common cause of failure in threaded fastener assemblies is fatigue. A properly tightened, rigid-joint fastener will not fail in fatigue.

All fasteners should be tightened within established torquing levels and the sequence of assembly is critical in avoiding misalignment or binding of parts.

The following charts provide the sequence for accepted torquing of multi-fastener installations and the industry accepted standard for torquing levels.

For multi-fastener installations, the sequential torquing should be applied in one-third increments of the torquing level in the sequence shown until the total torque level is achieved.

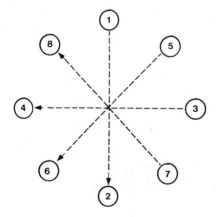

Even numbers of fasteners

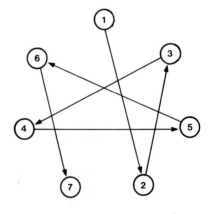

Odd number of fasteners

Maximum Tightening Torque (In-Lbs) for
Fasteners and Nuts of Various Materials

Screw Line	Aluminum or Copper	Brass	Monel	Silicon Bronze	Steel Low-Carbon
2-56	1.4	2.0	2.5	2.3	2.2
2-64	1.7	2.5	3.1	2.8	2.7
4-40	2.9	4.3	5.3	4.8	4.7
4.48	3.6	5.4	6.7	6.1	5.9
6-32	5.3	7.9	9.8	8.9	8.7
6-40	6.6	9.9	12.3	11.2	10.9
8-32	10.8	16.2	20.2	18.4	17.8
8-36	12.0	18.0	22.4	20.4	19.8
10-24	13.8	18.6	25.9	21.2	20.8
10-32	19.2	25.9	34.9	29.3	29.7
1/4-20	15				
1/4-28	22				

Torquing Levels (In-Lbs) for Tightening Nuts

Screw Size	Standard Bolts, Studs, and Screws With a Tensile Strength of 40,000 to 75,000 psi				High Strength Bolts, Studs and Screws With a Tensile Strength of 90,000 psi or Greater			
	Shear Nuts		Tension Nuts		Shear Nuts		Tension Nuts	
	Min.	Max.	Min.	Max.	Min.	Max.	Min.	Max.
2-56	.98	1.4	1.5	2.5	1.4	2.1	2.3	3.4
2-64	1.0	1.6	1.7	2.6	1.6	2.3	2.6	3.9
4-40	1.9	2.9	3.2	5.0	2.9	4.3	4.8	7.2
4-48	2.3	3.4	3.8	5.6	3.4	5.0	5.6	8.4
6-32	3.7	5.5	6.1	10.0	5.5	8.2	9.0	13.6
6-40	4.5	6.7	7.4	11.0	6.7	10.0	11.0	16.6
8-32	7.0	9.0	12.0	17.0	9.0	16.0	17.0	22.0
8-36	7.8	10.0	14.0	20.0	10.5	18.5	20.0	25.0
10-24	12.0	15.0	22.0	28.0	15.0	26.0	28.0	34.0
10-32	15.0	18.0	30.0	34.0	18.0	32.0	34.0	40.0
1/4-20	N/A	N/A	24.5	27.6	N/A	N/A	55.0	62.0
1/4-28	N/A	N/A	33.3	36.9	N/A	N/A	75.0	83.0

Tightening Torque Requirements for Threaded Semiconductor Devices, All Materials

Stud Size	Torque, In.-Lb
6-32	7-9
8-32	8-10
10-32	10-12
1/4-28	16-20
5/16-24	16-20
1/2-20	80-100

The reinstallation of a fastener that is sealant-locked into a tapped hole should always have the old sealant completely removed and new sealant applied.

Fastener installations should always be tested by applying 90% of the specified torque to the fastener.

Field reliability is often dependent on the proper installation of threaded fasteners, especially in equipment using rotating parts (fans, motors, pumps) or equipment installed in a vibrating environment such as aircraft, land vehicles, shipboard, or in a location near rotating equipment where the vibrations can be transmitted.

COUNTERSINKING AND COUNTERBORES

Countersinking

Never attempt to countersink screws in sheet metal which is too thin to take the full depth of the screw head, this will result in an extremly weak joint. When necessary to countersink in thin metal, the thinner top metal should be dimpled into the countersunk mating piece, as shown below:

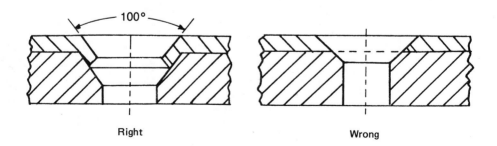

Right   Wrong

Minimum Sheet Thickness

The figure below shows the minimum thickness of sheet metal recommended for maximum joint strength countersinking. The countersink diameter (C) column should be used for one hole or "Make-from" pattern. Additional clearances may be required for multi-hole patterns, depending on design (i.e., material, thickness, etc.).

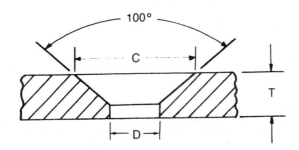

Screw Size	C Countersink Data	D Dia. (±0.005)	Recommended Min. T**	Minimum Csk. TA
4 (0.112)	0.227 to 0.237	0.120	0.071	0.050
6 (0.138)	0.281 to 0.291	0.147	0.090	0.063
8 (0.164)	0.334 to 0.344	0.173	0.112	0.080
10 (0.190)	0.387 to 0.397	0.199	0.125	0.090
0.250	0.509 to 0.519	0.261	0.160	0.112

** Minimum sheet thickness for maximum strength of joint (equal to 1-½ times the head height).
A Minimum sheet thickness for countersinking. Thinner sheets should be dimpled.

Minimum Sheet Thickness

Counterbores:

Counterbores for Socket Head Cap Screws

The allowable tolerance on the length under the head of hexagonal head cap screws is as follows: Up to 1 inch in length plus 0.000 minus 0.032; 1 to 2 inches inclusive, plus 0.000 minus 0.062; over 2 inches, plus 0.000 minus 0.093.

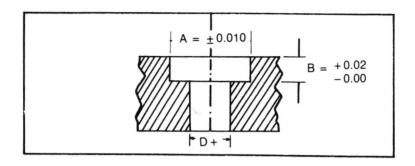

Screw Size	A	B
4	0.219	0.11
6	0.250	0.14
8	0.312	0.27
10	0.344	0.19
0.250	0.406	0.25

Screw Size	A	B
0.313	0.500	0.31
0.375	0.594	0.38
0.437	0.688	0.44
0.500	0.812	0.50

Counterbores for Socket Head Cap Screws

SCREW LENGTHS AND APPLICATIONS

Screw Lengths

Standard commercial lengths should be used whenever possible. When computing the required length, the requirements outlined should be fulfilled.

Engaged Lengths for Blind Threaded Holes

The length of engagement should be considered to mean the distance a screw extends into a tapped hole. It includes the chamber of the screw and the countersink of the tapped hole when applicable. Screws provide a minimum engaged length equal to one and one-half times the nominal diameter when high strength is required and when the screw is frequently disassembled. Where high strength is not a factor, a minimum engagement equivalent to one nominal diameter is required. (See Fig. 8-1.)

If the amount of looseness that is possible with Unified screw threads is critical in the maintenance of axial alignment, stability may be achieved by the use

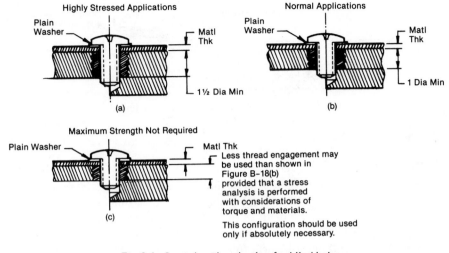

Highly Stressed Applications

Normal Applications

(a)

(b)

Maximum Strength Not Required

Less thread engagement may be used than shown in Figure B–18(b) provided that a stress analysis is performed with considerations of torque and materials.

This configuration should be used only if absolutely necessary.

(c)

Fig. 8-1. Screw lengths selection for blind holes.

of shoulder design in conjunction with a close fitting, plain cylindrical surface rather than by increasing the length of engagement. However, the eccentricity allowances for shoulder designs may necessitate the use of undersize threads on the screws, since Class 2 Unified threads are not provided with a positive clearance. A typical shoulder design is illustrated in Fig. 8-2.

Projected Lengths

The length of screws and bolts should provide a minimum thread projection of $1\frac{1}{2}$ threads. Maximum length should be limited by the nearest larger standard screw length. This rule applies except when such projection will result in corona discharge or when design requirements cannot be met. (See Fig. 8-3.)

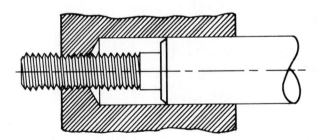

Fig. 8-2. Typical shoulder design.

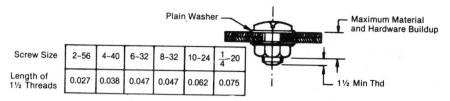

Screw Size	2-56	4-40	6-32	8-32	10-24	$\frac{1}{4}$-20
Length of 1½ Threads	0.027	0.038	0.047	0.047	0.062	0.075

Fig. 8-3. Screw length selection for through holes.

Internal Thread Lengths

Blind Holes

To insure positive assembly of a screw into a blind tapped hole, the minimum full thread length of the hole must be more than the maximum engaged length of the screw. A minimum of three threads should be added to the maximum engaged length to determine the minimum tap depth. An exception to this occurs when the depth must be held to an absolute minimum, in which case the minimum full thread depth is equal to the maximum engaged length of the screw, or one and one-half threads for bottom tap.

A dimension showing the depth of the tap drill should be specified when this is important. If this dimension is not important, the depth may vary at the discretion of the manufacturer.

The fabrication of unrelieved threads in a blind hole is often complicated by insufficient tap and chip clearance. (See Fig. 8-4). The specification of a com-

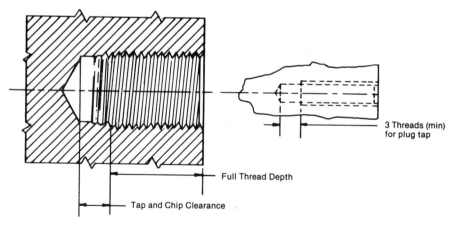

Fig. 8-4. Blind hole clearance.

patible depth of hole/thread length ratio, therefore, is an important factor in the design of such holes. Insufficient tool and chip clearance may result in faulty thread and decreased tool life, and consequently, in excessive fabrication costs. For these reasons, constants based upon the number of threads per inch to be used when calculating the depth of hole required for optimum internal thread fabrication are listed in Table 8-1.

The preferred clearance is based on a tap and chip clearance dimension equal to five times the thread pitch plus 0.10 inch. This value permits the use of a plug end tap in thread cutting.

Whenever possible, the preferred clearance should be utilized. Only in cases where space limitations prohibit the use of these values should the nonpreferred clearance be used. This is clearly understood by comparing the production cost of one operation (preferred clearance) with two operations (nonpreferred clearance). In calculating preferred clearance, the depth of the complete threads is established by design requirements. To this dimension is added the preferred clearance constant for the number of threads per inch (refer to Table 8-1). The sum of the two dimensions is the preferred depth of hole.

Nonpreferred Clearance

The nonpreferred clearance is based on a tap and chip clearance dimension equal to two times the thread pitch plus 0.10 inch. This value necessitates the use of a plug end tap followed by a bottoming end tap in thread cutting. If the design will not permit a preferred depth of hole as calculated, the nonpreferred clearance constant for the given number of threads per inch is added to the depth of the complete threads as established by the design requirements. The sum of the two dimensions is the nonpreferred depth of hole. Because of the two tapping operations involved, the nonpreferred is the more expensive.

Eccentricity

Designs requiring close control of eccentricity between threads and other integral features of the part should be avoided. If required, the eccentricity may be as little as 0.003 total indicator reading (0.0015 actual eccentricity) for cut threads. Because thread tolerances are greater than the normal eccentricity of turned or ground parts, closer control of eccentricity on mating threaded parts may be obtained by providing turned or ground pilot diameters. Figure 8-5 illustrates two applications of pilot diameters.

**Table 8-1. Recommended Tap and Chip Clearance
for Unrelieved, Blind, Tapped Holes**

Threads Per Inch	Pitch	Preferred Clearance*	Nonpreferred Clearance†
80	0.012	0.16	0.12
72	0.014	0.17	0.13
64	0.016	0.18	0.13
56	0.018	0.19	0.14
48	0.021	0.20	0.14
44	0.023	0.22	0.15
40	0.025	0.22	0.15
36	0.028	0.24	0.16
32	0.031	0.26	0.16
28	0.036	0.28	0.17
24	0.042	0.31	0.18
20	0.050	0.35	0.20
18	0.056	0.38	0.21
16	0.062	0.41	0.22
14	0.071	0.46	0.24
13	0.077	0.48	0.25
12	0.083	0.52	0.27
11	0.091	0.56	0.28
10	0.100	0.60	0.30
9	0.111	0.66	0.32
8	0.125	0.72	0.35
7	0.143	0.82	0.39
6	0.167	0.94	0.43
5	0.200	1.10	0.50
4-1/2	0.222	1.21	0.54
4	0.250	1.35	0.60

*Values in this column do not appear on engineering data, but are to be used in calculating the depth of hole required before tapping in all cases possible. These values are based on a figure of five times the thread pitch plus 0.10.

†Values in this column do not appear on engineering data, but are to be used in calculating the depth of hole required before tapping when space limitations prohibit the use of the values shown in the preferred clearance column. These values are based on a figure of two times the thread pitch plus 0.10.

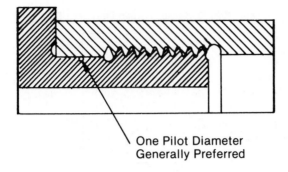

One Pilot Diameter
Generally Preferred

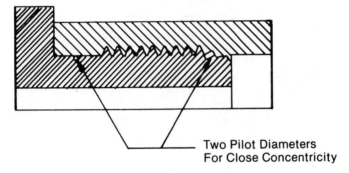

Two Pilot Diameters
For Close Concentricity

Fig. 8-5. Pilot diameters.

Undercuts (Reliefs)

Undercuts should be avoided unless absolutely necessary. They may be used only when thread runout on external and internal applications cannot be tolerated due to space limitations.

External Threads

Generally, an internally threaded part is counterbored to clear the incomplete threads of the mating part when it must assemble close to a shoulder. When the design does not permit the internally threaded part to be counterbored, the external thread may be undercut as shown in Table 8-2. External threads should not be undercut in sizes smaller than no. 4 (0.112 nominal diameter). Values for external thread undercuts are specified in Table 8-2.

Table 8-2. External Thread Undercut Data.

Threads Per Inch	A +0.016 −0.000	B		Threads Per Inch	A +0.016 −0.000	B		Threads Per Inch	A +0.016 −0.000	B
80	0.025	0.020		32	0.062	0.046		12	0.167	0.120
72	0.028	0.022		28	0.071	0.053		10	0.200	0.143
64	0.031	0.024		24	0.083	0.061		8	0.250	0.178
56	0.036	0.027		20	0.100	0.072		6	0.333	0.237
48	0.042	0.032		18	0.111	0.081		4	0.500	0.353
40	0.050	0.037		16	0.125	0.090				
36	0.056	0.041		14	0.143	0.103				

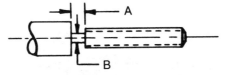

Fig. 8-6. Undercut Dia = Maximum Major Dia − B (Table 2) $^{+0.000}_{-0.005}$

Table 8-3. Internal Thread Undercut Data.

Threads Per Inch	A +0.016 −0.000				B
	2P	3P	4P	7P	
80	0.025	0.038	0.050	0.088	0.010
72	0.028	0.042	0.056	0.097	0.010
64	0.031	0.047	0.062	0.109	0.010
56	0.036	0.054	0.072	0.125	0.010
48	0.042	0.062	0.083	0.146	0.010
40	0.050	0.075	0.100	0.175	0.010
36	0.056	0.083	0.111	0.194	0.010
32	0.062	0.094	0.125	0.219	0.010
28	0.071	0.107	0.143	0.250	0.015
24	0.083	0.125	0.167	0.292	0.015
20	0.100	0.150	0.200	0.350	0.015
18	0.111	0.167	0.222	0.389	0.025
16	0.125	0.188	0.250	0.438	0.025
14	0.143	0.214	0.286	0.500	0.025
12	0.167	0.250	0.333	0.583	0.025
10	0.200	0.300	0.400	0.700	0.035
8	0.250	0.375	0.500	0.875	0.035
6	0.333	0.500	0.667	1.167	0.045
4	0.500	0.750	1.000	1.750	0.065

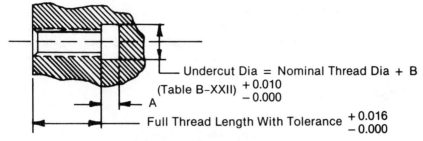

Figure 8-7.

Internal Threads

When the depth of a hole is restricted and thread runout cannot be tolerated, an undercut may be utilized. In the case of undercuts, the thread length should be specified as a nominal dimension with a plus 0.016 tolerance as shown in Table 8-3. The nominal length is equal to the nominal length of engagement of the screw. Since the thread is full its entire length, it is permissible for the screw to protrude into the undercut. The undercut width should be equivalent to a preferable length of four pitches, although three or a minimum of two may be used. Internal threads should no be undercut in sizes smaller than 0.31 nominal diameter. Values for widths are specified in Table 8-3.

RIVETS—DESIGN, METHODS, AND ASSEMBLY

Riveting is a low cost method of fastening when production procedures, can be incorporated. The assembly costs, however, are higher than for screws when quantities are such that production may not be used. The following information covers design data, methods of riveting, and assembly of parts using riveting techniques.

In selecting riveting as the means of manufacturing any given product, the designer must carefully consider the advantages and limitations of the riveting process.

Advantages

1. Either metallic or nonmetallic materials can be joined.
2. Dissimilar metals and assemblies having parts of nonuniform thickness can be fastened readily.
3. The rivet can be made out of a variety of materials.

4. No heat distortion or crystallization of the loose metal occurs during the riveting process.
5. Economy prevails because high production rates are possible, and rivets are less expensive than threaded fasteners.

Limitations

1. Parts joined by riveting cannot be disassembled easily.
2. Riveted joints are not normally watertight or airtight.
3. Riveting is relatively more expensive than spot welding.
4. Riveted joint strength is generally lower than that of bolter or welded assemblies.

Classes of Rivets

All rivets are driven in both the hot and cold condition by hammering or peening, or by the application of steady pressure or squeezing. Basically, the following three classes of rivets are most commonly used:

1. Solid rivet
2. Tubular rivet
3. Blind rivet

See Table 8-4 for a summary of rivet fasteners.

Hot Riveting

In hot riveting, the rivet is first brought to a high temperature which renders it soft and malleable. It is then quickly inserted through the parts to be joined and headed over by a riveting hammer while the other end is backed up with a tool called a buck or bucking bar. This method is usually applied to large rivets used for structural purposes.

Cold Riveting

In cold riveting, the rivet is driven at the normal ambient temperature. Cold-hammering methods are usually applied to small rivets. Cold squeezing in a riveting machine that accurately controls applied pressure is satisfactory for work which can be accommodated in a press of the portable or mounted type.

Solid-Shank Rivets

The solid-shank rivet is intended to permanently fasten and tightly join two or more parts where ready separation of the parts is not required. The solid-shank

Table 8-4. Summary of Rivet Fasteners

Description	Function	Precautions	Remarks
Solid Rivet Countersunk Head	Permanent fastener for joining two or more parts when a flush surface is required.	Flush riveting should be avoided wherever possible and, when it is necessary to flush both sides of a joint, the joint thickness must be at least 0.015 greater than the sum of the rivet head heights.	Requires countersinking hole prior to insertion.
Solid Rivet	Permanent fastener for general applications where head protection is not objectionable.		Least costly method of fastening when production methods can be incorporated.
Tubular Rivet	Permanent fastener for joining plastic, wood, and sheet metal in lightly loaded structures. Has shear and compressive strength equal to solid rivet.	Solid rivets should not be used in joints of soft or brittle materials such as laminated plastics. Tubular or blind rivets should be used in these applications.	Rivet should be clinched against thickest or strongest section.
Blind Rivet (Nonstructural)	Permanent fastener for assembly on nonstructural components, securing anchor nuts and fastening name, warning, and instruction plates. Head styles include: universal, 100 degree countersunk, and modified truss.	The tubular rivet should not be used to replace solid rivets or bolts of equal size in applications requiring the tensile and shear loads of solid rivets or bolts. Because of the method used for clinching, these rivets should not be expected to resist tensile loads. The head of the rivet should always bear against the thinnest or softest material.	

rivet is used in the assembly of parts which must carry structural or other heavy loads. There are two types of solid-shank rivets (protruding head and flush head) as listed in Table 8-5. Solid-shank rivet materials are listed in Table 8-6.

Protruding head rivets are generally preferred (over flush head) because they offer such advantages as high allowable joint strengths and do not require special machining or forming of the joined material to achieve flushness.

Flush head rivets require countersunk or dimpled surfaces, which tend to weaken the material. They should be used only where flushness is necessary to satisfy such design requirements as aerodynamic smoothness, appearance, mating part clearance, etc.

The following design considerations apply to solid-shank rivet applications:

1. Reinforcements or stiffeners should be located on one side of a web to simplify handling and tooling.

Table 8-5. Solid-Shank Rivet Installations and Applications

Description	Application
Countersunk	Used where a smooth surface is required on one side of an assembly. The top sheet must be thick enough to permit countersinking. APPLICABLE RIVET: 100° Flush Head
Dimple and Countersink	Used where a smooth surface is required on one side of an assembly. The top sheet must be dimpled since it is not thick enough for countersinking and the lower sheet must be thick enough for countersinking. APPLICABLE RIVET: 100° Flush Head
Double Dimple	Used where a smooth surface is required on one side of an assembly and neither sheet is thick enough for countersinking. APPLICABLE RIVET: 100° Flush Head
100° / 60° Double Flush	Used where surface flushness is required on both sides of an assembly. This installation formed by upsetting the shank of an MS20426 rivet into a 60° countersunk hole and subsequently shaving the upset head flush with rivet shavers. APPLICABLE RIVET: 100° Flush Head
Protruding	Used for structural and nonstructural assemblies where flushness requirements permit. APPLICABLE RIVET: Protruding (Universal) Head
NACA 60°	For restricted usage where a smooth surface is required on one side of an assembly and an MS20426 (double flush) rivet cannot be used. This rivet is formed by upsetting the shank of an MS20470 rivet into a 60° countersunk hole and subsequent shaving the upset head flush with rivet shavers. APPLICABLE RIVET: Protruding (Universal) Head

2. Straight runs of rivets of the same size and type are the easiest and most economical to produce because changing rivet size or type requires a new manufacturing set-up.

3. Areas to be riveted should have flat contacting surfaces. Shorter rivets minimize the tendency to distort or buckle.

4. Where possible, joints should be designed so that special tools are not required. This will reduce both tooling costs and assembly time. (For example, it is easier to drive one long rivet with a spacer between channel

Table 8-6. Solid-Shank Rivet Materials

Material	Remarks
1100 Aluminum Alloy	For riveting nonstressed or nonstructural assemblies; 3003 and 5052 or other soft-aluminum alloys; plastic and laminated plastic and joints where the rivet is welded. Recommended structural rivet for thin aluminum alloy sheets subject to low loads and for joining all aluminum alloys except 3003 and 5052. If used to join 3003 or 5052 to a structural material, the rivet head should be located against softer material.
2117 Aluminum Alloy	Used to join laminated plastics only if sheet thickness exceeds rivet diameter. These rivets are available in sizes of 0.094, 0.125, and 0.156 inch diameters only. If larger sizes and greater length characteristics are required, 2024 rivets should be used since they are approximately 35 percent stronger than 2117 rivets. The 2117 rivets are not heat treatable, but are driven in the annealed condition and work-hardened by driving.
5056 Aluminum Alloy	To prevent galvanic corrosion, this is the only type of rivet used for fastening magnesium alloys. May be used for aluminum-magnesium combinations. Used for high-strength rivets stronger than 2117. This is the recommended structural rivet for joining all aluminum alloys (except 3003 and 5052) where 2117 rivets are not suitable. If used to join 3003 and 5052 to a structural material, the rivet head should be located against softer material.
2024 Aluminum Alloy	May be used to rivet laminated plastics only if sheet thickness exceeds rivet diameter. Recommended diameters are 0.187 and 0.250 inch. On smaller sizes, 2117 rivets should be used since 2024 rivets in diameters smaller than 0.187 inch are difficult to drive. For high-strength applications, 2024 rivets shall be used since a stronger joint will result with fewer rivets required at lower cost and savings in weight. 2024 rivets must be heat treated. If not driven within 10 minutes after heat treating, they must be refrigerated until ready for use. These rivets remain in the soft or "SW" condition as long as they are refrigerated. Hardening occurs at room temperature and, therefore, these rivets must be driven in the heat treated or refrigerated condition. Otherwise, they become too hard to drive.
Monel	Monel rivets are used for high temperature applications. They are also used for riveting corrosion resistant steel carbon and alloy steel, titanium, and copper alloys. Monel rivets are magnetic at room temperature.

walls than to drive two small rivets.) Fillers or plugs for rivet support should be of a material with characteristics equal to or better than cloth-base phenolic.

5. Rivets are not recommended for castings. Where used with castings, however, the manufactured head should be positioned against the casting and the upset head against the noncast part.

6. If rivets are used to fasten soft materials (elastomers, plastics, etc.) that cannot resist high compression, back-up plates are recommended to help distribute the load over a large area instead of concentrating the bearing surface over a relatively small area. This prevents overcompression of the soft part and digging or tearing into the surface. The thickness of each back-up plate should be at least equal to $\frac{3}{4}$ of the manufactured head height of the rivet. Soft rivets should be used for these applications.

7. Rivets should not be designed to function as pivot joints.

8. Rivets should not be used as a means of attaching interchangeable items.

9. Cost is directly related to the number of rivets in a joint. For economy, the least number of rivets which will satisfy design requirements should be used.

Size Requirements

Generally, solid-shank rivet size is to be selected so that the shear strength of the rivet is equal to or slightly less than the allowable bearing strength of the weakest or thinnest sheet. The relationship of rivet diameter to sheet thickness is shown in Table 8-7.

Unless required by the design, the following considerations apply to the sheet thickness of all rivet joints:

1. The minimum recommended sheet thickness for protruding head rivets in inches is:

Rivet Diameter:	0.093	0.156	0.187	0.250
Minimum Sheet Thickness:	0.020	0.025	0.040	0.064

Table 8-7. Rivet Diameter vs. Sheet Thickness (Solid-Shank Rivets).

Joint Material	Single Shear t = Thinnest Sheet	Double Shear t = Middle Sheet	Flush Riveting t = C'sk Sheet
Aluminum Alloys	D Max = 3t	D Max = 1.5t	D Max = 2t
Other Alloys	D Max = 5.5t		

2. When riveting in light metal castings, the rivet diameter should not be greater than the thickness of that area of the casting being riveted.
3. The manufactured head should be positioned on the side of the thinnest sheet of the joint whenever possible.

Determining the length of a rivet requires consideration of total thickness of the member being joined, the difference between the rivet diameter and the hole size, and the type of driven head. This length is best determined by experiment, but, for purposes of specifying a length on a drawing, an allowance of $1\frac{1}{2}$ diameters may be used for the flat head rivet.

When deciding the length, too long a length is preferable to too short a length because too short a length may result in insufficient filling of the rivet hole and damage to the sheet. Table 8-8 relates suggested rivet length to the determining parameters.

The following consideration applies to rivet positioning: edge distance for rivets should be at least two rivet diameters plus 0.03 inch to avoid distortion or cracking of the sheet during rivet installation.

Table 8-8. Solid-Shank Rivet Length (Part I).

Rivet Data	Joint Thickness													
	0.032	0.063	0.094	0.126	0.157	0.188	0.219	0.251	0.282	0.313	0.344	0.376	0.407	
	Thru				Thru				Thru				Thru	
	0.031	0.062	0.093	0.125	0.156	0.187	0.218	0.250	0.281	0.312	0.343	0.375	0.406	0.437
	Rivet Length*													
1/16			3/16	1/4	1/4	5/16	5/16	3/8	3/8	7/16	7/16	1/2	1/2	
3/32		3/16	1/4	1/4	5/16	5/16	3/8	3/8	7/16	7/16	1/2	1/2	9/16	9/16
1/8		1/4	1/4	5/16	5/16	3/8	3/8	7/16	7/16	1/2	1/2	9/16	9/16	5/8
5/32	1/4	1/4	5/16	5/16	3/8	3/8	7/16	7/16	1/2	1/2	9/16	9/16	5/8	5/8
3/16	1/4	5/16	5/16	3/8	3/8	7/16	7/16	1/2	1/2	9/16	9/16	5/8	5/8	3/4
1/4	3/8	3/8	7/16	7/16	1/2	1/2	9/16	9/16	5/8	5/8	3/4	3/4	3/4	3/4

*Includes material allowance to permit formation of driven head.

Spacing

(See Figs. 8-8 and 8-9.) Rivet spacing is the center-to-center distance between rivets measured along a row of rivets; row spacing is the perpendicular distance between two adjacent rows of rivets. To prevent distortion, sufficient space between rivets should be provided for riveting tools. To facilitate manufacturing, rivet spacing should be as generous as practical. For normal structural applications, minimum rivet spacing for all applications should be four times the shank diameter. Spacing less than this minimum should be used only under special circumstances. With even and equal rivet spacing, minimum rivet spacing is equal to minimum row spacing; with equal but staggered rivet spacing, row spacing is limited to 0.9 times minimum rivet spacing. Any reduction in these values will bring the diagonal distance between rivets closer than the minimum rivet spacing.

Straight runs of equally spaced rivets of the same size and type are recommended to facilitate production and to permit the use of automatic riveting and indexing fixtures. Rivets may be staggered if necessary.

The minimum distance between the center line of a rivet and the bend tangent

Table 8-8. (*Continued*)

Rivet Data	Joint Thickness													
	0.438	0.469	0.501	0.532	0.564	0.595	0.626	0.657	0.688	0.719	0.751	0.782	0.813	0.844
	Thru				Thru				Thru				Thru	
	0.468	0.500	0.531	0.563	0.594	0.625	0.656	0.687	0.718	0.750	0.781	0.812	0.843	0.875
	Rivet Length*													
$\frac{1}{16}$														
$\frac{3}{32}$	$\frac{5}{8}$	$\frac{5}{8}$	$\frac{3}{4}$	$\frac{3}{4}$	$\frac{3}{4}$	$\frac{3}{4}$	$\frac{7}{8}$	$\frac{7}{8}$	$\frac{7}{8}$	$\frac{7}{8}$	1	1	1	1
$\frac{1}{8}$	$\frac{5}{8}$	$\frac{3}{4}$	$\frac{3}{4}$	$\frac{3}{4}$	$\frac{3}{4}$	$\frac{7}{8}$	$\frac{7}{8}$	$\frac{7}{8}$	$\frac{7}{8}$	1	1	1	1	
$\frac{5}{32}$	$\frac{3}{4}$	$\frac{3}{4}$	$\frac{3}{4}$	$\frac{3}{4}$	$\frac{7}{8}$	$\frac{7}{8}$	$\frac{7}{8}$	$\frac{7}{8}$	1	1	1	1		
$\frac{3}{16}$	$\frac{3}{4}$	$\frac{3}{4}$	$\frac{3}{4}$	$\frac{7}{8}$	$\frac{7}{8}$	$\frac{7}{8}$	$\frac{7}{8}$	1	1	1	1			
$\frac{1}{4}$	$\frac{7}{8}$	$\frac{7}{8}$	$\frac{7}{8}$	$\frac{7}{8}$	1	1	1	1						

*Includes material allowance to permit formation of driven head.

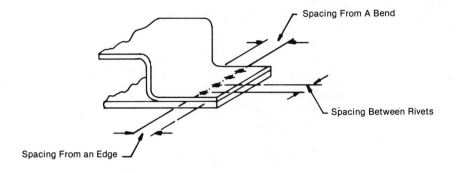

Fig. 8-8. Recommended rivet spacing.

Spacing From an Edge	Spacing Between Rivets	Spacing From a Bend
1-1/2D Minimum	3D Minimum	2D Minimum

Note: D = Rivet Diameter

line of a joggle is proportional to the size of the tool required to install the rivet. Clearance must be sufficient so that any required dimpling or countersinking tools and the rivet set tools will clear the bend tangent line of the joggle radius. (See Fig. 8-10.)

To improve the effectiveness of stiffener angles, attaching rivets should be located as near to the vertical leg of the angle as practical. Rivets attaching extruded and formed angle stiffeners (nondimpled) should be located one rivet diameter plus 0.06 inch from the tangent line of the fillet or bend radius of the angle. (See Fig. 8-11.)

The minimum structural clearance for rivet installations, after all access and tool clearance requirements have been met, is shown in Fig. 8-12.

To insure a proper grip and a tight strong joint, the rivet head should be at right angles to the rivet axis. Table 8-9 illustrates the maximum permissible inclination of driven or manufactured heads of standard 2117 and 2024 aluminum alloy rivets between the rivet center line and surface normal.

The maximum unsupported length of 2117 or 2024 aluminum alloy rivets through tubing, core material, honeycomb, etc., that is not capable of supporting rivet expansion should not exceed 1.5 times the rivet diameter. The maximum unsupported length of other rivets should not exceed the rivet diameter.

The maximum grip length of 2117 or 2024 alloy rivets should not exceed 5.3

Even and Equal Spacing

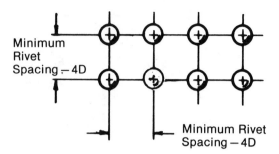

Minimum Rivet Spacing — 4D

Equal But Staggered Spacing

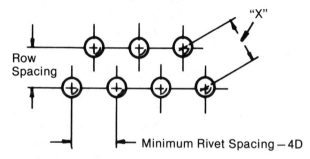

Minimum Rivet Spacing — 4D

Note: "X" is not rivet spacing, but it must never be less than the minimum rivet spacing.

Fig. 8-9. Rivet and row spacing.

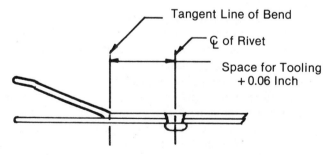

Fig. 8-10. Rivet adjacent to joggle.

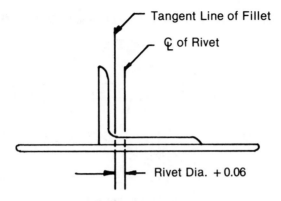

Fig. 8-11. Rivet adjacent to angle.

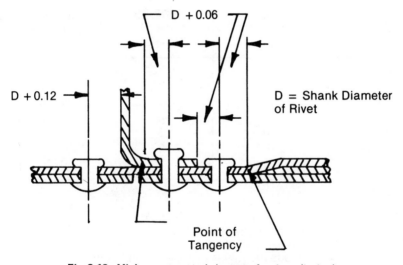

Fig. 8-12. Minimum structural clearance for rivets (inches).

times the rivet diameter. The maximum grip length of other rivets should not exceed 2.5 times the rivet diameter.

Tubular Rivets

The tubular rivet is essentially a solid rivet except that the lower section of the rivet shank is extruded or drilled to provide the clinch. Tubular rivets are made

Table 8-9. Head Angular Limits for 2117 and 2024 Alloy Rivets.

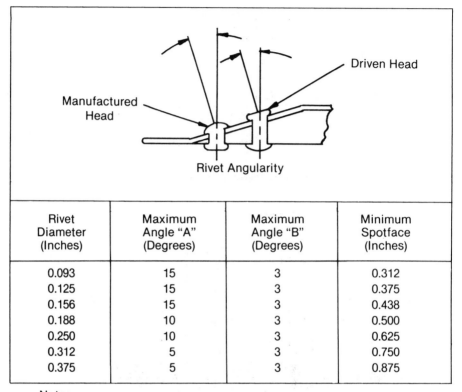

Rivet Diameter (Inches)	Maximum Angle "A" (Degrees)	Maximum Angle "B" (Degrees)	Minimum Spotface (Inches)
0.093	15	3	0.312
0.125	15	3	0.375
0.156	15	3	0.438
0.188	10	3	0.500
0.250	10	3	0.625
0.312	5	3	0.750
0.375	5	3	0.875

Notes:

1. Angles A and B equal the maximum angle at which rivets can be driven satisfactorily.
2. Where inclination between the rivet center line and surface normal exceeds the values of angles A and B, the minimum spotfacing listed shall be provided.

of steel, brass, and aluminum. Basic types of tubular rivets are shown in Table 8-10.

The tubular rivet is used in the assembly of parts that will be subjected to shear rather than tensile stresses. It may be used where the materials being joined are too brittle or soft to withstand the impact necessary to form a solid rivet. It is intended to fasten two or more parts permanently in lightly loaded structures where ready separation of the parts is not required. The principal use of tubular

Table 8-10. Basic Types of Tubular Rivets.

Description	Application
 Truss Head Flat Countersunk Head Oval Head Straight Hole Tapered Hole Semitubular	Most widely used type of small rivet. Depth of hole in rivet, measured along wall, does not exceed 112% of mean shank diameter. Hole may be straight or tapered. When properly specified and set in a prepared hole, this rivet becomes essentially a solid member, since hole depth is just enough to form clinch. Used whenever maximum shear strength is needed. Strength in shear or compression is comparable to solid rivet. Dimensions have been standardized by Tubular and Split Rivet Council: nominal body diameters, 0.032 to 0.310 in.; corresponding minimum lengths, 1/32 to 1/4 in.
 Full Tubular	Has a shank with hole depth more than 112% of mean shank diameter. Can be used to punch its own hole in fabric, some plastic sheet, and other soft materials, eliminating a preliminary punching or drilling operation. Shear strength is less than for a semitubular rivet.
 Bifurcated (Split)	Rivet body is sawed or punched to produce a pronged shaft that punches its own hole through fiber, wood, plastic, or metal. With a few exceptions, punched shanks are more suitable than sawed shanks for piercing operations on nonmetallic materials. Sawed or broached types serve for applications in nonmetallic materials such as leather or fabric. Sawing or broaching does not distort leg as much as punching does, but punching cold works the material and makes it stronger. However, size of rivet may affect this general rule.
 Compression	Consists of two elements: solid or blank rivet and deep-drilled tubular members. Pressed together, these form an interference fit. Because heads of both members can be produced to close tolerances, these rivets are commonly used when appearance from both sides of the work must be uniform and heads must flush to prevent accumulation of dirt or waste. Can be used in wood, brittle plastics, or other materials with little danger of splitting during setting.

rivets is in assembling parts fabricated from plastic, wood, and sheet metal whenever the full strength of a rivet is not required and where pressures required for solid rivets would buckle or otherwise damage the sheet.

The cost of tubular rivets is greater than the cost of solid-shank rivets. Installation costs, however, are less with tubular rivets so that the overall cost is less.

The design considerations specified for solid shank rivets apply for tubular rivets as well. The relationship of tubular rivet diameter to sheet thickness is shown in Table 8-11.

Determining the length of a tubular rivet requires consideration of the total thickness of the member being joined and the clinch allowance. Table 8-12 shows the recommended lengths for tubular rivets.

Table 8-11. Rivet Diameter vs. Sheet Thickness (Tubular Rivets).

Joint Material	Single Shear t = Thinnest Sheet	Double Shear t = Middle Sheet	Flush Riveting t = C'sk Sheet
Aluminum Alloys	D Max = 3t	D Max = 1.5t	D Max = 4t
Other Alloys	D Max = 5.5t		

Table 8-12. Tubular Rivet Length.

Dia		Tubular Rivet Length						
Domed Head	$\frac{1}{8}$	Joint Thickness	0.010–0.050	0.051–0.090	0.091–0.140	0.141–0.190	0.191–0.250	0.251–0.300	
		Length	0.130	0.190	0.240	0.290	0.350	0.400	
	$\frac{5}{32}$	Joint Thickness	0.010–0.080	0.081–0.140	0.141–0.200	0.201–0.250	0.251–0.290	0.291–0.350	
		Length	0.190	0.240	0.300	0.370	0.400	0.450	
	$\frac{3}{16}$	Joint Thickness	0.010–0.120	0.121–0.170	0.171–0.230	0.231–0.250	0.251–0.360	0.361–0.510	0.511–0.610
		Length	0.240	0.300	0.360	0.390	0.500	0.650	0.750
Countersunk Head	$\frac{1}{8}$	Joint Thickness	0.030–0.120	0.121–0.170	0.171–0.220	0.221–0.280	0.281–0.330		
		Length	0.226	0.276	0.326	0.386	0.436		
	$\frac{5}{32}$	Joint Thickness	0.040–0.110	0.111–0.160	0.161–0.220	0.221–0.290	0.291–0.320	0.321–0.370	
		Length	0.229	0.279	0.339	0.409	0.439	0.489	
	$\frac{3}{16}$	Joint Thickness	0.030–0.150	0.151–0.210	0.211–0.260	0.261–0.290	0.291–0.400	0.401–0.550	0.551–0.650
		Length	0.285	0.345	0.405	0.435	0.545	0.695	0.795

Table 8-13. Clearance Holes for Tubular Rivets.

Tubular Rivet Diameter	Single Hole	Clearance Holes		
		Centerline Tol. $\begin{array}{l}+0.000\\-0.003\end{array}$	Multiple Holes	
			Centerline Tol. $\begin{array}{l}+0.000\\-0.004\end{array}$	Centerline Tol. $\begin{array}{l}+0.000\\-0.005\end{array}$
0.060	0.067	0.067		
0.088	0.094		0.096	
0.098	0.102			0.109
0.120	0.128			0.136
0.146	0.152			0.159

Spacing requirements for tubular rivets are identical to the solid-shank rivets. (See Figures 8-8 and 8-9.) Recommended clearance hole diameters are shown in Table 8-13. Clearance holes are such that holes may be punched or drilled prior to setting the rivet.

When assemblies are composed of soft or brittle materials and a metal sheet, the rivet should be clinched against the metal side. In assemblies consisting of two or more metal sheets, the head of the rivet should bear against the thinnest sheet. When assemblies are composed of metal, the rivet should be clinched against the thickest section, as shown in Figure 8-13. When assemblies are composed of metal and an insulating material, the rivet should be clinched against the metal side, as shown in Figure 8-14.

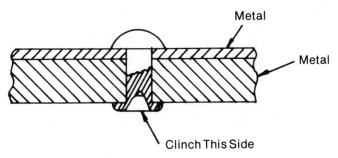

Metal

Metal

Clinch This Side

Fig. 8-13. Riveting metal sheets.

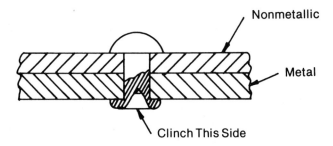

Nonmetallic

Metal

Clinch This Side

Fig. 8-14. Riveting metal to insulating material.

When designing for entry of rivet setting tools, less room is required for clinching rivets than for inserting them. (See Fig. 8-15.)

Blind Rivets

Blind rivets are intended for use where one side of an assembly is inaccessible or where space on one side is too limited to permit the use of a conventional fastener. This type may be assembled mechanically by means of a hand or power tool. Basically, blind rivets are classified by the method of setting, into pull-stem, drive pin, and explosive as follows (see also Fig. 8-16):

1. Pull-Stem Rivets. These can be subdivided into pull-through rivets, where a mandrel or stem is pulled completely out, leaving a hollow rivet; self-plugging rivets, where the stem is pulled into but not through the rivet body and the projecting end, and, in most cases, is removed in a separate

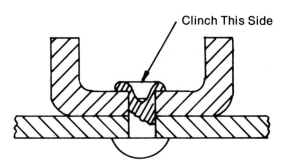

Clinch This Side

Fig. 8-15. Setting tool clearance.

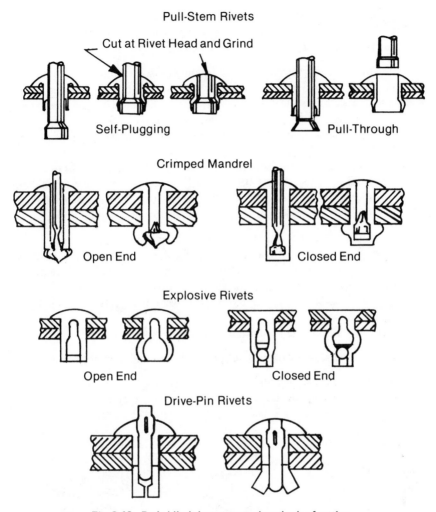

Fig. 8-16. Basic blind rivet types and methods of setting.

operation; and crimped-mandrel rivets, where a part of the mandrel remains as a plug in the rivet body.

2. Drive-Pin-Rivets. These are two-piece rivets consisting of a rivet body and separate pin installed from the head side of the rivet. The pin is hammered into the rivet body and flares out the slotted ends on the blind side.

3. Explosive Rivets. These rivets carry a chemical charge in the body. Application of heat to the rivet head by a setting iron or similar tool activates the charge to expand and set the blind end.

Among the principal considerations to be evaluated concerning pull-stem, drive-pin, and explosive rivets are cost, structural integrity of the joint, speed of assembly, clinching ability of the rivet, ease with which it may be removed after setting (if necessary), size range of available rivets, and adaptability of the particular type to the assembly in question. Some blind rivet applications are illustrated in Figure 8-17.

Where possible, blind rivets should be avoided in primary structures or heavily stressed locations. They should be used only where it is possible to hold the sheets together by clamps or other methods while the rivets are expanded. In addition, blind rivets should be used only under the following conditions:

1. Where it is difficult or impossible to upset solid-shank rivets.
2. Where their use is clearly advantageous from a production and cost standpoint in comparison with solid-shank rivets or other assembly techniques.
3. Where access to both sides of the work is limited or impossible.
4. Where the material to be riveted is such that blind rivets are less likely to damage the part.
5. Where they can be used to advantage in repair work.

The following general design considerations apply to blind rivet applications:

1. To obtain a proper grip and a tight, strong joint, the rivet axis of all types of blind rivets must be normal to the surfaces being joined.

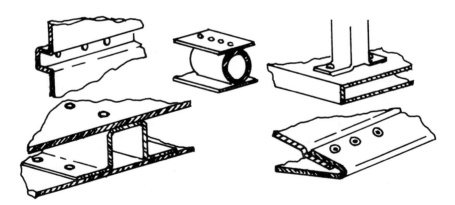

Fig. 8-17. Blind rivet applications.

2. Blind side structural clearances must be considered and will vary for each application. The amount of clearance required depends upon the type of rivet, grip and shank length, protruding stem (before or after installation), and distance of surfaces from the shank end.
3. For high-stress applications, provisions should be made for the shank upset to be visible for inspection.
4. Protruding head blind rivets are preferred over flush head rivets.

Positioning and Joint Design

The positioning of blind rivets and the attendant joint design are described in Table 8-14.

Minimum clearance on the open side (the side from which the rivet is inserted into the hole) is determined by rivet length plus head height (not flush head) and 0.060 inch clearance or the dimensions of the rivet gun, whichever is greater.

Table 8-14. Blind Rivet Design Criteria.

Description	Application
Edge Distance	Average recommended edge distance is two times the diameter of the rivet. In lightly loaded structures where the rivet performs only a holding function, this can be decreased to 1-1/2 diameters, and in heavily loaded structures, it may be necessary to increase it to as much as three diameters to develop required joint strength.
Rubber, Plastic, and Fabric Joints	Some plastics, such as reinforced molded fiberglass or polystyrene, which are reasonably rigid, present no problem for most small rivets. However, when the material is very flexible or is a fabric, set the rivet as shown at (a) or (b) with the upset head against the solid member. If this practice is not possible, use a back-up strip as shown at (c).
Sheet Metal	Some blind rivets are well adapted to assembly of sheet metal. Others that rely on shank deformation for holding action are not. If a choice can be made, it is always desirable to set the rivet against the thicker member. Backing it up with a washer or back-up strip in assemblies that are accessible from both sides is desirable for relatively thin sections, but not absolutely necessary.
Spacing	Rivet pitch should be three times the diameter of the rivet. Depending upon the nature of load, it may be desirable to decrease or increase this distance. However, it is generally considered that three diameters overcome the tendency of the material to fail and concentrate the load on the rivet.

Table 8-14. (*Continued*)

 Flush Joints	Generally, flush joints are made by countersinking one of the sections and using a rivet with a countersunk head (a). Some rivets are available in either a round or flat-top countersunk head, but not both. The rivet at (b) is capped. Another popular method of providing flush assembly and gaining additional bearing strength is to dimple the sheet by forming a conical projection on the back of the sheets with a die (c). A countersink-like recess is left on the front of the sheet which allows flush mounting of the rivet head. The projections in the two sheets are nested together to increase bearing strength.
 Clearance	An angular section is riveted to a straight member (a). If the angle projects above the rivet head, it may be difficult to apply the tool to upset the rivet. One method of solving this problem is to pull the rivet from the underside of the work. In riveting the U-channel (b) to the outer members, the rivet must be fed from the outer side. On this joint, allow enough back-up clearance to prevent interference. At (c), not only tool clearance but also rivet head clearance is critical. A large diameter head may interfere with the wall.
 Length	The amount of length needed for clinching action varies greatly and depends on material being fastened, necessary strength, and method of riveting. Most rivet manufacturers provide data on grip ranges of their rivets to simplify selection for the user.
 Honeycomb Sections	Inserts should be employed to strengthen the section and provide a strong joint. Otherwise, setting the rivet may deform the section and cause a structural weakness that may later result in failure.
 Riveted Joints	Riveted cleat or batten holds a butt joint (a). The simple lap joint (b) must have sufficient material beyond hole for strength. Excessive material beyond rivet hole (c) may curl up, vibrate, or cause interference problems, depending on the installation. The best solution is to trim the panel to leave an edge about twice the hole diameter in width. An alternate solution would be to relocate the rivet at a position halfway between the upper and lower panel edges.
 Making Use of Pull Up	By judicious positioning of rivets and parts that are to be assembled with rivets, the setting force can sometimes be used to pull together unlike parts. Rivets pull flat cleat into close over-all juxtaposition to the curved surface at (a) and (c), but the rivet at (b) does little to help mold the parts together.

Blind side minimum clearance is determined by the protrusion of the stem before it is drawn.

Thickness can be critical in the application of some types of blind rivets and inconsequential in others. This depends on the basic rivet design and on the setting technique. Some rivets may be set in materials as thin as 0.020 inch, if, after the rivet is set, the head is formed properly and shank expansion is closely controlled. Conventional practice, however, is to form the blind head against the thicker sheet of a combination, if it is accessible. When one component is of compressible material, such as wood or soft plastics, extra large head-diameter rivets provide excellent bearing.

Blind rivets should be located so that the surfaces are normal to the pin axis. Rivet axes may deviate from the surface normal provided that: (1) the manufactured head is not inclined more than $\pm 0°30'$, (2) the upset shank head is not inclined more than $15°$; and (3) the radius of curvature of a curved surface is not more than 10 times the rivet diameter. In some cases, spotfacing of the surface adjacent to the shank end may be possible, but spotfacing the surface common with the manufactured rivet head is preferred, since access of the tool head is essential for proper installation.

Materials

Conventional riveting materials and applications are shown in Table 8-15.

If possible, rivet patterns in simple lap joints should have two or more rows, preferably arranged in tandem. (For example, a two-row joint at one inch spacing has the same static strength as a one-row joint with a one-half inch spacing, but the fatigue life of a two-row joint is usually greater.) Staggered patterns, however, are better for tension field beams because the web tension loads place the rivets in preferred alignment relative to the direction of the tension load. In addition, staggered patterns produce lighter weight joints than tandem patterns. (See Fig. 8-18.)

Recommended practices for joint details of rivet spacing, edge distance, and rivet diameters are summarized in Table 8-16.

Loads on rivets should be in simple shear. To develop maximum strength in multiple-rivet designs, the holes must be accurately aligned by line reaming. Avoid tensile and cantilever loads on rivets. See Figs. 8-19 and 8-20.

Countersunk head rivets are not preferred due to the additional cost for countersinking and their tendency to loosen more rapidly than the universal head type rivets. This type should be used only where a flush surface is required. Do not use where maximum head height exceeds the minimum thickness of material being joined unless the thin material is dimpled. Reference dimensions for countersunk head rivets are shown in Table 8-17. The maximum countersunk diam-

Table 8-15. Conventional Rivet Materials.

Materials to be Riveted	Rivet Materials					
	Aluminum Alloy 2117-T4	Aluminum Alloy 2024-T4	Aluminum Alloy 5056	Aluminum Alloy 1100-F	Monel	A-286 CRES (3)
Aluminum alloy to aluminum alloy	Preferred	Preferred	Permissible but undesirable	(1)		
Aluminum alloy to cadmium-plated steel	(2)	(2)			Permissible but undesirable	
Aluminum alloy to CRES	(2)	(2)			Permissible but undesirable	
Aluminum alloy to titanium	(2)	(2)			Permissible but undesirable	
Cadmium-plated steel to cadmium-plated steel	Permissible but undesirable	Permissible but undesirable			Permissible but undesirable	Permissible but undesirable
CRES to CRES					Preferred	Preferred
CRES to titanium					Preferred	Permissible but undesirable
Titanium to titanium					Preferred	Permissible but undesirable
Magnesium alloy to aluminum alloy			Preferred			
Magnesium alloy to cadmium-plated steel			Preferred			
Magnesium alloy to magnesium alloy			Preferred			
Soft material (plastic, fabric, etc.)	Acceptable		Permissible but undesirable	Preferred		

Notes: (1) May be used for welding, where extra softness is required, or for plugging holes.

(2) May be used for weight saving.

(3) A286 rivets are more difficult to drive and should only be used in preference to monel where higher shear strength is a necessity.

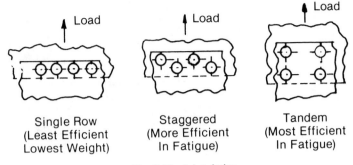

Fig. 8-18. Joint design.

eter has been calculated to allow the rivet head to vary a maximum of 0.010 inch above the surface and 0.002 inch below the surface. The excess material may then be removed to produce a flush surface.

CLEARANCE HOLES FOR PRODUCTION ASSEMBLIES

In designing mechanical parts that are to be assembled in quantity, consideration must be given to the interchangeability of parts avoiding the use of match drilling at assembly in order to achieve the most economical advantage. Predrilled parts when properly designed will assemble with no additional machining.

The following tables for clearance holes given here are a simple means of selecting the hole sizes needed for parts interchangeability when the proper design and tolerances have been established. These tables are acceptable for any standard linear or coordinate dimensioning system that does not have an excess of accumulated tolerances.

The tables are separated into categories for ease of access: (1) simple linear patterns, (2) complex linear patterns, (3) simple coordinate patterns, and (4) complex coordinate patterns.

Simple Patterns have no more than one dimension in each coordinate direction, and parts must be free to move in order for holes to align.

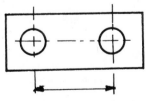

Simple linear pattern, two holes on a common center line.

Table 8-16. Design Tips for Small Rivets.

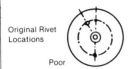   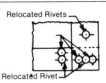

Original Rivet Locations

Equal

Relocated Rivets

Poor

Best

Relocated Rivet

Rivet Symmetry

Maintain symmetry of rivet pattern for multiple riveted joints. Space rivets equally to simplify machine indexing and to allow for riveting clearances.

Possible Jaw Interference

Poor

Better

Best (Head)

Channel Sections

Channels should be wide enough to provide clearance for rivet-setting tools. Clinching rivet against channel is best.

Clearance Hole

Poor

Better

Best

Angular Section

Keep rivet from under angle section to provide clearance for riveting or provide an access hole to simplify joining.

Poor

Best

Angular Shapes

Counterboring provides flat surface for uniform clinching.

Poor

Best

Edge Clearance

Rivet too close to edge of joint prevents solid positioning of riveting tool.

Tube

Poor

Poor

Poor

Flat

Rod

Better

Better

Better

Flat

Flats

Best

Best

Best

Clearance Hole for Tool

Rod and Tube Joints

Flat surface machined or formed on rod or tube permits solid seating of rivet. Keep rivets short to prevent buckling and distortion.

Table 8-16. (Continued)

 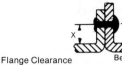

Poor   Flange Clearance   Best

Rivets close to flanges interfere with rivet settling. Distance X should equal radius of riveting machine jaws in open position, plus a tolerance, to allow sufficient clearance for jaws.

Poor   Hole Clearance   Best

Long rivets in oversized holes may buckle when set. Use recommended allowances for hole clearance.

  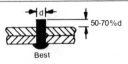

50-70%d

Poor   Poor   Best

Clinch Allowance

Rivets should be long enough to insure adequate holding. Recommended clinch allowances should be used. Allow 50 to 70% of rivet diameter for semitubular rivets.

Poor   Heavy and Thin Gauge Stock   Best

Rivet head should be on side of thin sheet. Upset rivet against heavy-gauge material. For dissimilar materials, clinch against stronger part or against a washer.

 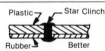

Plastic   Star Clinch   Washer

Poor   Rubber   Better   Washer   Best

Compressible Materials

Star set rivet or use cap in conjunction with large diameter head in soft materials. For maximum strength, use washers or burrs.

Washer

Poor   Better   Washer   Best   Washer

Weak Materials

Use washers to strengthen brittle materials or where counterboxes or countersinks cause weakening in thin sections.

Poor   Tight Joints   Best

Long rivets can buckle and have tendency to increase length through thermal expansion. Short rivets allow for swelling.

Table 8-16. (*Continued*)

Poor   Better   Best

Counterboring

Provide clearance for rivet-settling tools.

Poor   Best

Countersinking

Provide clearance for rivet-settling tools.

  Washer Shoulder

Poor   Good   Better   Best

Pivoted Joints

Several methods are satisfactory, but washers or shoulder rivets are generally preferred.

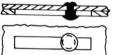

Sliding Joint

Use shoulder rivet if one part of assembly slides.

Poor   Best

Multiple Rivets

Space rivets to insure clearance for riveting tools, especially if rivets are set simultaneously.

Serrated
Head

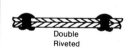 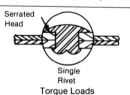 

Double
Riveted   Interlocking

Single
Rivet

Torque Loads

Use two rivets for joint if torque loads are expected. Special rivet with serrated surface under head, or interlocking joint design can also be used.

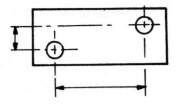

Simple coordinate patterns, two holes not on a common center line.

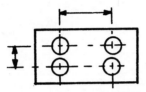

Three or four holes in a square or rectangular pattern.

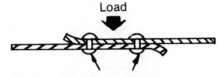

Load

Rivets Subjected to
Combination of Shear
And Bending Stress

Poor

Rivets Subjected to
Bending Stress Only

Load

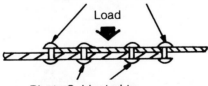

Rivets Subjected to
Shear Stress Only

Good

Fig. 8-19. Stresses on rivets.

Table 8-17. 100° Countersunk Head Rivets.

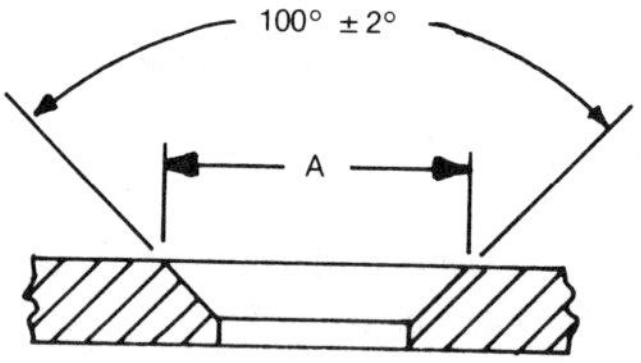

Flush Condition

Rivet Body Diameter	MS20426		MS20427	
	Countersink Diameter (A)	Minimum Material Thickness	Countersink Diameter (A)	Minimum Material Thickness
0.062	0.125 ± 0.005	0.032 ± 0.002	0.142 ± 0.005	0.040 ± 0.003
0.094	0.190 ± 0.005	0.050 ± 0.003	0.197 ± 0.005	0.050 ± 0.003
0.125	0.236 ± 0.005	0.063 ± 0.003	0.243 ± 0.005	0.063 ± 0.003
0.156	0.297 ± 0.005	0.080 ± 0.004	0.305 ± 0.005	0.080 ± 0.004
0.187	0.364 ± 0.005	0.090 ± 0.004	0.372 ± 0.005	0.090 ± 0.004

Protruded Condition

Rivet Body Diameter	MS20426		MS20427	
	Countersink Diameter (A)	Minimum Material Thickness	Countersink Diameter (A)	Minimum Material Thickness
0.062	0.104 ± 0.006	0.032 ± 0.002	0.120 ± 0.005	0.040 ± 0.003
0.094	0.165 ± 0.010	0.050 ± 0.003	0.175 ± 0.005	0.050 ± 0.003
0.125	0.211 ± 0.010	0.063 ± 0.003	0.220 ± 0.005	0.063 ± 0.003
0.156	0.272 ± 0.010	0.080 ± 0.004	0.282 ± 0.005	0.080 ± 0.004
0.187	0.339 ± 0.010	0.090 ± 0.004	0.350 ± 0.005	0.090 ± 0.004

Poor   Good   Best

Fig. 8-20. Loads on rivets.

Complex Patterns have two or more dimensions in one or both coordinate directions.

Complex linear patterns.

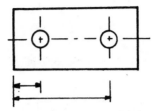

Two or more holes on a common center line dimensioned to a common base line. Part edges used as base lines (datums) will align.

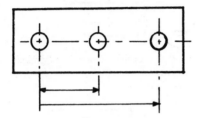

Three or more holes on a common center line, dimensioned to the center line of one of the holes.

Complex coordinate patterns.

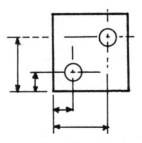

Two or more holes are dimensioned to a common base line in one or both coordinate directions. The part edges used as base lines will align within the limits of squareness held on the parts.

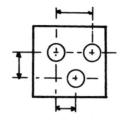

Three holes in a triangular pattern.

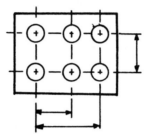

More than four holes in a square or rectangular pattern.

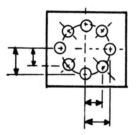

Any number of holes, dimensioned by coordinates, in a circular or irregular pattern.

The tables are defined as shown below. Four conditions are tabulated for each screw size, except the number 2 screw. Use of number 2 screws should be limited to use in linear patterns only due to the size and strength characteristics.

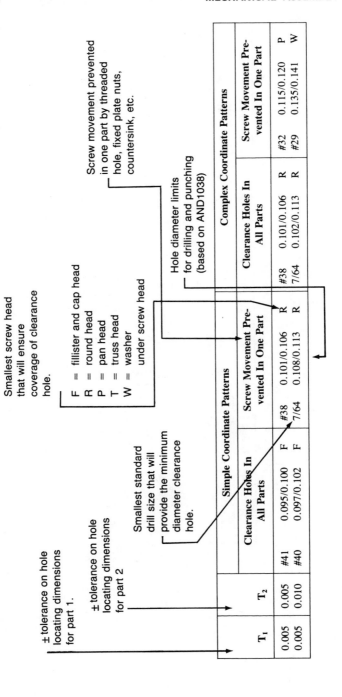

± tolerance on hole locating dimensions for part 1.

± tolerance on hole locating dimensions for part 2

Smallest standard drill size that will provide the minimum diameter clearance hole.

Smallest screw head that will ensure coverage of clearance hole.

F = fillister and cap head
R = round head
P = pan head
T = truss head
W = washer under screw head

Screw movement prevented in one part by threaded hole, fixed plate nuts, countersink, etc.

Hole diameter limits for drilling and punching (based on AND1038)

		Simple Coordinate Patterns		Complex Coordinate Patterns	
		Clearance Holes In All Parts	Screw Movement Prevented In One Part	Clearance Holes In All Parts	Screw Movement Prevented In One Part
T_1	T_2				
0.005	0.005	#41 0.095/0.100 F	#38 0.101/0.106 R	#38 0.101/0.106 R	#32 0.115/0.120 P
0.005	0.010	#40 0.097/0.102 F	7/64 0.108/0.113 R	7/64 0.102/0.113 R	#29 0.135/0.141 W

Clearance Holes For No. 2 Screw (max OD 0.086).

		Simple Linear Patterns		Complex Linear Patterns	
T_1	T_2	Clearance Holes In All Parts	Screw Movement Prevented In One Part	Clearance Holes In All Parts	Screw Movement Prevented In One Part
0.005	0.005	#42 0.093/0.098 F	#40 0.097/0.102 F	#40 0.097/0.102 F	7/64 0.108/0.113 R
0.005	0.010	#41 0.095/0.100 F	#38 0.101/0.106 R	#38 0.101/0.106 R	#31 0.119/0.124 W
0.005	0.020	#39 0.099/0.104 F	#33 0.112/0.117 P	#33 0.112/0.117 P	#28 0.139/0.145 W
0.005	0.030	#36 0.105/0.110 R	1/8 0.124/0.129 W	1/8 0.124/0.129 W	#22 0.156/0.162 W
0.010	0.010	#40 0.097/0.102 F	7/64 0.108/0.113 R	7/64 0.108/0.113 R	#30 0.127/0.133 W
0.010	0.020	#38 0.101/0.106 R	#31 0.119/0.124 W	#31 0.119/0.124 W	#30 0.127/0.133 W
0.010	0.030	7/64 0.108/0.113 R	#30 0.127/0.133 W	#30 0.127/0.133 W	
0.020	0.020	7/64 0.108/0.113 R	#30 0.127/0.133 W	#30 0.127/0.133 W	
0.020	0.030	#33 0.112/0.117 P	9/64 0.139/0.145 W	9/64 0.139/0.145 W	
0.030	0.030	#31 0.119/0.124 W	#26 0.146/0.152 W	#26 0.146/0.152 W	

Clearance Holes For No. 3 Screw (max OD 0.099).

Simple Linear Patterns

T_1	T_2	Clearance Holes In All Parts	Screw Movement Prevented In One Part
0.005	0.005	#36 0.105/0.110 F	#35 0.109/0.114 F
0.005	0.010	7/64 0.108/0.113 F	#32 0.115/0.120 F
0.005	0.020	#33 0.112/0.117 F	1/8 0.124/0.129 R
0.005	0.030	#31 0.119/0.124 R	#29 0.135/0.141 W
0.010	0.010	#35 0.109/0.114 F	#31 0.119/0.124 R
0.010	0.020	#32 0.115/0.120 F	#29 0.135/0.141 W
0.010	0.030	#31 0.119/0.124 R	9/64 0.139/0.145 W
0.020	0.020	#31 0.119/0.124 R	9/64 0.139/0.145 W
0.020	0.030	1/8 0.124/0.129 R	#25 0.149/0.155 W
0.030	0.030	#29 0.135/0.141 W	#20 0.160/0.166 W

Complex Linear Patterns

Clearance Holes In All Parts	Screw Movement Prevented In One Part
#35 0.109/0.114 F	#31 0.119/0.124 R
#32 0.115/0.120 F	#29 0.135/0.141 W
1/8 0.124/0.129 R	#25 0.149/0.155 W
#29 0.135/0.141 W	
#31 0.119/0.124 R	#28 0.139/0.145 W
#29 0.135/0.141 W	#20 0.160/0.166 W
9/64 0.139/0.145 W	
9/64 0.139/0.145 W	
#25 0.149/0.155 W	
#20 0.160/0.166 W	

Simple Coordinate Patterns

T_1	T_2	Clearance Holes In All Parts	Screw Movement Prevented In One Part
0.005	0.005	7/64 0.108/0.113 F	#32 0.115/0.120 F
0.005	0.010	#34 0.110/0.115 F	1/8 0.124/0.129 R
0.005	0.020	#31 0.119/0.124 R	#29 0.135/0.141 W
0.005	0.030	1/8 0.124/0.129 R	#25 0.149/0.155 W
0.010	0.010	#32 0.115/0.120 F	#30 0.127/0.133 R
0.010	0.020	1/8 0.124/0.129 R	#27 0.143/0.149 W
0.010	0.030	#29 0.135/0.141 W	#22 0.156/0.162 W
0.020	0.020	#29 0.135/0.141 W	#22 0.156/0.162 W
0.020	0.030	#29 0.135/0.141 W	
0.030	0.030	#27 0.143/0.149 W	

Complex Coordinate Patterns

Clearance Holes In All Parts	Screw Movement Prevented In One Part
#32 0.115/0.120 F	#29 0.135/0.141 W
1/8 0.124/0.129 R	#27 0.143/0.149 W
#29 0.135/0.141 W	
#25 0.149/0.155 W	
#30 0.127/0.133 R	#22 0.156/0.162 W
#27 0.143/0.149 W	
#22 0.156/0.162 W	
#22 0.156/0.162 W	

Clearance Holes For No. 4 Screw (max OD 0.112).

Simple Linear Patterns

T₁	T₂	Clearance Holes In All Parts			Screw Movements Prevented In One Part		
0.005	0.005	#31	0.119/124	F	1/8	0.124/0.129	F
0.005	0.010	1/8	0.124/0.129	F	#30	0.127/0.132	F
0.005	0.020	#30	0.127/0.132	F	#28	0.139/0.145	F
0.005	0.030	#29	0.135/0.141	F	#25	0.149/0.155	R
0.010	0.010	1/8	0.124/0.129	F	#29	0.135/0.141	F
0.010	0.020	#30	0.127/0.133	F	#27	0.143/0.149	R
0.010	0.030	#29	0.135/0.141	F	#23	0.153/0.159	R
0.020	0.020	#29	0.135/0.141	F	#23	0.153/0.159	R
0.020	0.030	#28	0.139/0.145	F	#19	0.165/0.171	P
0.030	0.030	#26	0.146/0.152	F	#17	0.172/0.178	T

Complex Linear Patterns

T₁	T₂	Clearance Holes In All Parts			Screw Movement Prevented In One Part		
0.005	0.005	1/8	0.124/0.129	F	#29	0.135/0.141	F
0.005	0.010	#30	0.127/0.132	F	#27	0.143/0.149	F
0.005	0.020	#28	0.139/0.145	F	#19	0.165/0.171	P
0.005	0.030	#25	0.149/0.155	R	#13	0.184/0.190	W
0.010	0.010	#29	0.135/0.141	F	#23	0.153/0.159	R
0.010	0.020	#27	0.143/0.149	R	#17	0.172/0.178	T
0.010	0.030	#23	0.153/0.159	R	#10	0.193/0.199	W
0.020	0.020	#23	0.153/0.159	R	#10	0.193/0.199	W
0.020	0.030	#19	0.165/0.171	P			
0.030	0.030	#17	0.172/0.178	T			

Simple Coordinate Patterns

T₁	T₂	Clearance Holes In All Parts			Screw Movement Prevented In One Part		
0.005	0.005	#31	0.119/0.123	F	#30	0.127/0.133	F
0.005	0.010	1/8	0.124/0.129	F	#29	0.135/0.141	F
0.005	0.020	#29	0.135/0.141	F	#25	0.149/0.155	R
0.005	0.030	#28	0.139/0.145	F	#19	0.165/0.171	P
0.010	0.010	#30	0.127/0.133	F	#27	0.143/0.149	F
0.010	0.020	#29	0.135/0.141	F	5/32	0.155/0.161	R
0.010	0.030	#27	0.143/0.149	F	#18	0.169/0.175	T
0.020	0.020	#27	0.143/0.149	F	#18	0.169/0.175	T
0.020	0.030	#25	0.149/0.155	R	#13	0.184/0.190	W
0.030	0.030	5/32	0.155/0.161	R	# 8	0.198/0.204	W

Complex Coordinate Patterns

T₁	T₂	Clearance Holes In All Parts			Screw Movement Prevented In One Part		
0.005	0.005	#30	0.127/0.133	F	#27	0.143/0.149	R
0.005	0.010	#29	0.135/0.141	F	#23	0.153/0.159	R
0.005	0.020	#25	0.149/0.155	R	#13	0.184/0.190	W
0.005	0.030	#19	0.165/0.171	P			
0.010	0.010	#27	0.143/0.149	F	#18	0.169/0.175	T
0.010	0.020	5/32	0.155/0.161	R	# 8	0.198/0.204	W
0.010	0.030	#18	0.169/0.175	T			
0.020	0.020	#18	0.169/0.175	T			
0.020	0.030	#13	0.184/0.190	W			
0.030	0.030	# 8	0.198/0.204	W			

Clearance Holes For No. 6 Screw (max OD 0.138).

Linear Patterns

		Simple Linear Patterns		Complex Linear Patterns	
T_1	T_2	Clearance Holes In All Parts	Screw Movement Prevented In One Part	Clearance Holes In All Parts	Screw Movement Prevented In One Part
0.005	0.005	#27 0.143/0.149 F	#25 0.149/0.155 F	#25 0.149/0.155 F	#21 0.158/0.164 F
0.005	0.010	#26 0.146/0.152 F	#23 0.153/0.159 F	#23 0.153/0.159 F	#18 0.169/0.175 F
0.005	0.020	#24 0.151/0.157 F	#19 0.165/0.171 F	#19 0.165/0.171 F	#12 0.188/0.194 R
0.005	0.030	#22 0.156/0.162 F	#16 0.176/0.182 F	#16 0.176/0.182 F	# 4 0.208/0.214 T
0.010	0.010	#25 0.149/0.155 F	#21 0.158/0.164 F	#21 0.158/0.164 F	#15 0.179/0.185 R
0.010	0.020	#23 0.153/0.159 F	#18 0.169/0.175 F	#18 0.169/0.175 F	# 8 0.198/0.204 P
0.010	0.030	#21 0.158/0.164 F	#15 0.179/0.185 R	#15 0.179/0.185 R	7/32 0.218/0.224 T
0.020	0.020	#21 0.158/0.164 F	#15 0.179/0.185 R	#15 0.179/0.185 R	7/32 0.218/0.224 T
0.020	0.030	#19 0.165/0.171 F	#12 0.188/0.194 R	#12 0.188/0.194 R	C 0.241/0.247 W
0.030	0.030	#18 0.169/0.175 F	# 8 0.198/0.204 P	# 8 0.198/0.204 P	

Coordinate Patterns

		Simple Coordinate Patterns		Complex Coordinate Patterns	
T_1	T_2	Clearance Holes In All Parts	Screw Movement Prevented In One Part	Clearance Holes In All Parts	Screw Movement Prevented In One Part
0.005	0.005	#26 0.146/0.152 F	#23 0.153/0.159 F	#23 0.153/0.159 F	#18 0.169/0.175 R
0.005	0.010	#25 0.149/0.155 F	#20 0.160/0.166 F	#20 0.160/0.166 F	#15 0.179/0.185 R
0.005	0.020	#22 0.156/0.162 F	#16 0.176/0.182 F	#16 0.176/0.182 F	# 3 0.212/0.218 T
0.005	0.030	#19 0.165/0.171 F	#12 0.188/0.194 R	#12 0.188/0.194 R	B 0.237/0.243 W
0.010	0.010	#23 0.153/0.159 F	#18 0.169/0.175 F	#18 0.169/0.175 F	# 9 0.195/0.201 P
0.010	0.020	#20 0.160/0.166 F	#14 0.181/0.187 R	#14 0.181/0.187 R	# 1 0.227/0.233 T
0.010	0.030	#18 0.169/0.175 F	# 9 0.195/0.201 P	# 9 0.195/0.201 P	F 0.256/0.263 W
0.020	0.020	#18 0.169/0.175 F	# 9 0.195/0.201 P	# 9 0.195/0.201 P	F 0.256/0.263 W
0.020	0.030	#16 0.176/0.182 R	# 3 0.212/0.218 T	# 3 0.212/0.218 T	
0.030	0.030	#14 0.181/0.187 R	# 1 0.227/0.233 T	# 1 0.227/0.233 T	

Clearance Holes For No. 8 Screw (max OD 0.164).

Linear Patterns

		Simple Linear Patterns		Complex Linear Patterns	
T_1	T_2	Clearance Holes In All Parts	Screw Movement Prevented In One Part	Clearance Holes In All Parts	Screw Movement Prevented In One Part
0.005	0.005	#18 0.169/0.175 F	#16 0.176/0.182 F	#16 0.176/0.182 F	#13 0.184/0.190 F
0.005	0.010	#17 0.172/0.178 F	#15 0.179/0.185 F	#15 0.179/0.185 F	# 9 0.195/0.210 F
0.005	0.020	#15 0.179/0.185 F	#11 0.190/0.196 F	#11 0.190/0.196 F	7/32 0.218/0.224 F
0.005	0.030	#13 0.184/0.190 F	# 7 0.200/0.206 F	# 7 0.200/0.206 F	B 0.237/0.243 P
0.010	0.010	#16 0.176/0.182 F	#13 0.184/0.190 F	#13 0.184/0.190 F	# 5 0.205/0.211 F
0.010	0.020	#15 0.179/0.185 F	# 9 0.195/0.201 F	# 9 0.195/0.201 F	# 1 0.227/0.233 R
0.010	0.030	#13 0.184/0.190 F	# 5 0.205/0.211 F	# 5 0.205/0.211 F	D 0.245/0.251 T
0.020	0.020	#13 0.184/0.190 F	# 5 0.205/0.211 F	# 5 0.205/0.211 F	D 0.245/0.251 T
0.020	0.030	#11 0.190/0.196 F	7/32 0.218/0.224 R	7/32 0.218/0.224 R	H 0.265/0.272 T
0.030	0.030	# 9 0.195/0.201 F	# 1 0.227/0.233 R	# 1 0.227/0.233 R	

Coordinate Patterns

		Simple Coordinate Patterns		Complex Coordinate Patterns	
T_1	T_2	Clearance Holes In All Parts	Screw Movement Prevented In One Part	Clearance Holes In All Parts	Screw Movement Prevented In One Part
0.005	0.005	11/64 0.171/0.177 F	#15 0.179/0.185 F	#15 0.179/0.185 F	#10 0.193/0.199 F
0.005	0.010	#16 0.176/0.182 F	3/16 0.187/0.193 F	3/16 0.187/0.193 F	# 5 0.205/0.211 F
0.005	0.020	#13 0.184/0.190 F	# 7 0.200/0.216 F	# 7 0.200/0.206 F	B 0.237/0.243 P
0.005	0.030	#11 0.190/0.196 F	7/32 0.218/0.214 F	7/32 0.218/0.214 F	G 0.260/0.267 T
0.010	0.010	#15 0.179/0.185 F	#10 0.193/0.199 F	#10 0.193/0.199 F	# 1 0.227/0.233 T
0.010	0.020	3/16 0.187/0.193 F	# 4 0.208/0.214 F	# 4 0.208/0.214 F	1/4 0.249/0.255 T
0.010	0.030	#10 0.193/0.199 F	# 1 0.227/0.233 R	# 1 0.227/0.233 R	
0.020	0.020	#10 0.193/0.199 F	# 1 0.227/0.233 R	# 1 0.227/0.233 R	
0.020	0.030	# 7 0.200/0.206 F	B 0.237/0.243 P	B 0.237/0.243 P	
0.030	0.030	# 4 0.208/0.214 R	1/4 0.249/0.255 T	1/4 0.249/0.255 T	

Clearance Holes For No. 10 Screw (max OD 0.190).

Linear Patterns

T_1	T_2	Simple Linear — Clearance Holes In All Parts	Simple Linear — Screw Movement Prevented In One Part	Complex Linear — Clearance Holes In All Parts	Complex Linear — Screw Movement Prevented In One Part
0.005	0.005	# 9 0.195/0.201 F	# 7 0.200/0.206 F	# 7 0.200/0.206 F	# 3 0.212/0.218 F
0.005	0.010	# 8 0.198/0.204 F	# 5 0.205/0.211 F	# 5 0.205/0.211 F	# 2 0.220/0.226 F
0.005	0.020	# 6 0.203/0.209 F	7/32 0.218/0.224 F	7/32 0.218/0.224 F	C 0.241/0.247 F
0.005	0.030	# 4 0.208/0.214 F	# 1 0.227/0.233 F	# 1 0.227/0.233 F	G 0.260/0.267 R
0.010	0.010	# 7 0.200/0.206 F	# 3 0.212/0.218 F	# 3 0.212/0.218 F	A 0.233/0.239 F
0.010	0.020	# 5 0.205/0.211 F	# 2 0.220/0.226 F	# 2 0.220/0.226 F	F 0.256/0.263 R
0.010	0.030	# 3 0.212/0.218 F	A 0.233/0.239 F	A 0.233/0.239 F	I 0.271/0.278 P
0.020	0.020	# 3 0.212/0.218 F	A 0.233/0.239 F	A 0.233/0.239 F	I 0.271/0.278 P
0.020	0.030	7/32 0.218/0.224 F	C 0.241/0.247 F	C 0.241/0.247 F	M 0.294/0.801 T
0.030	0.030	# 2 0.220/0.226 F	F 0.256/0.263 R	F 0.256/0.263 R	

Coordinate Patterns

T_1	T_2	Simple Coordinate — Clearance Holes In All Parts	Simple Coordinate — Screw Movement Prevented In One Part	Complex Coordinate — Clearance Holes In All Parts	Complex Coordinate — Screw Movement Prevented In One Part
0.005	0.005	# 8 0.198/0.204 F	# 5 0.205/0.211 F	# 5 0.205/0.211 F	# 2 0.220/0.226 R
0.005	0.010	13/64 0.202/0.208 F	# 3 0.212/0.218 F	# 3 0.212/0.218 F	A 0.233/0.239 F
0.005	0.020	# 4 0.208/0.214 F	# 1 0.227/0.233 F	# 1 0.227/0.233 F	H 0.265/0.272 R
0.005	0.030	7/32 0.218/0.244 F	C 0.241/0.247 F	C 0.241/0.247 F	L 0.289/0.296 T
0.010	0.010	# 5 0.205/0.211 F	# 2 0.220/0.226 F	# 2 0.220/0.226 F	1/4 0.249/0.255 R
0.010	0.020	# 3 0.212/0.218 F	A 0.233/0.239 F	A 0.233/0.239 F	J 0.276/0.283 T
0.010	0.030	7/32 0.218/0.224 F	1/4 0.249/0.255 R	1/4 0.249/0.255 R	5/16 0.311/0.318 T
0.020	0.020	7/32 0.218/0.224 F	1/4 0.249/0.255 R	1/4 0.249/0.255 R	5/16 0.311/0.318 T
0.020	0.030	# 1 0.227/0.233 F	H 0.265/0.272 R	H 0.265/0.272 R	
0.030	0.030	A 0.233/0.239 F	J 0.276/0.283 T	J 0.276/0.283 T	

Clearance Holes For 1/4 Screw (max OD 0.250).

Simple Linear Patterns

T_1	T_2	Clearance Holes In All Parts			Screw Movement Prevented In One Part		
0.005	0.005	F	0.256/0.263	F	G	0.260/0.267	F
0.005	0.010	G	0.260/0.267	F	H	0.265/0.272	F
0.005	0.020	H	0.265/0.272	F	J	0.276/0.283	F
0.005	0.030	I	0.271/0.278	F	L	0.289/0.296	F
0.010	0.010	G	0.260/0.267	F	I	0.271/0.278	F
0.010	0.020	17/64	0.271/0.278	F	K	0.280/0.287	F
0.010	0.030	I	0.271/0.278	F	M	0.294/0.301	F
0.020	0.020	I	0.271/0.278	F	M	0.294/0.301	F
0.020	0.030	J	0.276/0.283	F	N	0.301/0.308	F
0.030	0.030	K	0.280/0.287	F	5/16	0.311/0.318	F

Complex Linear Patterns

Clearance Holes In All Parts			Screw Movement Prevented In One Part		
G	0.260/0.267	F	I	0.271/0.278	F
H	0.265/0.272	F	K	0.280/0.287	F
J	0.276/0.283	F	N	0.301/0.308	F
L	0.289/0.296	F	P	0.322/0.329	R
I	0.271/0.278	F	M	0.294/0.301	F
K	0.280/0.287	F	5/16	0.311/0.318	F
M	0.294/0.301	F	Q	0.331/0.338	R
M	0.294/0.301	F	Q	0.331/0.338	R
N	0.301/0.308	F	T	0.357/0.364	R
5/16	0.311/0.318	F	3/8	0.374/0.381	T

Simple Coordinate Patterns

T_1	T_2	Clearance Holes In All Parts			Screw Movement Prevented In One Part		
0.005	0.005	G	0.260/0.267	F	H	0.265/0.272	F
0.005	0.010	H	0.265/0.272	F	I	0.271/0.278	F
0.005	0.020	I	0.271/0.278	F	L	0.289/0.296	F
0.005	0.030	J	0.276/0.283	F	N	0.301/0.308	F
0.010	0.010	H	0.265/0.272	F	K	0.280/0.287	F
0.010	0.020	I	0.271/0.278	F	M	0.294/0.301	F
0.010	0.030	K	0.280/0.287	F	5/16	0.311/0.318	F
0.020	0.020	K	0.280/0.287	F	5/16	0.311/0.318	F
0.020	0.030	L	0.289/0.296	F	P	0.322/0.329	R
0.030	0.030	M	0.294/0.301	F	R	0.338/0.345	R

Complex Coordinate Patterns

Clearance Holes In All Parts			Screw Movement Prevented In One Part		
H	0.265/0.272	F	K	0.280/0.287	F
I	0.271/0.278	F	L	0.288/0.296	F
L	0.289/0.296	F	P	0.322/0.329	R
N	0.301/0.308	F	S	0.347/0.354	R
K	0.280/0.287	F	5/16	0.311/0.318	F
M	0.294/0.301	F	R	0.338/0.345	R
5/16	0.311/0.318	F	U	0.367/0.374	T
5/16	0.311/0.318	F	U	0.367/0.374	T
P	0.322/0.329	R	X	0.396/0.403	T
R	0.338/0.345	R			

Example

Assuming the location dimensions on parts 1 and 2 are toleranced at ±0.020 and a no. 8 screw is to be used. Determine the clearance hole size by referring to the table for no. 6 screws.

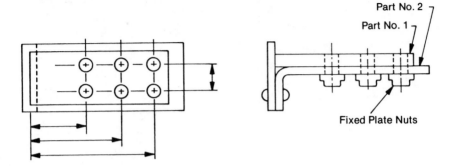

Part No. 2

Part No. 1

Fixed Plate Nuts

Solution

This is a complex coordinate pattern because there are six holes and because edge alignment is required. Since fixed plate nuts are used on part 2, only the clearance holes for part 1, are or interest. Therefore, on the table for no. 6 screws, in the column "Complex Coordinate Patterns—Screw Movement Prevented in One Part," opposite 0.020-0.020, it is found in the data that a F0.256/0.263W drill will give the required clearance hole in part 1.

REFERENCES

1. Assembly Engineering Master Catalog
   Hitchcock Publishing Co.
   Wheaton, Illinois

2. Avex Rivets
   Marina Del Rey, California

3. Machinery's Handbook
   The Industrial Press
   New York, N.Y.

4. Milford Rivet & Machine Co.
   Sunnyvale, California

5. USM "Pop" Rivets
   Temple City, California

9
PLASTIC-COATED ELECTRONIC EQUIPMENT

INTRODUCTION

The purpose of this chapter is to acquaint electrical and mechanical engineers with the design concepts, materials, and fabrication techniques used in producing plastic-embedded electronic modules.

Definitions

Encapsulating, *potting*, *embedding*, and *impregnating* are terms used to describe the process by which electronic parts are protected. These related terms are used throughout the industry with the following accepted definitions:

Encapsulating. Coating or enveloping the components with protective materials. May be accomplished by dipping, spraying, or spreading. Finished parts usually retain the original shape of the components.

Potting. Pouring or injecting protective materials around components within a shell. The shell remains with the component and becomes the "skin" of the part.

Embedding. Similar to potting except that the shell, a mold, is removed.

Impregnating. Similar to encapsulating except that a vacuum is used to obtain a void-free penetration of the protecting material. Usually done to transformer coils.

336

Embedded Module

Embedded and encapsulated electronic equipment differ from conventional equipment because of the addition of an enclosing structure, a supporting and surrounding plastic. (See Fig. 9-1.) This embedded configuration greatly increases the strength of the part and allows it to meet more severe environmental conditions. The protection provided by this process applies to any form of component assembly such as soldered, welded, conductive-paint, vacuum-deposited, and functional crystal construction.

Heat

The plastic conducts heat away from each component at a rate of from 2 to nearly 20 times that in air. However, the heat must still be transferred away from the plastic surface. Heat sinks may be embedded in the module to transfer heat to the mounting structure in addition to providing a means of support for the module.

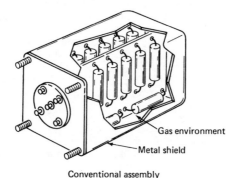

Gas environment

Metal shield

Conventional assembly

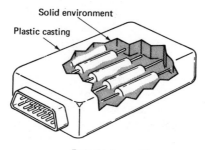

Solid environment

Plastic casting

Embedded assembly

Fig. 9-1. Comparison of conventional and embedded module.

Interconnection

Interconnections within an embedded module may be broken by the differential thermal expansion rates of the conductors and the plastic. Fine wires found in toroidal winds are susceptible to this type of damage. Extreme care must be exercised to avoid this condition during the selection of materials.

PLASTICS

Fundamentals

There are two major divisions in the plastics used in the electronics industry:

1. The *thermoplastic materials* are solid at ordinary temperatures and viscous liquids at elevated temperatures. Polyethylene, vinyls, acetates, and nylon are examples of thermoplastic materials.
2. The *thermosetting materials* begin as liquids or low-temperature–melting solids, and, with heat, they form permanent solids unaffected by subsequent elevated temperatures. Examples of this class of materials are phenolics, urethanes, melamines, phthalates, alkyds, and epoxies.

Because thermosetting plastics have favorable dielectric properties, mechanical strength, and resistance to elevated temperatures, they are the most widely used plastics in the electronics industry.

Epoxy Constituents

The constituents of epoxy-based compounds are the resin and the hardener (catalyst). Fillers or pigments are added when required. Physically, the resins and hardeners vary from water-clear liquids to low-temperature–melting solids. When a small amount of liquid hardener is added to the liquid resin, the resin polymerizes. Solid hardeners and resins, in sintered pellet or powder form, polymerize after melting.

Epoxy Cure Cycle

The polymerization period, called the *cure cycle*, may vary from minutes to days depending on the materials and the ratio of materials. Although many resin formulations cure at room temperature, exothermic heat liberated by the chemical action may harm temperature-sensitive electronic components. It is generally true that epoxies cured at higher temperatures will withstand higher operating temperatures after cure. However, the temperature limits of components in the nonoperating condition often prevent the use of the high-temperature–cure epoxies. The pot life indicates to the user how much working time is available before the epoxy becomes difficult to mold. The curing action of most epoxies may be arrested by refrigerating the epoxy. Uncontrolled acceleration of cure results in unwanted excess exothermic heat and entrapment of the bubbles.

Epoxy Properties

A general range of mechanical and electrical properties of epoxies is shown in Table 9-1. Care should be exercised when examining Table 9-1 to avoid assum-

Table 9-1. Epoxy Properties.

Property	General Range
Pot life at room temperature	15 min to 1 yr
Viscosity	300 cP to pastelike and proportionally higher upon addition of fillers
Cure time	30 min to 4 days
Temperature rise from exothermic heat	20°C to 150°C
Flexural deformation temperature	50°C to 180°C
Specific gravity	0.7 for sphere (microballoons) filled to 2.5 for oxide-and-metal-filled
Coefficient of thermal expansion	3 to 7 $\times$ 10^{-5} in./in. °C
Shrinkage during cure (10-in. mold)	0.37% to 2.5%
Moisture absorption (% weight)	0.07% to 0.44%
Heat resistance (% weight loss)	0.02% to 6.0%
Thermal shock	−55°C to 130°C
Tensile strength	5000 to 10,000 psi
Compressive strength	13,000 to 25,000 psi
Coefficient of thermal conductivity	10^{-4} to 10^{-3} $\dfrac{\text{cal.cm}}{\text{cm}^2 \text{ °C sec}}$
Dielectric strength	200 to 500 volts/mil (10^{-3} in.)
Volume resistivity	10^9 to 10^{15} ohm cm
Dielectric constant	3 to 5 at frequencies 60 to 10^7 cycles
Dissipation (power) factor	0.01 to 0.14 at frequencies 60 to 10^7 cycles

Table 9-2. Fillers.[a]

	Increases							Decreases			
Filler	Thermal Conduction	Compressive Strength	Machinability	Electrical Conductivity	Dielectric Constant	Impact Resistance	Dielectric Strength	Thermal Expansion	Specific Gravity	Shrinkage	Cost
Sand	X	X					X	X		X	X
Silica	X	X					X	X		X	X
Mica					X	X	X			X	X
Quartz	X	X				X	X	X		X	
Alumina	X	X			X		X	X		X	
Powdered metals	X	X	X	X			X	X		X	X
Glass spheres									X		

[a]X = most common usage.

ing that any one epoxy formulation can contain the best value for each property. For example, the two properties of low density (specific gravity) and high thermal conduction are difficult to obtain in one formulation.

Fillers and Pigments

Fillers and pigments may be added to the resin to promote better electrical and mechanical properties and to provide colors. Most fillers and pigments may be obtained in prefilled resins from the manufacturer. Table 9-2 shows representative types of fillers and the resulting change in the properties of the epoxy system.

ASSEMBLY TECHNIQUES

General

Embedded modules generally fall into categories: (1) those with pins or leads projecting from the molded surface, and (2) those with connection devices flush or nearly flush with the molded surface. These two categories of modules require slightly different mold designs.

Mold

A mold is any hollow form in which components are cast or embedded. High-production molds are usually made of hard materials and are easily taken apart for loading and part removal. Multipart molds are good for many repeated castings; however, they are expensive because of machining and inspection costs. A bad feature of any mold is that only one part may be cast at a time. If the protective material requires hours to cure, many expensive molds are required to produce quantities of parts.

A nonmold production technique utilizes the potting cup. (See Fig. 9-2.) The cup is made prior to assembly by injection or thermal compression molding. On assembly the electronic components are placed into the cups, the potting material is added, and the potting cup remains a part of the assembly.

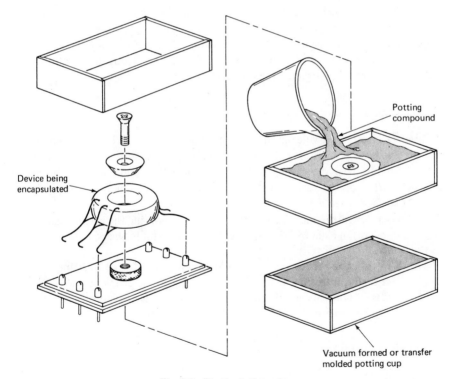

Device being encapsulated

Potting compound

Vacuum formed or transfer molded potting cup

Fig. 9-2. Plastic shell potting

Mold Material

The most versatile mold material is the room-temperature-vulcanizing (RTV) silicone-rubber compounds. The properties of this material may be summarized as follows:

1. Cures to a solid rubbery mass at room temperature.
2. May be obtained in compositions that will cure in from 30 minutes to 24 hours.
3. Remains solid at temperatures from $-70°F$ to $500°F$.
4. Accurately reproduces fine details.
5. Will not adhere to smooth surfaces.
6. Will make many molds from one pattern.

Mold Pattern

The mold pattern is the male die over which the hollow mold forms are made. The patterns are mostly made from metal, usually aluminum, to obtain a mirrored finish and close dimensional tolerance control. Mold patterns are made to the exact size and shape required in the final part. A tapped thread is required to keep the pattern snugly against the container.

Making a Mold

Figure 9-3 illustrates a vulcanized mold. Evacuation of the RTV compound just after it has been poured over the pattern eliminates any internal bubbles in the mold. Bubbles near the casting surface tend to expand under exothermic heat and evacuation of the casting material. This results in indentation pockets on the cast surface. (See Fig. 9-4A.)

Sharp corners in the mold are to be avoided. In pouring the casting material, sharp corners often will not fill completely and will result in the condition shown in Fig. 9-4B.

The RTV silicone-rubber mold is generally good for 10 to 12 casting operations. Eventually, intrusion of the RTV particles into the plastic surface occurs, which results in poor surface finish for the cast part. Removing the part from the mold also becomes difficult because of this intrusion.

Protruding Lead Molds

A one-piece mold for protruding lead types of modules is shown in Fig. 9-5. To insure a flat surface on which the module may rest when placed on the inter-

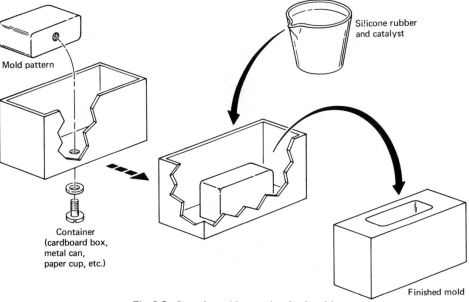

Mold pattern

Silicone rubber
and catalyst

Container
(cardboard box,
metal can,
paper cup, etc.)

Finished mold

Fig. 9-3. Steps in making a vulcanized mold.

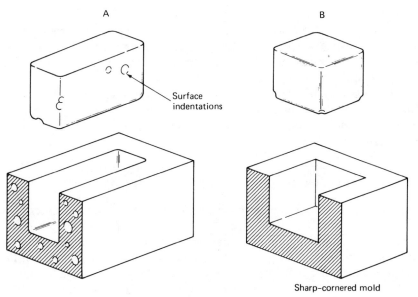

A

B

Surface
indentations

Sharp-cornered mold

Fig. 9-4. Vulcanized mold irregularities.

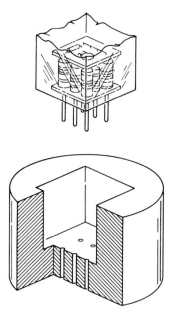

Fig. 9-5. One-piece protruding lead mold.

connection board, one surface of the mold contains holes for the leads. If the side of the module containing the leads is placed in the open end of the mold, an irregular meniscus forms on the leads and the sides of the module. (See Fig. 9-6.)

Lubricants may be coated on the leads to reduce insertion friction forces. There are, however, some lubricants that should not be used. Silicone-based greases and lubricants that may make it difficult to obtain a good solder or weld joint should be avoided.

Flushed Connector Molds

The mold design for modules containing flush or nearly flush connectors is similar to the protruding lead type. The difference occurs in the manner in which the mold pattern and mold are made. The mold pattern generally contains the connector. A one-piece mold can be used unless the geometry of the part is complex or entrapping air is a problem. (See Figs. 9-7 and 9-8.) One problem encountered from embedding connectors is the possibility of getting casting material into the mating portions of the connector. Thus, care should

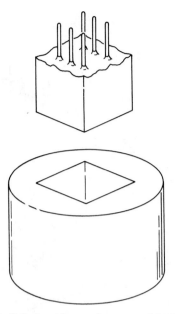

Fig. 9-6. Irregular meniscus on cast parts.

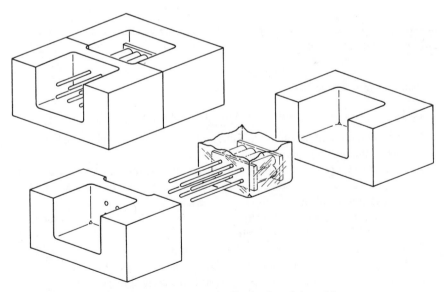

Fig. 9-7. Two-piece protruding lead module mold.

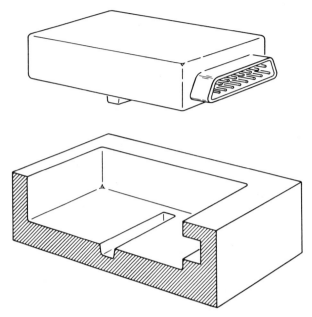

Fig. 9-8. Flush connector module mold.

be taken to protect the connector such as applying a release material to allow stripping the connector clean.

Embedding and Casting

Procedure for casting components in epoxy (see Figs. 9-9 and 9-10):

1. Preheat mold and assembly to same temperature as or slightly higher than that of the potting compound.
2. Pour epoxy compound slowly along the side of the mold to the desired cast level. Allow to set 20–30 minutes, and repour to compensate for settling and shrinkage.
3. Place entire system (mold) into vacuum oven at 140°F, and place a vacuum of 25 to 27 inches for 10–15 minutes.
4. Release vacuum, and allow to bake for 1 hour.
5. Remove mold from the oven, and allow to cool at room temperature.
6. Remove module from mold, and forward to electrical inspection.

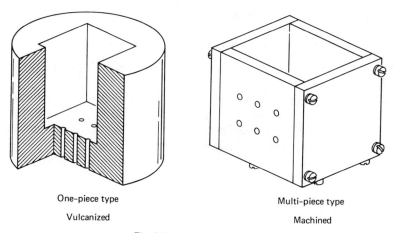

One-piece type   Multi-piece type

Vulcanized   Machined

Fig. 9-9. Prototype molds.

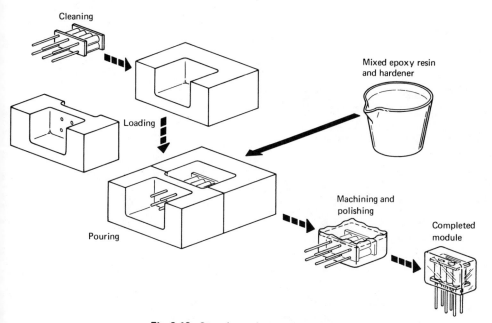

Fig. 9-10. Steps in casting components.

DESIGN RECOMMENDATIONS

The following considerations are recommended during the design:

1. Keep in mind that the part will be molded. The embedding, potting, or casting potential must be partially "designed-in."
2. Orient the components to minimize air entrapment in one or two axes.
3. Strain-sensitive components should be encapsulated and cured in flexible compound prior to placement in the mold. The epoxy provides a hard second protective layer in a two-step molding process.
4. Use attachment devices that have a knurled surface on which the plastic materials will adhere.
5. Examine connection devices (connectors and headers) for possible leakage of the plastic through and around the insert face.

10
ADHESIVE
BONDING

Preparation and Use

With each new advance in the electronics industry comes the need for better methods of joining one or more parts together making them able to withstand a variety of requirements (i.e., environmental, electrical, mechanical, etc.). Three of the most common methods of bringing together different materials are welding, fastening, and bonding adhesives.

This chapter will cover adhesives and will give an overview of the purpose and use of adhesives.

As stated above, the environment in which an adhesive is to be used determines which classification of adhesives it will fall into: structural or machinery. Structural adhesives are generally used for non-operating conditions such as joining four corners of a steel box together, joining plastic laminates together, etc. Machinery adhesives, on the other hand, refer to moving parts or operating parts. Examples of machinery adhesives would be those used for gear trains, attachment of accelerometers, and other torque related requirements.

In order for an adhesive to be suitable to be a given environment, it must fulfill three basic requirements. One is the *flow* or the adhesive's ability to fill spaces between the surfaces of two parts. Two is *wetting*. This is the adhesive's ability to join intimately with the attaching substrates. And lastly is *curing*. Curing is the chemical reaction taking place in the adhesive which permits it to harden, giving it its strength and rigidity.

An important factor in determining which adhesive to use in a given situation is the substrate being bonded. For example, if you wanted to attach a ceramic chip to a metal substrate you could use epoxy, silicone, or anaerobics, among others.

A new method of bonding ceramics to metals is presently under investigation. It is referred to as *reaction bonding*.

The metal substrate undergoes a corrosion-like reaction which is catalyzed by the ceramic chip. This reaction stimulates the growth of tendrils that entwine or weave into the molecular structure of the ceramic. The bond takes a few minutes to form, although to optimize strength the materials can be held together under high temperature conditions for several hours.

Reaction bonding will occur between all metals and most ceramics, as well as glass and some gemstones.

Some adhesives, referred to as convenience adhesives, have several common characteristics which enable them to be grouped in one classification even though they may be chemically quite different. Generally, they:

Do not require energy for cure,
Do not require mixing,
Cure within 5 minutes,
Achieve high strengths,
Bond a multitude of substrates,
Can be automated easily.

Anaerobics, cyanoacrylates and rubber modified acrylic adhesives fit easily into this category.

See Tables 10-1 to 10-4 for selection of adhesives and convenience adhesive properties respectively.

Prior to the application of any adhesive or sealant, it is of utmost importance that the surface to be bonded is clean. For effective bonding to occur, the adhesive must wet the surface of each substrate being joined. It is necessary that a chemical bond form between the surface of the adhesive and the substrate (see Figure 10-1) to satisfy these conditions, the substrate's surfaces must be clean of dust, grease, oil, etc. and chemically receptive to the adhesive chosen.

Surface preparation generally includes four types of cleaning: (A) detergent, (B) solvent, (C) chemical, and (D) mechanical. The amount of cleaning required depends upon the bonding strength required.

Detergent Cleaning

Detergents, soaps, and caustic sodas are reasonably inexpensive methods of preparing surfaces for adhesives but in the cleaning process, may in some instances, cause further contamination of the surface being cleaned.

Table 10-1. Adhesive Selection.

	Vinyl Plastic	Phenolic Plastic	Rubber	Glass	Metals
Glass, Ceramic	E, N	A, C, H, I, M, N	C, D, F, I, M, N	B, H, J, M, N	
Metals	E, N	A, C, H, I, J, K, M, N	D, C, H, I, J, K, M, N	F, H, J, K, L, M	F, G, H, I, J, K, N
Phenolic Plastic	N	C, J, K, N			
Rubber	E, N	D, H, N	C, D, F, I, M, N		
Vinyl Plastic	E, N				

Thermoplastic
A   Acrylic
B   Cellulose Nitrate

Elastomeric
C   Natural Rubber
D   Reclaim Rubber
E   Buna-N

Special Purpose
F   Cyanoacrylates

Thermosetting
G   Phenol Foramaldehyde (phenolic)
H   Epoxy

Resin Blends
I   Phenolic-vinyl
J   Phenolic-polyvinyl Butyral
K   Phenolic-Polyvinyl Formal
L   Phenolic-nylon
M   Phenolic-neoprene
N   Phenolic-butadiene-acrylonitrite rubber

Solvent Cleaning

Solvents should be used both prior to and after abrasion cleaning and should precede any chemical cleaning. This prevents contaminants from becoming embedded in the substrate rendering the solvent useless.

Chemical Cleaning

Chemical cleaning methods include degreasing, alkaline, pickling, and chemical surface alteration compounds.

DEGREASING is the first chemical cleaning step and entails dipping or spraying substrate surface to be bonded with a solvent.

ALKALINE CLEANING usually comes after degreasing. In this process, the parts to be bonded are dipped in a continuously agitated alkaline bath.

PICKLING involves immersing a ferrous metal in an aqueous acid solution to remove a loose layer of oxide from its surface.

CHEMICAL SURFACE ALTERATION when deemed necessary would be the final step in the surface preparation. The metal adherend is dipped into a chemi-

Table 10-2. One-Part Adhesive Properties.

Property	Cyanoacrylate	Anaerobic	Acrylic
Liquid Properties			
Viscosity Range	1–8000 cps	1,000–10,000	20,000–2,000,000
Flash Point	180° F	200° F	125
Color	Clear	Amber	Dark Amber
Shelf Life	1 Year	6 mos–1 Year	6 mos–1 Year
Toxicity	Low	Low	Moderate
Gas Filling Capability	.020"	.040"	.030"
Strength Properties			
Shear Strength			
Steel	3,000	3,000–7,000	4,500
Oily Steel	500	1,800	3,000
Aluminum	2,500	3,000	3,500
Impact Ft. Lb.	5–10	1–50	10–20
Peel Plw	1–5	2–65	25–65
Rubber	Excellent	Fair	Good
Plastics	Excellent	Good	Very Good
Ceramic	Excellent	Excellent	Excellent
Durability Properties			
Temperature Limit	100° F	450° F	250° F
Solvent	Good	Good/Excellent	Very Good
Humidity	Fair	Good	Good
Weathering	Fair	Excellent	Very Good

Table 10-3. Performance Properties.

Property	Cyanoacrylate	Anaerobic	Acrylic
Speed	1	2	3
Safety	3	1	2
Versatility (surfaces bonded)	1	3	2
Automation	1	2	3
Gap filling	3	1	2
Shear strength	2	1	3
Tensile strength	2	1	3
Impact strength	3	1	2
Peel strength	3	1	2
Fatigue strength	3	1	2
Temperature resistance	180° F	400° F	250° F
Solvent resistance	3	1	2
Weather resistance	3	1	2
Temperature shock	3	1	2
Water resistance	3	1	2
Humidity resistance	3	1	2

Notes: 1–preferred; 2–fair; 3–acceptable

Table 10-4. One-Part Adhesive Material Selection.

Property	Cyanoacrylate	Anaerobic	Acrylic
Steel	3	1	2
Oily steel	2	3	1
Aluminum	2	1	3
Plated metal	3	2	1
Glass	3	1	2
Ceramic	3	1	2
Rubber	1	3	2
Phenolic	3	2	1
Acrylic	1	3	2
Polycarbonate	1	3	1
Epoxy (glass reinforced)	2	3	1
Polyester (glass reinforced)	2	3	1
Acetal	1	3	2
Polyimide	3	1	2

Notes: 1–preferred; 2–fair; 3–acceptable

cally active solution. This solution either dissolves part of the surface or transforms it, making it more chemically active and more receptive to bonding with adhesives.

Mechanical Cleaning

Mechanical cleaning involves the use of sandblasting, hand brushing, chipping, scraping and tumbling among other methods. This method of cleaning produces varying degrees of surface texture, making it more responsive to certain types of adhesives.

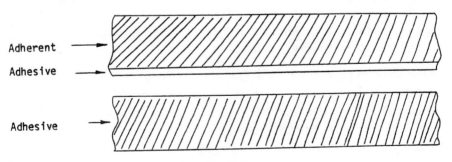

Adherent

Adhesive

Adhesive

Fig. 10-1.

Suggested Procedures for Treatment of Metal Faying Surfaces

Wrought Aluminum and Aluminum Alloys (Except Sandwich Cores)

Degrease surfaces with an organic solvent, immerse for 10 minutes at 150° to 160° F. (65.6 to 71.1° C.) in a chromic acid solution of approximately the following composition by weight:

30 parts clean, demineralized, oil free water
10 parts concentratred sulfuric acid (1.84 specific gravity (sp gr))
1 part sodium dichromate ($Na_2Cr_2O_7 \cdot 2H_2O$–technical grade)

Rinse in demineralized water at a temperature no higher than 150° F. (65.6° C.) and with a pH no greater than 7.5. Observe immediately for water breaks on the surfaces as evidenced by the water not cascading from the parts or specimens in a smooth, continuous sheet. If water breaks occur, repeat above treatment. Dry the rinsed parts or specimens for 30 minutes, or until thoroughly dry, at a temperature no greater than 150° F. (65.6° C.), and repeat the water-break test or subject to instrumentation test.

Stainless Steel (Except Sandwich Cores)

Degrease and proceed as specified above except that immersion will be for 15 minutes at 150° to 160° F. (65.6 to 71.1 C.) in a strong acid-dichromate solution of the following composition:

96.6 percent by volume concentrated sulfuric acid (1.84 sp gr)
3.4 percent by volume saturated water solution of $Na_2Cr_2O_7 \cdot 2H_2O$.

Sandwich Cores (Aluminum Alloys)

Degrease and proceed as specified above except that immersion will be for 15 minutes at 150° to 160° F. (65.6° to 71.1° C.) in a mild acid-detergent solution of the following composition by weight:

92.9 percent clean, demineralized, oil free water
1.0 percent concentrated sulfuric acid (1.84 sp gr)
6.0 percent sodium dichromate ($Na_2Cr_2O_7 \cdot 2H_2O$-technical grade)
0.1 percent surface active detergent

Sandwich Cores (Titanium and 302 Type Stainless Steel)

Degrease and proceed as specified above except that immersion will be for 15 minutes at 150° to 160° F. (65.6° to 71.1° C.) in a detergent solution of the following composition by weight:

94.4 percent clean, demineralized, oil free water

2.0 percent sodium metasilicate (Na_2SiO_2)

3.6 percent surface active detergent

Water Contact Angle Test

A drop of distilled or demineralized water placed on a flat, dry, thoroughly prepared metal surface will spread rapidly and uniformly over the surface.

If the water drop does not so disperse itself, the metal surface is not properly prepared for subsequent adhesive bonding and must be reprocessed. With proper instrumentation, the angle of contact of a drop of water with a base surface can be measured. A low contact angle (10 degrees or less) indicates a surface satisfactorily prepared for adhesive bonding. A satisfactory contact angle measuring apparatus consists principally of a telemicroscope with a crosshair in a rotating eyepiece and with a prism on the objective end. The process specification will include an effective quality control provision based on contact angle measurements and instrumentation described herein.

Treatment of Titanium Prior to Adhesive Bonding

The following treatment is suggested:

1. Methyl-ethyl-ketone wipe.
2. Trichloroethylene vapor degrease.
3. Pickle in the following water solution at room temperature for 30 seconds:
   Nitric acid—15 percent by volume of 70 percent nitric acid solution.
   Hydrofluoric acid—3 percent by volume of 50 percent hydrofluoric acid solution.
4. Rinse in tap water at room temperature for 2 minutes.
5. Immerse in the following water solution at room temperature for 2 minutes:
   Trisodium phosphate—6.68 ounces per gallon of solution.
   Potassium fluoride—2.7 ounces per gallon of solution.
   Sodium fluoride—1.2 ounces per gallon of solution.
   Hydrofluoric acid (50 percent solution) 2.6 percent by volume of solution.
6. Rinse in tap water at room temperature.
7. Soak in 150° F. tap water for 15 minutes.
8. Spray with distilled water and air-dry.

NOTE: If heavy heat-treat scale present, molten salt descale prior to above procedure.

Treatment of Magnesium Alloy, AZ-31-H24 Prior to Adhesive Bonding

The following treatment is suggested:

1. Degrease panels for 10 minutes at room temperature in trichloroethylene, then treat for 10 minutes at 160° to 190° F. in the following solution:
   95.3 percent by weight water
   2.2 percent by weight Na_2SiO_3 (metal)
   1.1 percent by weight $Na_4P_2O_7$ (pyrophosphate)
   1.1 percent by weight NaOH
   0.3 percent by weight Nacconol NR.
2. Rinse panels with cold distilled H_2O and then treat for 10 minutes at 155° ± 5° F. in a solution of 20 percent CrO_3.
3. Panels are then given a final rinse in warm distilled H_2O and air-dried.
4. Prepared surfaces should be primed with adhesive or bonded as soon as possible; elapsed time not to exceed 8 hours."

Extract from MIL-A-9067C.

REFERENCES

Science Digest, August, 1983, New York, New York
Marshall Claude Michael, 1982, El Monte, California
MIL A-9067-C

11
REWORK AND REPAIR

REWORK

Rework is the performance of an operation needed to:

1. Correct an improperly performed previous operation.
2. Replace a defective component.
3. Correct or compensate for an unexpected quality characteristic.
4. Effect a change due to design change, shortage, customer requirements, etc.

SOLDER-REMOVAL

Never strike the printed-circuit board or assembly against a hard object to jar solder loose. Remove excess solder with the soldering-iron tip by careful shaking, brushing, or wicking.

Wicking may be performed by placing the end of a piece of shielding or ground braid (after fluxing with an approved flux) against the heated joint. The molten solder will flow away from the joint and into the shielding.

REPAIR OF PRINTED-CIRCUIT-BOARD ASSEMBLIES

Pattern Repair

Damaged Conductor

Nicks, scratches, gouges, or lifting of foil and platings.

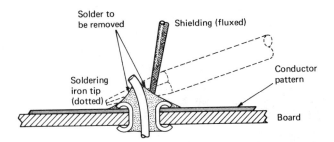

Fig. 11-1. Solder removal by wicking.

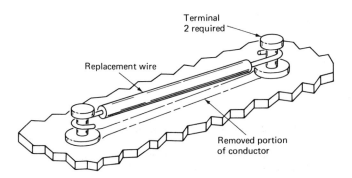

Fig. 11-2. Damaged conductor replacement.

1. Cut through the conductor to the base laminate at the edge of each affected pad and remove the damaged pattern (see Fig. 11-2). Care shall be taken to prevent cutting of laminating fibers on the board surface.
2. Install a funnel terminal in each hole, with the terminal post on the component side of the board.
3. Solder an insulated wire of 22- or 24-gauge AWG solid copper wire between the terminal posts as required. Also solder to the post any component leads that were originally intended for the affected plated-through holes. The wire should be woven beneath and around components for protection, if possible, but this is not a requirement.

Damaged or Defective Conductors

A damaged conductor may be a complete break (A), or scratches (B), nicks (C), or pinholes (D) which reduce the cross-sectional area of the conductor by more than 25 percent. (See Fig. 11-3.)

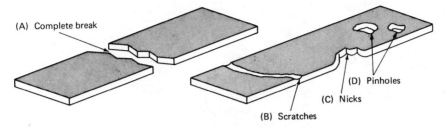

(A) Complete break

(D) Pinholes

(C) Nicks

(B) Scratches

Fig. 11-3. Damaged conductor.

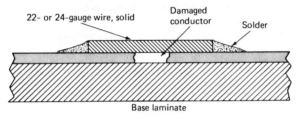

22- or 24-gauge wire, solid

Damaged conductor

Solder

Base laminate

Fig. 11-4.

Recommended repair for complete break (Fig. 11-4)

1. Clean both sides of break on the conductor about $\frac{1}{4}$ inch back from break with an eraser and then with Freon TMC.
2. Cut piece of 22- or 24-gauge solid copper wire a minimum of $\frac{1}{4}$ inch longer than the break.
3. Hold wire on centerline of conductor, across the break, and solder in place.

Recommended repair for scratches and pinholes that reduce the conductor width by 25 percent (Fig. 11-5)

1. Clean surfaces to be repaired thoroughly with a soft-pencil eraser and then Freon TMC.

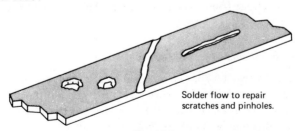

Solder flow to repair scratches and pinholes.

Fig. 11-5.

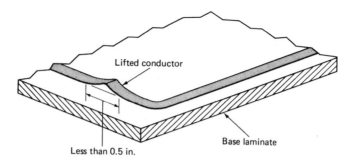

Less than 0.5 in.

Fig. 11-6. Lifted copper circuit.

2. Reflow solder over scratches and pinholes to fill any voids on the conductor. (*Note:* Use a soldering-iron tip that closely matches the width of the conductor to be repaired, melt solder on the tip, and use brushlike strokes over the faults to obtain best results.)

Lifted Conductors

A lifted conductor exists when a portion of the conductor pattern is lifted from the base laminate or substrate, but not broken. The length of the lifted conductor that is repairable shall not exceed one-half the length of the conductor path between two terminal areas or 0.5 inch, whichever is smaller. (See Fig. 11-6.) Lifted conductors are repaired by one of the following methods:

- Method A, epoxy under conductor (preferred). Clean area to be repaired, spread Hysol 2038 or equivalent under entire length of lifted conductor and press conductor into contact with substrate and hold until set. Clean off excess epoxy. Then cure or bake per manufacturer's instructions.
- Method B, epoxy over conductor (not preferred). Clean area to be repaired; press conductor into contact with substrate; then apply Hysol 2038 or equivalent to surface of lifted conductor and to a distance of at least 0.125 inch in all directions from the damaged area. Then cure or bake per manufacturer's instructions.

Lifted, Damaged, or Missing Terminal Areas

The unshaded area in Fig. 11-7 is the terminal and/or land area under question. In those areas, a 0.0015-inch feeler gauge can penetrate a distance equal to no more than one-half the distance from the interface of the hole to the nearest edge of the terminal (annular ring) for not more than 180° of the periphery. When feeler-gauge penetration does not exceed limits just described, the terminal

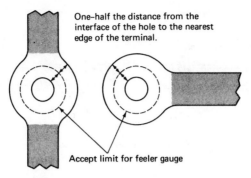

One–half the distance from the interface of the hole to the nearest edge of the terminal.

Accept limit for feeler gauge

Fig. 11-7. Printed-circuit terminal pads.

area shall be acceptable and not considered as lifted. However, if the feeler-gauge penetration is greater than described or if the defect can be visually determined, repair may be made in accordance with the following procedures.

1. For partially lifted terminal areas. Clean immediate area of solder using a vacuum-type solder remover. Clean area to be repaired with Freon TMC. Insert Hysol 2038 or equivalent epoxy resin under the copper with a camel's-hair brush or other suitable applicator. Press terminal area down with a blunt instrument and hold until set. Clean off excess epoxy. Cure or bake according to manufacturer's instructions.

2. For a completely lifted terminal area. Remove any solder with a vacuum-type solder remover. Clean area to be repaired with Freon TMC. Insert Hysol 2038 or equivalent under copper with a suitable applicator. Adhere to board by pressing with a blunt instrument or by applying the flat side of a clean soldering iron for 5 seconds. After adhering or curing, check for good bond. Clean the repaired areas with solvent and remove excess epoxy. Install a fused or funnel eyelet of sufficient size to receive the component lead. The outer diameter of the eyelet should be within 0.010 inch of the inner diameter of the hole. Solder eyelet, and clean area with Freon TMC.

COMPONENT REMOVAL

1. Cut the leads as close to the board as possible on component side. (See Fig. 11-8.)

2. Heat and remove solder and portion of lead remaining from dip side of board. *Do not overheat joint, and avoid touching other components with soldering iron.*

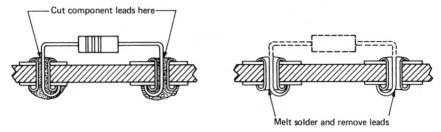

Fig. 11-8. Component removal.

3. Mount replacement component, and solder on dip side of board. *Note:* When using a soldering iron to replace semiconductors, each lead joint must be made within 3 seconds, and time must be allowed between each connection for the part to cool. Use heat shunts between joint and part wherever possible.

COMPONENT REPAIR

Mounting Defects–Uncoated Components

Resistors or similar items having cold solder joints, lead misalignment, excess solder at the joint, poor electrical connection, or that are improperly functioning. (No coating on the board or components.)

1. Using a hand-soldering iron of wattage appropriate for the repair, remove the affected components. Heating time per joint is 7 seconds maximum. *Caution:* Never apply heavy pressure to a joint with a soldering iron, since the pad may be lifted.
2. Remove excess solder from the component lead, pad, and hole. If necessary, heat a piece of metallic braid in contact with the hole to draw out the solder, or use a "solder sucker."
3. Resolder or replace the component, using Sn63 solder. Heat-sink all diodes and transistors during removal and replacement.

Mounting Defects–Coated Components

Resistors or similar items having cold solder joints, lead misalignment, excess solder at the joint, poor electrical connection, or for similar parts that are improperly functioning. Polyurethane coating has been applied to the board and components.

1. Use a $47\frac{1}{2}$ watt, pencil-type, hand-soldering iron with a $\frac{1}{8}$-inch chisel tip to melt the solder joint. This will also crystallize the polyurethane coating without burning the board. Heat for 7 seconds maximum. If additional time is required, let the solder joint cool to room temperature before re-applying heat for an additional 7 seconds (maximum).
2. Remove crystallized coating material from the joint area. Remove excess solder from the component lead, pad, and hole. If necessary, heat a piece of metallic braid in contact with the hole to draw out the solder, or use a "solder sucker."
3. Resolder or replace the component using Sn63 solder. Heat-sink all diodes and transistors during removal and replacement.
4. Replace the coating according to applicable print requirements.

Coating Repair

Holes, gaps, breaks, or scratches through the polyurethane board-coating to the base laminate; bubbles in the coating.

1. Inspect the board under ultraviolet light and mark defects.
2. Remove the top of each bubble with a scalpel.
3. Fill each bubble crater, hole, scratch, or similar flaw with polyurethane. Overlap the defect on all edges by $\frac{1}{16}$ inch to insure sealing. Most repairs may be made with a small, artist's brush.

Mislocated Holes

Misplaced holes may be ignored if the electrical or mechanical function of the board is not impaired. Holes that require filling shall be processed as follows. (This method does not apply to a defect if any hole dimension exceeds 0.5 inch.)

1. Bevel the hole edges as shown in Fig. 11-9.

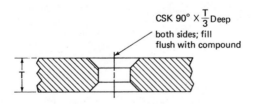

Fig. 11-9. Repair of mislocated holes.

2. Fill the prepared hole with epoxy potting compound. Spread the resin approximately $\frac{1}{16}$ inch beyond the edges of the defect. Cure the resin for $1\frac{1}{2}$ hours at $150° \pm 5°F$. The cured resin shall be flush with the parent laminate within $-0.005 + 0.015$ inch. However, in the ground frame and contact areas, this thickness shall not exceed the detail drawing dimension.

Edge Defects

Board-edge damage to laminate only. This method shall not be used if any defect dimension exceeds 0.5 inch, after preparation per No. 1 below.

1. Remove damaged material until board thickness at the defect edges measures within applicable print requirements.
2. Fill the damaged area to the original board dimensions with epoxy potting material. Spread potting $\frac{1}{16}$ inch beyond defect edges onto the undamaged board surfaces.
3. Cure.

Warped Laminate, Minor

For boards that are warped in the conductor area. Warpage slightly exceeds maximum requirements.

1. Place the board between two flat plates of sufficient weight to flatten the board.
2. Heat the board for 4 hours at $150° \pm 5°F$.

Warped Laminate, Major

For those boards that are warped in the connector contact area.

1. Place the board contact area between two aluminum plates.
2. Apply enough pressure, using clamps, to straighten the board.
3. Heat the board for 1 hour at $150° \pm 5°F$.
4. Increase the pressure slightly and heat for 3 hours at $150° \pm 5°F$.

12
PRINTED CIRCUIT PROCESSING AND ASSEMBLY

Basic to considerations of Processing and Assembly of printed circuit boards is the fact that the circuit and its components will ultimately be soldered. At this point, all the decisions made in design regarding layout, material selection, and processing should add up to a complete, reliable circuit. In pursuing this objective, low-cost uniformity and high reliability are prime goals.

There is no single determining factor or set of factors that define the materials or process to use. The volume of circuits required, and the availability of engineering time, talent, plant space and investment dollars are preliminary guides used in the selection procedure. Where in-house capability is not necessary or advantageous, manufacture of circuits can be detailed in specifications to job shops. Specific products, such as solderable organic coatings and soldering resists, can be called out on the purchasing specification. The facilities of these companies allow engineering time and manufacturing space to be devoted entirely to the assembly and manufacture of the basic product being produced.

If a specialist job shop is selected to produce printed circuits they will assist in determining the following factors:

1. Selection of Base Material. Laminate manufacturers provide excellent assistance in this area. The operating environment (temperature and humidity), the electrical qualities (dielectric and resistance values), the process to be used, and the material and manufacturing costs must be considered. Although there are many grades of laminate, the most widely used are: XXP-Phenolic, XXXP-Phenolic, G-10 Epoxy Glass, G-11 Epoxy Glass.

2. Material Characteristics. These include thickness of the copperclad, (2 or 3 oz. copper, one- or two-sided clad) and the thickness of the laminate needed to meet the strength and weight requirements.
3. Use of eyelets or plate-through holes.
4. Use of solder resist on the circuits.

If the circuits are to be manufactured in-house, additional decisions must be made in regard to material selection and technique of processing. The specific type of circuit selected, such as gold or solder plated, determines the general process used. With each decision, the number of production alternatives are reduced. Regardless of the method, cleanliness is vital and quality control paramount.

CLEANING COPPERCLAD

The need for cleaning the copper surface is dictated by the requirements of the production steps. Some of the methods of cleaning are degreasing in cold or vapor degreasing solvents, cleaning in alkaline cleaners, tampico scrubbing, blasting, and chemical cleaning.

Improved Resist Adhesion

When cleaning copperclad for improved adhesion of resists (such as with cold top enamels), tampico scrubbing and/or chemical cleaning is recommended. Non-passivating chemical brighteners with controlled etching will clean without etching away valuable copper. The circuit is dipped into the brightener and pumice is sprinkled on the circuit surface and brushed. The oxides are chemically reduced and the pumice abrades the surface giving a finish that is receptive to the resist.

Cleaning for Plating

There are several reasons for plating of circuit: (1) For use as the resist in the etching of the circuit; (2) To achieve through-connections of holes in the circuit for better soldering; (3) To increase surface wearability by plating with rhodium, nickel-rhodium, gold, etc.; (4) To maintain the solderability of the circuit surfaces.

The condition of the laminate determines the steps to be used in the preparing of the surface for plating. If oil or grease is present it is good practice to clean with alkali or hot or cold solvent. Here too the practice of pumice brushing is sometimes used; chemical cleaning is considered best—with or without the pumice. A non-passivating chemical brightener can give maximum cleanliness to copper surfaces. If the brightener has strong chelating and solubilizing action, it will insure the removal of copper complexes. This action on the metallic oxides, carbonates, and salts insures their removal during water rinsing.

Although water rinsing minimizes carryover of the brightener to the plating baths, many processes include an additional chemical dip. When circuits are to be solder plated, a dip in a mild solution of fluoroboric acid is sometimes used to insure plating bath compatability. Also, when plating with cyanide gold, a short dip in a 2 oz. solution of cyanide is used to insure no acid carry-over. The procedure used is determined by the processing controls and techniques.

It is desirable in cleaning copperclad to do it with a minimum of attack on the copper, maintaining the physical and electrical constants of the laminate. Chemical brighteners that have controlled etching action yield low operating costs. A typical cleaning process is shown in Fig. 12-1.

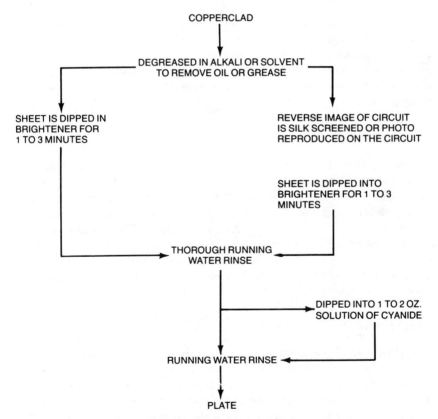

Fig. 12-1. Typical copperclad cleaning sequence.

The chemically cleaned surfaces should have maximum wettability (low surface tension) and offer excellent plating adhesion. Adhesion may be evaluated by the coagulation test on solder plate; with gold, nickel-rhodium, and silver, the 180° mandrel bend and oven exposure tests can be used.

DEFINITION OF CIRCUIT PATTERN

There are several methods of defining the circuit pattern: the photo process, silk screen, offset printing, dry screen, plating, and various combinations. The demand for fineness of line definition, volume or production, compatability with other steps in the process, cost, and equipment available determine the method to be selected. The most popular methods are described below.

Photo Resist Process

Where fine lines and detail are needed, the copperclad is coated with a photosensitive emulsion such as Eastman Kodak's KPR by whirling, spraying, or dipping. A negative of the circuit pattern is superimposed on the coated copperclad and, securely held in a vacuum printing frame, it is exposed to arc lights. The exposed sheet is developed in solvent to remove the portion of the emulsion that is not exposed.

Silk Screen Process

This process is used where there is a requirement for long production runs and where exacting fine-detail definition is not needed. In this process a positive film is made of the circuit drawing. The image is then transferred to the sensitized polyvinyl on the screen by exposure to arc lights. Water is used to rinse off unexposed film and the circuit pattern is defined. Where long runs are required or where dimensions are critical, a stainless steel screen is used. The inks used in this process are generally a balance between adhesion (for good line definition with no undercutting) and case of removal after the etching or plating operations are completed.

Plating as a Resist

Solder and gold plating are used as resists in the etching of circuits; gold is seldom used due to high costs. Photo or silk screen silk screen resist is applied on the plated sheet to define the circuit for etching. (The resist is applied prior to the solder plate.) After solder plating the resist is removed and the solder serves as the direct resist to the etchant. Although the plating resist process requires additional steps compared to other methods, it is used because it provides the solderability value of solder or gold and enables the plating of holes at the same time that the circuit is defined, or because it deposits wear-resistant metals for contact surfaces.

ETCHING

The type of etchant is determined by the resist method used. Where gold or solder plate is used for circuitry definition, the etchant must attack (etch) the copper but not the placed metal; etchants such as ammonium persulfate and chromic-sulfuric acid mixtures are used in etching these circuits. However, when the silk screen or photo method is used on the copper surface to define circuitry, ferric chloride or ammonium persulfate is generally used.

Since still-tank etching is generally too slow a method, air-agitated tanks and paddle etchers are used. The best production is achieved in spray-type etching machines. Automated etchers permit control of etching, rinsing, and drying and the etching process is tied into subsequent operations. Control of the etching baths is important. As the etchant is used, its action slows down (temperature control is generally used to maintain etch rate). Metal complexes form that are difficult to remove. Controls such as the copper colorimeter used in testing ferric chloride baths are worthwhile accessories.

After etching the circuits must be thoroughly rinsed. This is a critical phase in the manufacture of circuits. Since the chemicals used as the etching agents are water soluble, their residues become less water soluble as they age and react. As a result metal complexes form which make rinsing difficult. Cleaning agents, with the ability to chelate and solubilize these complexes and render them water soluble, become valuable in this state of circuit processing. A typical process in after-etching rinsing is illustrated in Fig. 12-2.

FERRIC CHLORIDE SPRAY ETCH

SPRAY WATER RINSE

REMOVAL OF RESIST

SPRAY OR DIP WITH BRIGHTENER

COOL SPRAY WATER RINSE

Fig. 12-2. Post-etch rinsing procedure.

Etching resist is removed prior to the use of the chelating and solubilizing cleaning material because a clean solderable condition can be gained on circuitry and pads while insuring complete etching contaminate removal. Whatever the etchant, complete and thorough rinsing is needed. As a quality control check, resistivity measurement of the circuit itself, made at specific humidity and temperate values, is a sensitive measure of circuit contamination. Any residue or reaction of the etchant with the plate metal resists must be controlled because it has a direct bearing on the ultimate solderability of the circuitry and pads.

RESIST REMOVAL

Removal of the resists from printed circuits is not difficult. However, removal of these resists without attacking or harming the laminate is a problem.

Epoxy-glass boards will show exposed fibres when immersed for too long a time or in too harsh a removal agent. If the resist has been flashdried or if a bake cycle was used, removal is slower and the dwell time in the remover must be increased. A line of removers has been developed where the choice between speed of action and complete safety can easily be made.

Removal Techniques

Plain steel tanks into which the circuits are dipped are widely used. To shorten the time of circuit exposure to the remover, spray systems are also used. For small production quantities, trays or dishes are adequate. Brushing following immersion in the remover or forceful spraying of cool water should be used.

The time of exposure of the circuits to the remover should be measured and controlled. Where control is difficult to maintain the safest type of remover should be selected. A typical system for removing photosensitive resist is shown in Fig. 12-3.

If the resist is not completely removed it will adversely affect soldering and subsequent operations. Sometimes the dye may be leached from the resist, giving the appearance of complete removal while small spots or dots of the resist may still remain. The condition of the circuit surface is part of the mechanism of good and reliable soldering. The interfacial tension characteristics of the circuit surface directly affects solder flux wetting and flow.

CLEANING CIRCUITS

Because of the direct relationship between circuit surface and soldering, definite steps must be taken to insure cleanliness.

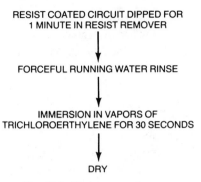

Fig. 12-3. Resist removal process.

Copper Circuits

After etching and removal of photo or silk screen resist, the circuit is sprayed or dipped into the chemical brightener. Not only are the copper oxides removed, but the salts and other copper-iron complexes are solubilized for ease of removal by water rinsing. Etchant baths such as ferric chloride build up a sludge. This sludge consists mainly of extremely fire particles of iron and copper complexes that cling to the circuit surface. The chemical brightener solubilizes these particles and chelates them so the circuits can be easily water rinsed. For small production, dipping is adequate. Fig. 12-4 shows a typical cleaning procedure.

Gold-plated Circuits

Although gold is a noble metal it has a highly absorptive surface and there is ample field evidence to indicate the need for cleanliness. Contamination can origi-

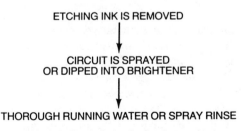

Fig. 12-4. Typical circuit cleaning method.

nate from extraneous fumes and plating or etching residues. Cyanide-based gold-plating baths can leave hard-to-remove residues, particularly when holes have been plated. Chemical cleaning/brightening, with its ability to solubilize and chelate these residues, does an excellent job.

Solder-plated Circuits

Circuits are generally solder plated in order to use the plating as an etchant resist and to plate the holes for through-conductors. When sulfuric-chromic acid mixtures are used in etching solder-plated circuits, the solution coats the solder with a film of lead sulfate-lead chromate, and this prevents further action of the etching solution on the solder. This film becomes a barrier to good solder wetting.

Chemical processing by dip or spray can help to break this "skin," but for maximum cleanliness the circuits should be dipped into the chemical bath and then sprinkled with a mild abrasive and scrubbed. Where plated holes are present ultrasonic baths should be used. The cavitation of the bath insures penetration and cleaning action in the holes of the circuits.

SOLDERABLE SURFACES

One of the basic functions of printed circuits is to aid reduction of soldering time. Therefore appropriate attention should be given to the solderability of the surfaces. There is no best surface. Some of the factors which should be considered are: (1) The basic process by which the circuits are to be made (gold or solder plate as rests become the final circuit surface); (2) Length of time that the circuits are to be stored awaiting assembly and soldering; (3) Equipment available; (4) Cost; (5) Type of solderiing flux to be used.

Electroplated Metals

When the circuit is processed using solder or gold plating as an etching resist, these metals are automatically present on the circuit and offer solderable surfaces. Their cleanliness after plating and/or etching determine their basic solderability. The cleaning materials used and the temperature of the plating solutions can have serious effects on the base laminate and the type of plate deposited. A flash immersion of gold will not give as good long-term solderability in storage as electroplated 0.0001 to 0.0092 gold. Electroplated metals are also used in combination with solderable organic coatings to insure resistance to temperature and moisture, and thus maintain their solderability.

Immersion Metals

These metals are most often used on the copper circuits after they have been etched, etc. They are easy to apply and offer fair protection from oxidation. By nature they are porous; oxidation (particularly in the presence of humidity) of the base copper will occur through the immersion plate. To protect against humidity, organic coatings are used over the immersion plate. A typical immersion method is shown in Fig. 12-5.

Solderable Organic Coatings

Plain organic coatings, although excellent in exclusion of oxidation, are difficult to solder through. The development of solderable protective coatings achieved the needed balance of protection and solderability. Solder-assisting protective coatings can be applied by spraying, dipping, or roller coating.

A solderable protective coating gives greater protection against humidity than immersion plating but it is more sensitive to temperature. The coating can be used in combination with immersion or plated metals, offering double protection plus assistance. Where plated holes are present, the solder assistance from the coating coats the holes and reduces the oxides that form as the circuit comes up to temperature during soldering. This insures good capillarity of the solder in

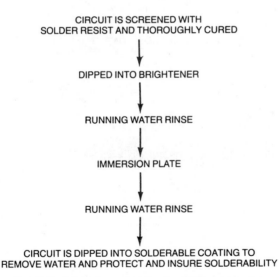

Fig. 12-5. Processing and protection of immersion plate.

the holes. These organic coatings should not be used when water-soluble fluxes are used for soldering. Typical procedures are shown in Fig. 12-6.

Roll Soldering

Owing to the excellent solderability of solder-coated surfaces, equipment has been designed to roller-coat solder onto printed circuits. Because the solder is rolled-on in the molten state, alloying occurs with the copper giving an excellent bond and a dense metallic coating. This equipment consists of a solder pot in which high carbon steel rollers (pretinned) are immersed. Pressure rollers are used to hold the circuit in intimate contact with the tinned rollers. These pressure rollers have individual spring adjustments and are generally chromium plated. The circuits are fluxed and fed through the rollers. The temperatures of the solder bath and the speed of travel of the circuit are critical in order to minimize solder tears on the trailing edges of pads and circuitry.

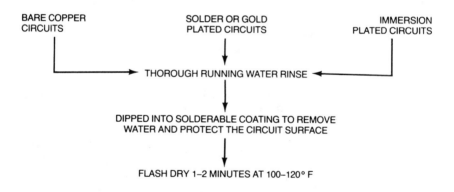

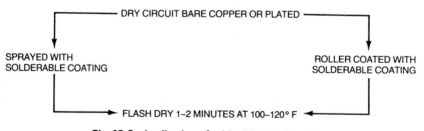

Fig. 12-6. Application of solderable organic coating.

After solder rolling, the flux is removed in a flux remover. Fig. 12-7 shows the roller coating process.

Bare Copper

Since cleanliness is the first step to good soldering, circuits are left with the copper circuitry and pads unprotected. Just before soldering they are cleaned, dipped, or automatically brushed with brightener, thoroughly rinsed and dried, and then the components are inserted.

Pre-Fluxing of Circuits

Application of soldering fluxes to the printed circuit boards is also done. This pre-fluxing should not be considered a replacement for liquid fluxing prior to soldering. Resin and rosin fluxes are used followed by a short drying cycle. In general, the coating of circuits with solderable organic coatings has replaced the pre- or dry-fluxing methods.

SOLDERING OF CIRCUITS

The soldering of circuits involves two prime steps after the insertion of the components, the application of soldering flux and the soldering operation itself.

Soldering Flux

Fluxes are used to reduce the interfacial tension between the solder and the metal surface to be soldered. Proper surface tension of the flux greatly assists

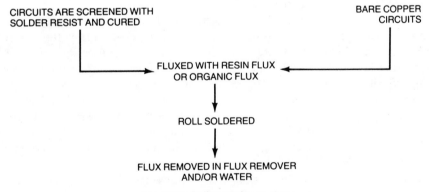

Fig. 12-7. Roller coating process.

in the removal of surface oxides. The flux exerts a physical surface tension, lowering action between the metals, much as a wetting agent reduces the surface tension throughout the alloying action, called "staying power," is what enables flow of the solder through the plated holes and eyelets to form good solder joints. Three basic types of flux are used in the soldering of printed circuits:

1. Rosin Flux: Rosin flux is plain water-white (WW) rosin with a suitable dissolving solvent. This flux is the least corrosive or conductive but its limited wetting properties and relatively short staying power have downgraded its use.
2. Resin Flux: Resin flux results from processing and addition of other chemicals to the base rosin. Resin fluxes are available that give maximum wetting action (good flow-out) and which are adapted to flux application equipment.
3. Organic Flux: Organic flux is water soluble. The inherent nature of organic fluxes is such that flux residues is often of utmost importance, organic fluxes are available that are designed with control of residue and ease of removal in mind. Non-charring residues and elimination of ingredients that react and become less water soluble after soldering are built-in features.

Flux Selection

In choosing a flux, the first consideration is the electrical value that is being sought in the assembled product. Next, two questions need to be answered. Are the residues of the flux to remain or are they to be removed? How is the flux to be applied to the circuits?

It is possible by means of a thorough, controlled process to use any of the three fluxes. If eyelets, funnelettes, etc., are to be used, organic fluxes have a built-in danger since they are conductive and corrosive and must be removed. If the residues are trapped in the holes (between eyelet and laminate), removal is difficult if not impossible. Organic flux, in this instance, would not be recommended since its use would adversely affect the electrical values of the assembled part. Some brands of resin and rosin fluxes can be left on the circuit after soldering because in most applications they will not affect the electrical values.

If plans call for removal of flux residues, organic fluxes can be considered. Military specifications call for all flux residues to be removed.

Method of flux application is important because it determines the volume of flux required and the possibility of close residue control.

Flux Application

Fluxes can be applied by brushing, dipping, spraying, bubbling, or foaming. The number of circuits being soldered is the first determinant in the method selected. When small volume is involved, hand brushing or dipping into a dish containing flux is inexpensive, but it is important to deliver the same quantity and quality of flux to the circuit. Automatic fluxing units have been designed that ensure close control of these variables.

1. *Spraying.* Spraying may be accomplished by mounting a spray gun over the path taken by an assembled circuit. The ratio of solids to solvent in the flux is always constant: timing devices control the time of the flux spray burst. Although templates are often used, overspray remains a disadvantage of this method. A modified spray technique makes use of a stainless steel screen that revolves in a bath of flux. An air blast through a stainless tube (concentric with the screen) transfers the flux from the screen to the circuit. Overspray is minimized and lead length of components is not critical.
2. *Brushing.* This method is slow and complete coverage of all component leads is difficult. Automatic units utilizing revolving brushes or combination pump-brush applicators give better results than the hand method.
3. *Foam Fluxing.* Air is introduced into the flux through a porous stone causing the flux to foam or bubble. The bursting of the bubble gives the wetting action of a spray without the problem of overspray. It is important that the bubbles are uniform for thorough wetting and that surface tension during soldering is compatible. Penetration, wetting, and control of deposited flux are advantages of this method.

Component Leads

Soldering of circuits is not complete without mentioning the importance of component leads. RETMA, the Naval Quality Assurance Group, and manufacturers themselves are developing component solderability standards. The soldering mechanism involves the alloying of the component lead metal with the solder and the circuit pad to form a continuous path for electrical energy. Cleanliness, plating techniques, protection, and storage considerations apply equally to leads as well as to circuits.

SELECTIVE SOLDERING

Solder resists are widely used as a selective soldering measure. Solder resist was originally used to prevent bridging between close pads and conductors. Today, solder resist is also used to (1) improve soldering by permitting better wetting in pad areas and minimizing icicling in automated soldering equipment, (2) improve insulation resistance and protect the circuitry, (3) reduce weight of the soldered unit (20% average weight reduction), (4) reduce solder consumption (as much as 40%), and (5) minimize copper contamination in the solder pot.

Applications

Solder resist is used on plain copper and on tinned- and gold-plated circuits. It is applied prior to roll soldering and before immersion plating so that only surfaces to be soldered are roll or immersion coated.

Screening

A definite amount of resist must be applied through the screen to the board surface to afford resistance to the heat of soldering and to the solvency of the fluxes during the soldering operation. Silk screen (or stainless steel) of 10XX and 12XX mesh size are generally used.

Curing

Curing is the most important step in the use of solder resist. Thorough cure is needed to obtain maximum electrical characteristics and to polymerize the resin so that it can withstand the soldering flux/solvents and the solder temperature.

A combination oven with radiation and convection heat gives the best cure in the shortest time. The radiation phase, 1 min. with the surface of the circuit at 260° F, brings the circuit to heat and starts the resin polymerization. The 2–4 min. period in the convection phase holds this temperature and removes the solvent fumes. Convection and gas-fired infrared ovens are also used. Longer times are required in the convection ovens and the control of the cure is more difficult to maintain from batch to batch. Gas-fired infrared ovens perform well when tied into a circulating air system to remove volatiles.

Use of solder resist in military applications where weight savings are critical is steadily growing. In the design of etched coils, antennas, etc., solder resist is also finding increased use. When the resist is correctly cured, it is possible to use a process such as shown in Fig. 12-8.

OTHER CIRCUIT MANUFACTURING TECHNIQUES

There are two other methods of circuit manufacture that have specific applications and in some instances wide areas of usability.

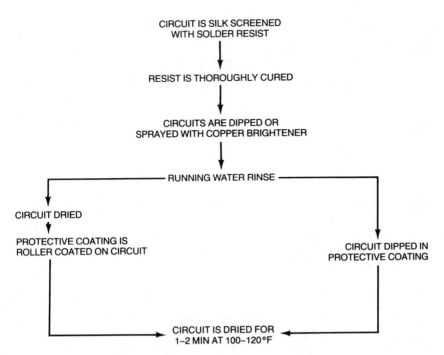

Fig. 12-8. Typical process for protecting cured, solder-resisted printed circuits

Plated Circuits

A plated circuit, different from conventional plated circuit, occurs when the starting point is plain laminate rather than copperclad. Figure 12-9 illustrates fabrication techniques when circuits are processed beginning at the laminate stage.

Stamped/Embossed Circuits

Elimination of the etching and chemical exposure of the laminate was one of the goals in developing this method. Using dies, the circuitry is defined and bonded. Several methods and variations give different characteristics to the finished circuit.

Molded Circuits

The molded circuit technique produces a circuit base via molding, then deposits the conductors in the molded (depressed) areas. This method has the advantage

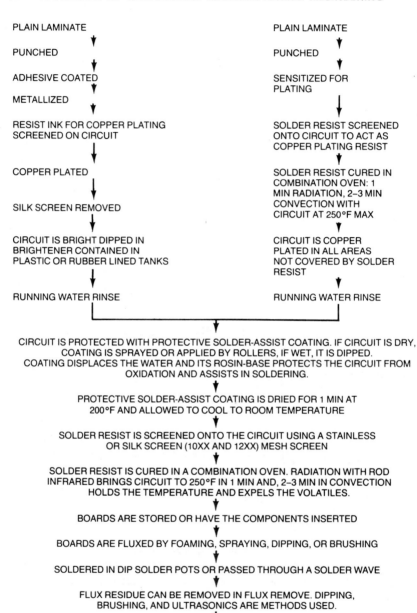

Fig. 12-9. Various steps in the production of printed circuits using the plated circuits technique beginning with plain laminate.

of producing holes directly during the molding process. Variations in method give specific advantages to the user.

Flexible Circuits

The flexible circuit method permits shaping the circuitry into various forms to fit individual applications. This type of circuit is made by conventional etching on flexible stock.

Ceramic Circuits

The original circuits used in the development of the proximity fuse were made on a ceramic base. In the process, silver inks are screened and fired onto the ceramic. Other metals and techniques are also sometimes used. Ceramic circuits, aided by the development of high-fired glass ceramics, are widening the applications of printed circuits.

Other types of circuits are also in use in varying quantities— and there will surely be more in the future. New requirements of miniaturization and ruggedness are already stimulating new decision concepts. In these methods of circuit manufacture, the same problems of cleanliness of surface, protection for solderability, etc., are important. For example, in the plating of copperpyrophosphate solder resist functions as a plating resist; it can be considered truly an additive process because it is not removed.

SOLIDS CONTROL

Whether flux residues are to be left on the soldered circuit or removed, the balance of the solids and volatile agents in the flux is important. Special hydrometers are available that read gravity between 0.85 and 0.90 with an expanded easy-to-read scale. Specific gravity solids control charts on resin fluxes and organic fluxes are also helpful. A small amount of the dry residue improves quality and, where flux removal systems are in use, the solids control pays for itself in flux remover savings and in time.

SOLDERING PRINTED CIRCUIT ASSEMBLIES

Soldering is an alloying operation. It is the formation of an alloy of the solder and the metal being soldered. Alloys formed with solder have a much greater tensile strength than that of the solder itself. Therefore, the thinner or more closely the solder lays down or wets the surface, the stronger is the alloy that is formed. The key, therefore, is the surface tension of the solder and the interfacial tension between it and the metal being soldered. The condition of the surfaces to be alloyed (circuit and component) after etching, plating, cleaning, etc., has a direct bearing on interfacial tension qualities. Contaminants, whether from processing or from oxides formed in storage, increase this interfacial tension and reduce the wetting action and flow of the solder. A proper flux lowers the interfacial tension of the solder and keeps it lowered to achieve a strongly marked affinity between the solid metal and the liquid metal.

Lowering of the interfacial tension with a suitable soldering flux, which has the correct surface tension and staying power, causes the phenomenon of wetting of the liquid to the solids as indicated by the formation of intermetallic compounds or solid solutions at the interface. When the right flux is used and contaminated surfaces cleaned, the atoms of the metallic surface are so completely unopposed in their natural bent for immediate chemical satisfaction that the powerful cohesive forces in the liquid solder are reduced, permitting maximum wetting flow. Dewetting (tearing and balling action), and failure of the solder to "take", are evidences of the opposite effects.

Surfaces to be soldered must be prepared in such a way that with a good liquid flux, the molten solder has greater affinity to alloy with the metal being soldered than its molecular attraction (cohesion) for itself. Fluxes have been designed with this phenomenon in mind.

In addition to overcoming the powerful effect of the cohesive forces within the liquid solder, the temperature of the metal being soldered must be brought up to the temperature of the solder in order for alloying to take place. In soldering operations a temperature of 100–150° F above the eutetic point of the solder should be used. Generally, 63% tin/37% lead or 60% tin/40% lead are the compositions of the solder alloys used. Their ranges from the eutetic to the final melting pot are quite short (60/40 solder, 361–379° F), enabling use of lower temperatures and thus producing stronger joints. For good quality control, the solder should be analyzed regularly.

If the metal being alloyed with the solder does not reach the proper temperature, poor joints and unreliable connections (cold joints) result. In automated circuit soldering lines, pre-heat stations minimize the temperature difference between circuit and component surfaces and the solder. In still-pot dipping of circuits, dwell time is important; the travel speed must be correct in-line wave soldering.

Soldering of printed circuits has promoted the need for control of solder temperature, particularly in high volume soldering operations, and has hastened the development of wave soldering, which permits delivery of solder at a constant temperature.

SOLDERING TECHNIQUES

Dip Pots

For small production, dip pots are widely used. The pot should be of adequate size for the circuit to be soldered. Since only the surface of the solder contacts the circuit, pre-heating the assembled circuit with a hot plate or infrared bank minimizes the temperature difference between the assembled circuit and the solder, and reduces the cooling action on the surface of the solder.

Wave Soldering

Outstanding advances have been made in the wave-soldering method. Fluxing, preheating, soldering, and removing flux residues are accomplished in sequence. Wave soldering not only provides the technique for high volume production, but ᵖives assured levels of quality.

Solder Pot Coverings

Control of gross on the surface of solder is important in wave soldering. Additives are available that keep the solder bright and dross-free. Pot coverings must not only exclude oxygen from the solder surface but must also be able to reduce the oxides that form the dross. For this reason, the bath life of pot coverings is limited. When installing soldering equipment, provision for adequate ventilation in the soldering area should be made. Most states require 100 lineal feet of air per minute.

REMOVAL OF FLUX RESIDUES

Flux residues are removed from the soldered printed circuit assembly not only when there is a possibility of conductance from the flux residue but also when the circuits are to be overcoated or encapsulated, when the residues might interfere with the mechanical operation of the unit, or when the neatness and cleanliness of the unit are important. Flux removal is required under most military procedures. Definite processes are called out in MIL-S-6872A.

Water Soluble Flux Removal

Water soluble fluxes are often selected because their residues can be removed in water systems. Removal procedure is as follows: (1) spraying with water at 100-120° F for 10–20 sec.; (2) two to three water baths at 100–120°F with dwell times of 30 sec. to 1 min. each; (3) spraying with water at 100–120°F for 10–20 sec.; (4) dry with forced air.

Resin and Rosin Flux Removal

Flux removers suitable for use with resin and rosin fluxes depend on the nature of the circuit components, inks, safety requirements, and equipment available. Common methods include: (1) bristle-brush dipped into remover and surface scrubbed clean; (2) assembled circuits dipped into a series of baths containing flux remover; (3) application by automatic equipment with rotating brushes and/or spray; (4) ultrasonic devices may be used when permissible-speed of removal is increased by scrubbing action of cavitation.

In designing or designation flux removers, requirements of safety to the assembled circuit, workers, and plant should be considered. Highly flammable and toxic materials should be used only after suitable precautions are taken.

STORAGE OF CIRCUITS

As indicated in the discussion of solderable surfaces, keeping surfaces of circuits clean and solderable is a problem that requires constant attention. When the time interval is long between manufacture of the circuit and its assembly and soldering, this is even more of a problem. Sulfite-free cartons to protect circuits in storage have long been an industry practice. Plastic bags are also used; in some instances the bags are sealed to exclude air and moisture. To insure that bags are not opened and circuits exposed to the atmosphere, the quantity in each package is geared to the rate of use of the circuits. Where protective packaging is not done (in many instances purposely), the bare copper circuits are cleaned just prior to assembly. Critical packaging, storage, and time lag factors are thus eliminated.

13
SAFETY

GENERAL SAFETY RULES

A good worker is a well-informed worker. Be sure you know the proper way to perform any job given to you. If you have any doubt, ask your supervisor.

You are not expected under any circumstances to take unnecessary chances, or to work under hazardous conditions without the maximum available safeguards.

Remember there are three actions that may be taken in regard to a hazard:

1. *Eradicate it* (do away with it completely).
2. *Isolate it* (fence it off, put a guard on it).
3. *Learn to live with it* (develop an understanding of the situation so that trouble will not develop from your being around it).

One of these actions must be taken with every hazard.

Report to your supervisor any condition or practice that might produce a loss.

Call other workers' (especially new workers') attention to practices that will lead to accidents.

Walk—do not run—in halls or on stairs. Keep your hand on the handrail, not in your pocket.

Obey warning tags and signs.

Horseplay and practical jokes are dangerous.

Consumption of alcoholic beverages during working hours can cause accidents.

In case you are injured, get first aid immediately.

Report all incidents of personal injury to your supervisor at once.

EYES

Many workers have been blinded while performing a grinding or chipping task, even though it only took a few seconds.

Always wear eye protection when handling chemicals, drilling, chipping, grinding, sandblasting, pounding overhead, welding, or using a wire brush or a portable grinder. In fact, when in the vicinity of these activities, you should *always* wear eye protection.

When you are performing welding operations, do not remove your eye protection until the last puddle has completely cooled. In addition, the entire area around the welding operation should be shielded to prevent flash burns to the eyes of other workers.

When working in high winds or bright sunlight, wear eye protection to prevent foreign objects from causing eye damage.

Even when a job is short, take the necessary action to prevent the loss of an eye.

LIFTING

Do not lift awkward or especially heavy materials by yourself. Get help.

Inspect the object you are going to lift for sharp corners, nails, or snags, which may cause injury.

Crouch as close to the load as possible. Bend your legs, not your back. Keep your back straight.

Get a firm grip on the object, straighten your legs, and rise to a vertical position.

When setting objects down, reverse the procedure, making sure your leg muscles do the work and not your back.

If special trucks, racks, hoists, or other devices are provided, use them. They are there to prevent injuries as well as to make work faster and more efficient.

MACHINERY

Do not start or operate any machine, crane, industrial truck, elevator, or other piece of equipment unless you have been both trained and authorized to do so.

All machinery operators must be alert to malfunctions or changes in characteristics in machinery operations, and they must report these conditions to their supervisor. It is far more economical to catch problems early than to wait until a machine breaks down, thus holding up production while repairs are made.

Before starting any machinery, make sure that everyone else is in the clear and that guards and protective devices are both in place and in working condition. Do not operate the machine if all conditions are not proper.

Stop all machinery before oiling or making adjustments.

After stopping a machine for repair or adjustment, use lockout devices or tags to insure that it will not be turned on until *you* have removed *your* tag.

Do not wear gloves, jewelry, neckties, long sleeves, or loose clothing when operating machinery.

Guards on machinery are provided for your protection. They should be removed only when the machine is shut down and replaced immediately after the corrective action is completed.

If a guard is removed and mislaid or if no guard has been provided for a hazardous point of operation, the supervisor should be notified before the machine is turned on.

HAND TOOLS

Defective tools produce loss. Use only tools that are substantial and well-maintained. Exchange worn and defective tools immediately.

Use the right size and type of tool for the job. There are special hammers, wrenches, pliers, screwdrivers, chisels, and saws for many types of work. The right type of tool makes the job easier and more efficient.

Carry tools in a toolbox, bag, or toolbelt, not in pockets or trouser belt. Edged or pointed tools should be carried in scabbards.

Keep tools clean. Grease and dirt cause slips and produce losses.

Tools lying near machines, and on benches, floors, and ladders cause losses and often get lost. Put them back in your kit when you're through with them. This is especially important for sharp, pointed, or heavy tools.

Keep the heads dressed on striking tools. Mushroomed and burred heads can cause serious loss.

HEALTH

If you become ill on the job, do not continue to work. Report to your supervisor; then, get medical care.

Fatigue, caused by lack of sleep or food, decreases efficiency and contributes to many losses. Be sure to get enough sleep, and eat well-balanced meals.

Wash your hands before eating or smoking. Do not wash them in gasoline or solvent: these cause skin ailments and internal poisoning.

ELECTRICAL

Do not use electrical equipment or activated circuits if your hands are wet or if you are standing on wet ground.

Immediately turn off the power in the case of overheating, or of sparking or smoking motors, wiring, or other electrical equipment. Report the condition to your supervisor.

Have a qualified electrician repair defective electrical equipment and make new electrical installations.

Electrical cords should be inspected for breaks in the insulation, kinks, and exposed strands before use. Protect the cord from oil, hot or rough surfaces, and chemicals. Keep cords out of aisles and other traveled paths.

Do not use standard electrical tools where there are flammable vapors or gases. A spark may cause an explosion or fire.

Ground wires must be intact at all times. All portable electrical hand tools should be grounded.

PROTECTIVE DEVICES

Wear suitable gloves or handpads when handling chemicals or sharp, heavy, rough, or hot materials.

Wear a hardhat when working in the vicinity of overhead hazards. If you are working near electric wires or trolley cables, exchange the metal hat for a fiberglass or plastic hat.

Wear a safety belt with lifeline when working over pits, in tanks, or wherever there is danger of falling.

Wear the proper type of respirator for dusty operations and spray painting, and when around toxic chemicals or in areas of dense smoke.

Wear good sturdy clothing that fits well and is comfortable.

Wear good, well-soled shoes. When working with heavy materials, wear safety shoes.

LADDERS AND SCAFFOLDS

Select a ladder with the right kind of safety feet for the surface, and secure it in place.

Before using a ladder, check it for weak or damaged rails and loose or broken rungs. If found defective, take it out of service, and turn it over to your supervisor for destruction or proper repair.

When going up or down a ladder, face the ladder and use both hands on the siderails.

Do not overreach from a ladder. Stay within an arm's length on a ladder. Never try to shift a ladder while you're on it.

Do not leave tools on a stepladder, scaffold, or other elevated place from which they might fall.

Never work from the top step of a ladder.

Never use a metal ladder when working on or near electrical equipment.

Before climbing a ladder be sure that your hands and feet are free from mud or grease.

For overhead work, use the proper type of ladder. Do not use makeshifts.

Wooden ladders should not be painted except with clear wood preservative. Painting hides cracks and flaws that cause falls.

A scaffold must be a substantial, well-braced structure, built of selected knotfree lumber. It must be built to withstand four times the load it is expected to carry.

Adequate, substantial footing must be provided for ladders, especially when they rest on earth, sand, or loose material.

Handrails and intermediate rails must be provided on all open sides of the scaffold. Toe boards must also be provided on the open sides of the scaffold.

Permanent ladders or stairs should be provided for access to all scaffolds.

Do not get upon or down from the scaffold by climbing the bracing.

FIRE PREVENTION

Keep firefighting equipment and fire exits clean and ready for immediate use.

Observe no-smoking regulations where posted. They are for your protection.

Solvents and flammable liquids are major causes of fire. Use only the amounts necessary to do the job. Keep supplies of these flammables in self-closing containers, and do not use them around sparks, flame, or excessive heat.

Oily rags, old paint cans, and rubbish are fire hazards. Dispose of these in the containers provided as soon as you can. Be sure the lids are in place on all containers, and replace lids even if you didn't leave them off.

If you note a fire hazard and cannot correct it yourself, report it to your supervisor at once.

Learn to operate the various types of fire extinguishers provided.

In case of fire, the rule is *Sound the alarm, then fight the fire.*

SAFETY CHECKLIST

Be aware of the following items while you are performing your duties. If they are incorrect, they should be corrected. A suggestion to your supervisor is one method of getting corrective action.

1. Have guards been removed from machines, or have they been rendered ineffective?
2. Do people work under suspended loads?
3. Are open hatches, shafts, trenches, and excavations properly guarded?
4. Do workers ride loads?
5. Do workers ride materials hoists?
6. Is equipment operated properly and by qualified operators?
7. Is the lighting adequate to do the job properly?
8. Is the ventilating system removing smoke and gases?
9. Are worn or faulty machines taken out of service until repaired?
10. Are there any locations where workers can be struck by, strike against, or come into injurious contact with, an object?
11. Can a worker be caught in, on, or between objects?

12. Do any operations require severe pushing, pulling, or lifting?
13. Can material that is being worked on slip from stands, cradles, or platforms and fall on workers?
14. Are scrap and rubbish kept cleaned up in the work area throughout the day?
15. Is there an unusual amount of wasted product in a particular part of the operation?
16. Are quality-control requirements adhered to?

HANDLING HAZARDOUS AND FLAMMABLE MATERIALS

Hazardous and flammable materials include, but are not limited to, such items as radioactive materials, explosives, acids, toxic substances, volatile liquids, unstable chemicals, caustics, cyanides, flammable gases, or any item whose deterioration or mishandling would be a potential hazard to persons and property.

Measures required to prevent injury to personnel in the handling of hazardous and flammable materials are presented in this section. In addition, it points out the existence of certain hazards that require observing established precautions.

Radioactive Materials

The hazards associated with radioactive materials require specialized attention beyond the scope of this section. In view of the various legal and health problems involved with radioactive substances, only qualified personnel shall be allowed to handle these materials.

Personal Hygiene

Materials producing harmful vapors, fumes, dusts, or mists shall be handled only in areas with adequate ventilation. Contaminated protective clothing shall be properly cleaned and disinfected. Since even small amounts of acids, cyanides, caustics, solvents, other liquids, and chemicals can cause irritation if allowed to remain on the skin, frequent cleansing of exposed parts is important. The use of solvents in cleansing the skin should be avoided, but, if necessary, isopropyl alcohol should be used, followed by an emollient.

Protective clothing shall be provided when handling dangerous liquids and chemicals, such as acids, caustics, cyanides, solvents, epoxies, and epoxy strippers, as follows:

1. Protective shield or chemical goggles
2. Rubber or plastic gloves or fingercots, and apron

3. Rubber boots, where applicable
4. Head cover
5. Approved respirator if hazardous concentrations of dust, mist, or vapors are present. Use only safety-approved organic-vapor gas masks in areas where no oxygen deficiency will occur.

Additional protective equipment may be required if unusual handling situations occur.

Supervisors should inform employees of the hazards and safe handling methods associated with the materials being used, and shall be certain all employees know how to contact the emergency and medical departments in the event of an emergency.

Emergency First Aid

In the event of any accidental exposure, contact the medical department immediately for treatment, instructions, and assistance.

Safe Handling Procedures

Personnel shall be thoroughly trained and familiar with the handling of hazardous and flammable materials.

Work assignments involving the use of hazardous materials shall be performed only by persons familiar with the hazards involved and the necessary precautions.

Storage

Hazardous materials shall be stored in compliance with fire protection and safety requirements at all times. Containers are labeled indicating the contents and necessary precautions. Hazardous materials shall be stored in specially identified cabinets or lockers. Incompatible materials shall not be stored together. Bulk storage containers that will leak or easily break shall not be used.

Flammable material shall be stored in approved safety containers. Constant vigilance shall be practiced to insure adequate separation of flammables from open flames or spark-producing equipment.

Disposal

Hazardous materials shall be disposed of as determined by the safety engineer. Do not place dangerous materials in common trash containers nor pour down sink drains, unless specifically authorized. Disposal in public sewer systems is prohibited except by permission of the applicable state, county, or municipal agency.

Liquid Nitrogen, Oxygen, and Helium

Liquid nitrogen, oxygen, and helium are characterized by extremely low temperatures and high vapor pressures. The primary hazards from the inert liquid gases, nitrogen, and helium are:

1. Freezing of body tissue in event of direct contact
2. Overpressurization of storage containers due to improper venting of boil-off vapors
3. Oxygen depletion in unventilated room. In addition to these hazards, liquid oxygen may form an explosive gel if contaminated with hydrocarbon fuel; the resulting vapors increase fire hazards by impregnating combustible items such as clothing.

Safety precautions in handling liquefied gases shall be as follows:

1. Use only approved containers and storage vessels.
2. Personnel engaged in operations involving liquified gases shall wear loose, well-insulated gloves or mittens and such eye protection as safety goggles or a faceshield.
3. Prevent splashing onto exposed skin, as painful burns may result.
4. Insure that the boil-off vapors are vented to a safe location with sufficient dilution.
5. When handling liquid oxygen, prevent contamination and saturation of clothing with oxygen vapors, which are easily ignited.
6. Smoking restrictions are necessary, since oxygen-rich vapors, which are created adjacent to the surface of a chilled liquefied-gas container, form liquid-oxygen vapors.

Compressed Gases

Many types of gases are compressed and placed in gas cylinders (K-bottles) for industrial uses. In general, these gases have a hazard associated with explosion, fire, and pressure. If overheated, or subjected to serious damage, an opening may be forced in the cylinder, and the escaping gases would create a jet-propelled cylinder, which could be highly dangerous.

Compressed gases shall be handled only by experienced and properly instructed persons. When not being moved, cylinders shall be secured at all times with bench clamps. *Caution: Never attempt to connect any hardware to a gas cylin-*

der for the purpose of withdrawing the gas unless you are thoroughly familiar with the particular gas involved. Each type of gas shall be treated in a different manner, and no single general listing of safety procedures will apply to all cases.

The safety requirements for storage of pressurized-gas cylinders of nitrogen, helium, oxygen, hydrogen, acetylene, etc. shall be as follows:

1. Do not store near sources of high heat.
2. Store in a protected, secured manner to prevent accidental damage or falling.
3. Use valve-protection caps when not in use.

Fluorocarbons

Excess heat (greater than 450°F) on large quantities of these materials may generate toxic vapors. Standard, hot-wire-stripping and coating-marking machines create no health hazard due to the small heated area and short contact time.

Avoid prolonged heating of these resins except under proper ventilation (for example, by not forming, sintering, high-speed machining, welding, etc.). When these materials are heated over 450°F in furnaces or ovens, the inside oven air shall be purged through a suitable exhaust system before removal of parts, and personnel shall avoid excessive breathing of vapors when furnace or oven doors are opened. There shall be no smoking in areas where fluorocarbons are being handled. Normally, slow-speed turning (to prevent overheating) is all that is required to eliminate any vapor hazard. Chips and scraps shall be carefully collected in separate containers and disposed of by means other than burning.

Acids and Caustics

Use extreme care to avoid all splashes and acid vapors. Do not move carboys unless securely stoppered. Open acid containers in adequately ventilated areas only. Do not wash or clean empty containers unless specifically directed to do so. Do not force acids or caustics from containers by using compressed air or other compressed gases. Do not pour water into a caustic or acid. Add the caustic or acid slowly to cool water, and stir the mixture constantly with a suitable resistant implement.

Acids

Acids in either concentrated or dilute liquid or even in crystalline form may cause severe burns, give off poisonous vapors, cause fires, and, under proper conditions, generate explosions.

Acids are not flammable in themselves, but can cause fires upon contact with combustible materials. Fires of this nature require "dry chemical" or carbon dioxide extinguishers.

Acids shall not be stored near caustics, solvents, water, sources of heat, or combustible materials such as fuels, oils, packing materials, metallic or organic powders, carbides, lubricants or thinners, and oxidizing agents such as perchlorates or permanganates.

Caustics, Cyanides, and Alkalies

Caustics or alkalies are dangerous to handle in any form because of their corrosive action on all body tissues. Signs or symptoms are not evident immediately, and injury may result before one realizes that the chemical is in contact with the body. Caustics and cyanides are poisonous if swallowed. In combination with acids, cyanides give off deadly hydrogen cyanide gas.

Most soluble caustics, cyanides, and fluxes are neither flammable nor do they support combustion.

Do not mix cyanides (either in solid form or in solution) with acids. Do not place empty acid and cyanide containers near each other during transit, storage, or disposal. For solvents:

1. Minimize inhalation of vapors.
2. Avoid skin contact.
3. Electrically ground all spray equipment including the gun, and do not use near open flames.
4. Utilize personal protective clothing where need is indicated.

Caustics and cyanides shall not be stored where contact with acids or water can occur. Storage-area floors shall be kept clean and dry. Bottled ammonia shall not be stored in direct sunlight.

Bonding, Coating, and Encapsulating Materials

Bonding, coating, and encapsulating materials are hazardous due to their ability to sensitize and irritate the skin of handler. In some instances, these materials irritate or sensitize the respiratory system. The following safety precautions should be taken:

1. Avoid skin or eye contact.
2. Limit mixing to as few employees as possible.
3. Keep work area and tools clean.
4. Use local exhaust facilities where needed, and avoid excessive inhalation of vapors.

5. Shop operations involving peroxides should be separate from those involving epoxy or polyurethanes.
6. Do not use stiff-bristle brush to apply epoxy strippers.

These materials require the following special attention:

1. Store at room temperature.
2. Isolate from all other organic substances.
3. Smoking or open flames are not to be permitted.

Epoxy Strippers

Epoxy strippers are, in varying degrees, very strong primary skin irritants and are sometimes strong respiratory irritants. Acute skin burns due to splatter or other means of contact, coupled with nose and throat irritation, are the primary signs of hazard.

Most, if not all, strippers are nonflammable. (To be certain, read the label.) Beware of toxic and highly irritating products formed when strippers are thermally decomposed.

Solvents are hazardous due to their ability to enter the bloodstream rapidly through the lungs. Solvents may also cause dermatitis because of their ability to dehydrate and defat the skin. All pure solvents, except the chlorinated hydrocarbons, are flammable in varying degrees. Repeated overexposure to chlorinated hydrocarbons results in an accumulation of effects.

Storage of solvents shall be isolated from acids and oxidizers. Volatile solvents shall be stored out of direct sunlight, preferably in a cool area.

Solvent fires require "dry chemical" or carbon dioxide extinguishers. Do not use water, as it will merely spread the flames. *Caution: Beware of phosgene gas produced by thermal decomposition of chlorinated hydrocarbons.*

Fluxes and Solders

Fluxes may irritate or sensitize the skin, eyes, and respiratory tissues at room temperature, and may produce irritating and harmful fumes when heated. High-temperature silver solders and silver brazing alloys containing cadmium can release toxic cadmium oxide fumes considerably below the melting point of the alloy. Toxic lead fumes are produced whenever an alloy containing lead is heated above $750°F$.

Magnesium

Finely divided magnesium presents a serious fire hazard. If ignited, magnesium will burn with intense heat, and only special fire-extinguishing powder is effective for control purposes. Wherever possible, avoid the generation of dust particles. Equipment used for machining magnesium shall not be used for other metals without previous cleaning. Remove metal in as large a particle size as practical. Dry-dust collection is prohibited by state law.

Mercury

Mercury is used in manometers, gauges, and vacuum pumps used in manufacturing operations. The toxic vapors of mercury are evolved at room temperature and may arise from scattered droplets resulting from accidental spillage of mercury. Droplets shall be cleaned up as soon as possible after an accidental spill in a manner to avoid further dispersion.

Beryllium and Beryllium Compounds

This section covers the medical (health) and safety standards, and those steps that must be taken to maintain adequate safeguards in the handling of beryllium or alloys containing over 5 percent beryllium. Since the health hazards are directly related to the inhalation of respirable sized particles or fumes, provisions shall be made to control them within safe limits. These limits, as established by the American Conference of Governmental Industrial Hygienists, the USAEC, based on the data presented by the AMA in the *Archives of Industrial Health*, shall be used and are as follows:

1. In-plant air shall not exceed 2 micrograms per cubic meter (2 $\mu g/m^3$) of beryllium and its compounds, averaged over an 8-hour day.
2. The safety engineer or the designate shall sample the air in the working area at least twice each month. The sampling shall be done while work activities are in progress.
3. All processing of beryllium shall be performed only within areas approved by the safety engineer.

Working Area

The area in which beryllium is processed shall be maintained separately, and shall be equipped with such devices as to preclude ready access to the area; such access shall be restricted to personnel on official business. Access records shall be maintained by the supervisor. Entrance to the working area shall be through a change room, which is designed to serve as:

1. An "air lock" to assure continuous negative pressure within the work area.
2. A demarcation point between the area where street clothes may be worn and the area where protective garments must be worn.

The working area shall be equipped with its own ventilation equipment, which shall have provisions for safe removal of accumulated wastes. The air from the dust collectors shall not be recirculated. Personnel servicing the dust collectors shall wear approved respirators.

The air-supply system shall provide a minimum of 10 changes of air per hour in the working area, and shall be at a negative pressure with respect to surrounding nonsensitive areas.

The ventilation system shall be equipped with an alarm to signify equipment failure.

Floors, walls, and ceiling shall be of such construction that they may be readily and repeatedly cleaned by wet or vacuum methods.

Additional ventilation shall be provided at all points of potential beryllium release. Hoods and ducts shall be designed as simple and small as possible, with maximum visibility, and dampers and blast gates shall be of such configuration that operating personnel cannot make haphazard adjustments.

Enclosures and hoods shall have openings so designated that air velocity through the openings shall be 125 feet per minute minimum.

No circulating fans are to be permitted in the subject area.

Discharges of all ventilating and dust-collecting equipment shall be equipped with absolute-type filters installed after the blowers.

Individual, approved respirators shall be available near each work station. These respirators shall be available for personnel involved in dry machining operations. Dry machining includes grinding, honing, lapping, turning, and abrasive blasting without liquid coolant.

Note: Many solvents, or combinations thereof, react with beryllium with varying degrees of violence. Unless otherwise specified on the engineering drawing, the solvents used in the beryllium working area shall be restricted to those listed below and shall be used singly. (Solvents shall not be mixed.)

TT-I-735 Isopropyl alcohol

TT-M-261, Methyl ethyl ketone (shall not be used on cemented assemblies)

960171, 1-1-1 Trichlorotrifluorethane

960315, 1-1-1 Trichloroethane

Personnel Education

All personnel, including supervisors, shall be made aware of the hazards of beryllium exposure, the precautions necessary to avoid personal contamination, and the corrective action to be taken in case of accidental contamination.

Food or drink shall not be brought into the work or protective garment change area.

Personnel shall vacate the work area upon failure of the ventilation system.

Entrance to the work area shall be permitted only to authorized personnel wearing the protective clothing provided.

Personnel shall remove protective clothing and shall wash hands upon leaving the restricted area.

No smoking shall be permitted in either the working area or change area.

Any person who becomes contaminated shall report immediately to first aid office.

Any cut, abrasion, or suspected contamination must be reported to the first aid office immediately.

Personnel shall not be exposed to a concentration of beryllium and its compounds exceeding 25 $\mu g/m^3$ for any period of time, however short.

Housekeeping

General and emergency cleaning shall be done using a vacuum cleaning system that does not recirculate the air, or by wet mopping. Brooms and brushes are prohibited.

A periodic cleaning schedule of all equipment in the area shall be maintained and followed.

Covered containers shall be provided for all scrap and other beryllium wastes.

DESIGN CONSIDERATIONS FOR PERSONNEL SAFETY

Design of any equipment must embody features to protect personnel from electrical and mechanical hazards; also, from those dangers which may arise from fire, elevated operating temperatures and toxic fumes.

There are various methods of incorporating adequate safeguards for personnel, many of these methods being implicit in routine design procedures. However, certain procedures, design practices and related information are of much importance as to warrant special attention.

Operating Personnel

Operating personnel must be safeguarded from dangerous voltages, excessive temperatures, and mechanical hazards which may cause physical injury during either normal operation or malfunctioning of the equipment.

In design, attention must be given to the protection of both operating and maintenance personnel.

Operating personnel must not be exposed to any mechanical or electrical hazards, nor should operation of the equipment necessitate any unusual precautions, in particular, all parts accessible during normal operations should be reliably grounded.

Maximum safeguards must be provided inside the equipment to protect maintenance personnel while working on energized circuits.

The design must minimize the possibility of the operator's clothing becoming entangled in the equipment. Handles and knobs should be so arranged that clothing does not catch, corners should be rounded, and potentials greater than 70 volts must be physically shielded or removed by the use of interlock switches.

Some of these precautions are particularly important. In military service it may become necessary for the operator to rapidly manipulate or abandon the equipment. There is also the ever-present possibility that despite safety regulations, operating personnel may attempt to service equipment in a non-approved manner.

Maintenance Personnel

Safeguarding maintenance personnel is more difficult; tests and repairs must often be made wtih much of the apparatus exposed. It may be necessary to short out interlock switches and to remove covers which shield high voltages or moving machinery.

Every effort should be made in the design to protect maintenance personnel against contact with dangerous voltages in unexpected places. Controls for adjustment and points of access for lubrication should be located away from high voltage and moving parts. Danger signs imprinted next to dangerous parts or on protective covers should be used to alert maintenance personnel.

Electric Shock

Potentials exceeding 70 volts are considered to be possible electric shock hazards. Research reveals that most deaths result from contact with the relatively low potentials ranging from 70 to 500 volts, although under extraordinary circumstances, even lower potentials can cause injury.

Many severe injuries are caused not by electric shock directly, but by the reflex action and consequent body impact with nearby objects.

Some contact with electric potentials can be expected where maintenance personnel are, by the very nature of their duties, exposed to live terminals. Both shocks and burns, however, can be minimized by greater care in design and by a better understanding of electrical characteristics.

Three factors determine the severity of electric shock: (1) quantity of current flowing through the body, (2) path of current through the body, and (3) duration of time current flows through the body.

The most important variable is current. Amperage depends not only on voltage, but also on resistance of the circuit through the body, which in turn depends on whether points of contact are wet or dry. In cases on record, potentials below 10 volts have proved fatal when points of contact have pierced the skin.

Sufficient current passing through any part of the body will cause severe burns and hemorrhages. However, relatively small currents can cause death if the path includes the heart or lungs. Electric burns are usually of two types, those produced by heat of the arc which occurs when the body touches a high-voltage circuit, and those caused by passage of electric current through skin and tissue. (See Table 13-1).

Power Lines

Designers are often inclined to confine their safety considerations to high-voltage apparatus. However, it is important that considerable attention be devoted to the hazards of power lines. Fires, severe shocks, and serious burns are known to result from personnel contacting, short-circuiting, or grounding the incoming lines.

Main Power Switch

All equipment should be furnished with a clearly labeled main power switch which will remove all power from the equipment by opening all leads from the primary power service connections.

Main power switches should be equipped with safety devices that provide protection against possible heavy arcing. Barriers shielding fuses and conducting metal parts, and devices that prevent opening the switch box with the switch closed, should be provided as protection for personnel. Switches incorporating such safeguards are standardized, commercially obtainable equipment.

Table 13-1. Probable Effects of Shock.

Current Values (Milliamperes)	Effects
0-1	Perception
1-4	Surprise
4-21	Reflex action
21-40	Muscular inhibition
40-up	Respiratory block

Since electronic equipment must often be serviced with the power on, a switch enabling maintenance personnel to bypass the interlock system should be mounted inside the equipment. The switch should be so located that reclosing of the access door or cover automatically restores interlock protection. A panel-mounted visual indicator, such as a neon lamp and a suitable nameplate to warn personnel when interlock protection is removed should always be provided.

A "battle-short" switch or terminals for connection of an external switch should also be provided to render all interlocks inoperative. It differs from the switch used to bypass the interlock system for maintenance purposes in that the panel-mounted or remote controlled "battle-short" switch is designated for emergency use only. The circuit consists of a single switch, wired in parallel with the interlock system. Closing the "battle-short" switch places a short circuit across all interlock switches, thus assuring incoming power regardless of accidental opening of interlock switches.

Interlocks

Interlock switches are used to remove power during maintenance and repair operations. Each cover and door providing access to potentials greater than 70 volts should be equipped with interlocks. (See Table 13-2).

An interlock switch is ordinarily wired in series with one of the primary service leads to the power supply unit. It is usually actuated by the movable access cover breaking the circuit when the enclosure is entered. When more than one interlock switch is used, they should be wired in series. Thus, one switch may be installed on the access door of an operating subassembly and another one on the dust cover of the power supply.

The selection of a type of interlock switch must be based upon its reliable operation. Although it contains moving parts, the self-aligning switch appears most reliable. (See Figure 13-1.)

Where access to rotating or oscillating parts is required for servicing, it may be desirable to equip the protective covers or housings with safety switches or interlocks.

Table 13-2.

Voltage Range	Suitable Protective Measures
70–350 volts	Interlocks alone
350–500 volts	Barriers and interlocks
–500 volts and up	Enclosures, warnings, and interlocks

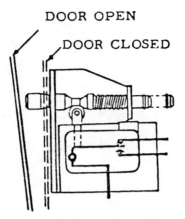

DOOR OPEN

DOOR CLOSED

Fig. 13-1. Door interlock switch.

Fusing

All leads from the primary service lines should be protected by fuses. Fusing of circuits should be such that rupture or removal of a fuse will not cause malfunction or damage to other elements in the circuit.

Fuses should be connected to the load side of the main power switch. Holders for branch-line fuses should be such that when correctly wired, fuses can be changed without the hazard of accidental shock. At least one of the fuse-holder connections should be normally inaccessible to bodily contact and this terminal should be connected to the supply. The accessible terminal should be connected to the load. Figure 13-2 shows the correct manner of wiring the instrument type of fuse holder to prevent accidental contact with the energized terminal.

Grounding

Various grounding techniques are used to protect personnel from dangerous voltages in equipment. All enclosures, exposed parts and chassis should be maintained at ground potential.

Specifications for the reduction of electrical noise interference should be consulted to determine the maximum permissible resistance of a grounding system. Reliable grounding systems should be incorporated in all electronic equipment. Enclosures and chassis should not be used as electric conductors to complete a circuit because of possible inter-circuit interference.

A terminal spot welded to the chassis provides a reliable ground connector.

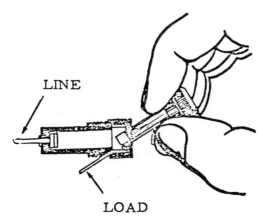

LINE

LOAD

Fig. 13-2. Fuse holder wiring.

(See Figure 13-3.) For aluminum chassis where welding is not feasible, a terminal properly secured by a machine screw, lockwasher and nut is safisfactory. (See Figure 13-4.)

A grounding lug should not be included as part of a "pile-up" that includes any material subject to cold-flow. The machine screw used should be of sufficient size so that eventual relaxation will not result in a poor connection. A lockwasher is necessary to maintain a secure connection. All nonconductive finishes of the contacting surfaces should be removed prior to inserting the screw. In no event should riveted elements be used for grounding since these cannot be depended upon for reliable electrical connections.

The common ground of each chassis should connect to a through-bolt, mounted on the enclosure and clearly marked "enclosure ground", which in turn should connect to an external safety ground strap. (See Figure 13-5.)

For the best design the external ground conductor should be fabricated from suitably plated, flexible copper strap, capable of carrying at least twice the current required for the equipment.

Fig. 13-3. Spot welded lug.

Fig. 13-4. Bolted lug.

Electronic test equipment must be furnished with a grounding pigtail at the end of the line cord. Signal generators, vacuum tube voltmeters, amplifiers, oscilloscopes, and tube testers are among the devices to be grounded. These leads are to be used for safety grounding purposes. In the event that a fault inside the portable instrument should connect a dangerous voltage to the metal housing, the dangerous current is bypassed to ground without endangering the operator.

Discharging Devices

Since high-grade filter capacitors can store lethal charges over relatively long periods of time, adequate discharging devices must be incorporated in all medium- and high-voltage power supplies. Such devices should be used whenever the time constant of capacitors and associated circuitry exceeds five seconds; they should be positive acting, reliable, and should be automatically actuated whenever the enclosure is opened.

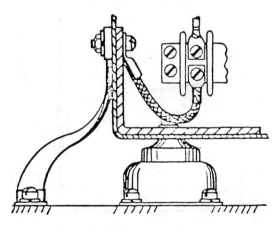

Fig. 13-5. Cabinet grounding system.

Shorting bars should be actuated either by mechanical release or by an electrical solenoid when the cover is opened. (See Figure 13-6.)

Good insurance is provided by the automatic charge-draining action of a bleeder resistor permanently connected across the output terminals of a DC power supply. Although bleeder current is an added load on the power supply, the systems should be designed to carry this slightly additional load. Bleeder resistance should be the lowest value, without presenting excessive loading, through which the capacitors can discharge quickly after the power is switched off.

However, in circuits where large high-voltage capacitors must be operated without adequate bleeding such as high-voltage radar apparatus, capacitors must be discharged by automatic devices. For high-voltage capacitors, discharging devices should be equipped with large resistors rated at 200 watts, 10,000 ohms, to limit discharge current and the possibility of damage.

Panel-Mounted Parts

Panel-mounted parts, especially jacks, are occasionally employed in power circuits for the insertion of meters, output lines, test apparatus, and other supplementary equipment. These items should be connected to the grounded leg of the monitored circuit rather than in the ungrounded, highvoltage line.

High-voltage meters should be recessed and shatter-proof windows used. (See Figure 13-7.)

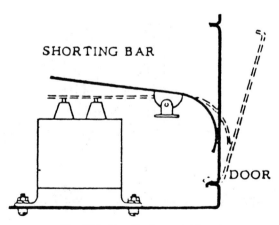

SHORTING BAR

DOOR

Fig. 13-6. Shorting bar actuation.

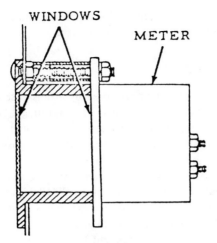

Fig. 13-7. High voltage panel meter.

Parts Safety

Electronic parts and circuits should be designed to minimize arcing in switches, relays, and other make-or-break apparatus. Fast-action switches are usually employed. Switches used in DC circuits employ magnetic arc blowouts and capacitors across the contacts.

Only explosion-proof switching devices should be employed where there is any possibility that equipment will be operated in an atmosphere of explosive gas or vapor. The design of explosion-proof equipment is covered by military specifications.

Protective devices should be incorporated in the design for all parts carrying hazardous voltages. Wherever possible, such components should be mounted beneath the chassis and ventilation requirements always considered.

When it is impractical to mount parts below the chassis to reduce the hazard to maintenance personnel when replacing above-chassis parts, protective housings having ventilating holes or louvers should be provided. If housings cannot be used, exposed terminals of the parts should be oriented away from the direction of easy contact.

All reasonable precautions should be taken to minimize fire, high temperature and toxic hazards. In particular, any capacitors, inductors, or motors involving fire hazards should be enclosed by a noncombustible material having minimum openings. As stated previously, elevated operating temperatures and ventilation requirements are primary considerations in personnel protection. Since many equipments are installed in confined spaces, materials that may produce toxic fumes must not be employed. Finished equipment should be carefully checked for verification of protective features in the design.

Shields and Guards

Safety enclosures covers should be anchored by means of screws or screwdriver-operated locks and should be plainly marked by warnings. On shipboard equipment, hinged covers, doors and withdrawable chassis should be counter-balanced or provided with a means to retain them in their open position, preventing accidental closing due to ship movement.

In their normal installed positions, chassis should be securely retained in enclosures. Stops should be provided on chassis slides to prevent inadvertent removal. Provision for firmly holding the chassis handles while releasing the equipment from the cabinet should also be incorporated. In the tilt-up position, a secure latch should support the equipment firmly despite conditions of shock, vibration, or inclination.

Suitable handles or similar provisions should be furnished for removing chassis from enclosures. Bails or other suitable means should be provided to protect parts when the chassis is removed and inverted for servicing. These serve also to protect the hands as the chassis is placed on the service bench.

Terminal boards carrying hazardous voltages about 500 volts should be protected by means of a cover provided with access holes for the insertion of test probes. Terminal numbers should be plainly marked on the external side of the cover. With this arrangement it is possible to check circuits that are energized.

Housing cabinets, or covers may require perforations to provide air circulation. The area of a perforation should be limited to that of a $\frac{1}{2}$-inch square or round hole. High-voltage components within should be set back far enough to prevent accidental contact. If this cannot be done, the size of the openings should be reduced.

To minimize the possibility of physical injury, all enclosure edges and corners should be rounded to maximum practical radii. This is especially important for front-top edges, front-side edges and enclosure, door and panel corners.

Thin edges should be avoided and chassis construction should be such that the chassis may be carried without danger of cutting the hands.

To prevent hazardous protrusions on panel surfaces, flathead screws should be used wherever sufficient panel thickness is available; otherwise, panhead screws should be used.

All accessible surfaces should be smooth. Surfaces that cannot be reasonably machined to a smooth finish should be covered or coated to prevent the possibility of skin abrasion. Small projections in areas where the rapid removal of plug-in units may cause injury to the hands must not be left uncovered.

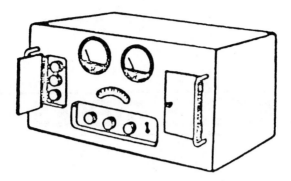

Fig. 13-8. Recessed controls.

Recessed mountings are recommended for small projecting parts such as toggle switches and small knobs located on front panels.

Care should be taken when designing equipment to prevent personnel from accidentally contacting rotating or oscillating parts such as gears, couplings, levers, cams, latches, or heavy solenoid equipment. Moving parts should be enclosed or shielded by protective guards wherever possible. Where such protection is not possible, warning signs must be furnished.

The cathode-ray tube is a special hazard in view of the high voltages that must be applied and the physical damage that may result from implosion. If the tube is accidentally nicked or scratched, resultant imploding may not occur until days later.

The face of such a tube must be safeguarded by a shatterproof glass or heavy plastic shield firmly attached to the panel. Signs warning personnel that the neck of the tube is easily broken and must be handled with caution should be posted inside the equipment.

Where access is provided to rotating, oscillating, or any other hazardous mechanisms, the cover or apparatus should bear a warning such as:

CAUTION–KEEP CLEAR OF ROTATING PARTS

Warnings

Warnings signs marked CAUTION–HIGH VOLTAGE, or CAUTION *XXX* VOLTS, should be placed in prominent positions on safety covers, access doors, and inside equipment wherever danger may be encountered. These signs should be durable, easily read, and placed so that dust or other foreign deposits will not obscure the warnings. Since signs are not physical barriers, they should be relied upon only if no other method of protection is feasible.

14

REFERENCE TABLES

Table 14-1. Temperature Conversion.

To use the table, look for the temperature reading you have in the middle column. If the reading you have is in °C, read the Fahrenheit equivalent in the right-hand column. If the reading you have is in °F, read the Centigrade equivalent in the left-hand column.

−80 to 34			35 to 77			78 to 290		
°C	Reading	°F	°C	Reading	°F	°C	Reading	°F
−62	−80	−112	1.7	35	95.0	25.6	78	172.4
−57	−70	− 94	2.2	36	96.8	26.1	79	174.2
−51	−60	− 76	2.8	37	98.6	26.7	80	176.0
−46	−50	− 58	3.3	38	100.4	27.2	81	177.8
−40	−40	− 40	3.9	39	102.2	27.8	82	179.6
−34	−30	− 22	4.4	40	104.0	28.3	83	181.4
−29	−20	− 4	5.0	41	105.8	28.9	84	183.2
−23	−10	14	5.6	42	107.6	29.4	85	185.0
−17.8	0	32	6.1	43	109.4	30.0	86	186.8
−17.2	1	33.8	6.7	44	111.2	30.6	87	188.6
−16.7	2	35.6	7.2	45	113.0	31.1	88	190.4
−16.1	3	37.4	7.8	46	114.8	31.7	89	192.2
−15.6	4	39.2	8.3	47	116.6	32.2	90	194.0
−15.0	5	41.0	8.9	48	118.4	32.8	91	195.8
−14.4	6	42.8	9.4	49	120.2	33.3	92	197.6
−13.9	7	44.6	10.0	50	122.0	33.9	93	199.4
−13.3	8	46.4	10.6	51	123.8	34.4	94	201.2
−12.8	9	48.2	11.1	52	125.6	35.0	95	203.0
−12.2	10	50.0	11.7	53	127.4	35.6	96	204.8
−11.7	11	51.8	12.2	54	129.2	36.1	97	206.6

Table 14-1. (*Continued*)

-80 to 34			35 to 77			78 to 290		
°C	Reading	°F	°C	Reading	°F	°C	Reading	°F
−11.1	12	53.6	12.8	55	131.0	36.7	98	208.4
−10.6	13	55.4	13.3	56	132.8	37.2	99	210.2
−10.0	14	57.2	13.9	57	134.6	37.8	100	212.0
− 9.4	15	59.0	14.4	58	136.4	43.0	110	230.0
− 8.9	16	60.8	15.0	59	138.2	49.0	120	248.0
− 8.3	17	62.6	15.6	60	140.0	54.0	130	266.0
− 7.8	18	64.4	16.1	61	141.8	60.0	140	284.0
− 7.2	19	66.2	16.7	62	143.6	66.0	150	302.0
− 6.7	20	68.0	17.2	63	145.4	71.0	160	320.0
− 6.1	21	69.8	17.8	64	147.2	77.0	170	338.0
− 5.6	22	71.6	18.3	65	149.0	82.0	180	356.0
− 5.0	23	73.4	18.9	66	150.8	88.0	190	374.0
− 4.4	24	75.2	19.4	67	152.6	93.0	200	392.0
− 3.9	25	77.0	20.0	68	154.4	99.0	210	410.0
− 3.3	26	78.8	20.6	69	156.2	100.0	212	413.6
− 2.8	27	80.6	21.1	70	158.0	104.0	220	428.0
− 2.2	28	82.4	21.7	71	159.8	110.0	230	446.0
− 1.7	29	84.2	22.2	72	161.6	116.0	240	464.0
− 1.1	30	86.0	22.8	73	163.4	121.0	250	482.0
− 0.6	31	87.8	23.3	74	165.2	127.0	260	500.0
0.0	32	89.6	23.9	75	167.0	132.0	270	518.0
0.6	33	91.4	24.4	76	168.8	138.0	280	536.0
1.1	34	93.2	25.0	77	170.6	143.0	290	554.0

Table 14-2. Decimal Equivalents.

Fraction						
–/2	–/4	–/8	–/16	–/32	–/64	Decimal Equivalent
0	0	0	0	0	0	0.000000
					1	0.015625
				1/32	2	0.03125
					3	0.046875
			1/16	2	4	0.0625
					5	0.078125
				3/32	6	0.09375
					7	0.109375
		1/8		2	4	8	0.125
					9.	0.140625
				5/32	10	0.15625
					11	0.171875
			3/16	6	12	0.1875
					13	0.203125
				7/32	14	0.21875
					15	0.234375
	1/4	2	4	8	16	0.250
					17	0.265625
				9/32	18	0.28125
					19	0.296875
			5/16	10	20	0.3125
					21	0.328125
				11/32	22	0.34375
					23	0.359375
		3/8	6	12	24	0.375
					25	0.390625
				13/32	26	0.40625
					27	0.421875
			7/16	14	28	0.4375
					29	0.453125
				15/32	30	0.46875
					31	0.484375
1	2	4	8	16	32	0.500

Table 14-2. (*Continued*)

			Fraction			
$-/2$	$-/4$	$-/8$	$-/16$	$-/32$	$-/64$	Decimal Equivalent
					33	0.515625
				17/32	34	0.53125
					35	0.546875
			9/16	18	36	0.5625
					37	0.578125
				19/32	38	0.59375
					39	0.609375
		5/8	10	20	40	0.625
					41	0.640625
				21/32	42	0.65625
					43	0.671875
			11/16	22	44	0.6875
					45	0.703125
				23/32	46	0.71875
					47	0.734375
	3/4	6	12	24	48	0.750
					49	0.765625
				25/32	50	0.78125
					51	0.796875
			13/16	26	52	0.8125
					53	0.828125
				27/32	54	0.84375
					55	0.859375
		7/8	14	28	56	0.875
					57	0.890625
				29/32	58	0.90625
					59	0.921875
			15/16	30	60	0.9375
					61	0.953125
				31/32	62	0.96875
					63	0.984375
2	4	8	16	32	64	1.000000

Table 14-3. Capacitor and Resistor Color Code.

| | Ceramics RMA Standard JAN Specification Capacitors | | | | | Resistors RMA and JAN | | | |
| | Tolerance | | Temperature Coefficient Parts per Million per °Centigrade | T.C. for Extended Range TC HI CAP | | | | Significant Figures | Color |
Decimal Multiplier	Capacity 10 MMF or Less	Capacity More than 10 MMF		Significant Figure	Multiplier	Multiplier	Tolerance		
1	±20 MMF (???)	±20%	NPO	0	−1	1	—	0	Black
10	±01 MMF	±1%	N33	—	−10	10	—	1	Brown
100	—	±2%	N75	1	−100	100	—	2	Red
1,000	—	±2.5% (MA)	N150	1.5	−1,000	1,000	—	3	Orange
10,000 (RMA)	—	—	N220	2.2	−10,000	10,000	—	4	Yellow
—	±0.5 MMF	±5%	N330	3.3	+1	100,000	—	5	Green
—	—	—	N470	4.7	+10	10^6	—	6	Blue
—	—	—	N750	7.5	+1,000	10^7	—	7	Violet
0.01	±0.25 MMF	—	+30	—	+10,000	10^8	—	8	Grey
0.1	±1.0 MMF	±10%	N330 ± 500	—	+100,000	10^9	—	9	White
—	—	—	+100 (JAN)	—	—	0.1	±5%	—	Gold
—	—	—	Bypass & Coupling (RA)	—	—	0.01	±10%	—	Silver
—	—	—	—	—	—	—	±20%	—	No Color

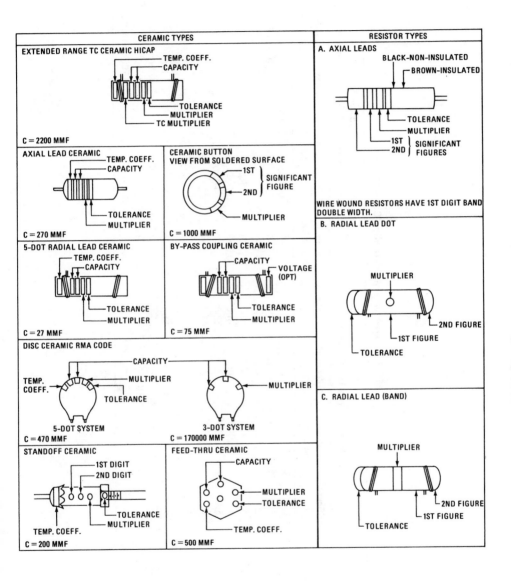

Molded Paper and Molded Mica Color Code

Color	Significant Figures	Molded Paper Tubular Capacitors			Molded Mica RMA Standard JAN Specification Capacitors			
		Decimal Multiplier	Tolerance ±%	Voltage Volts	Decimal Multiplier	Tolerance ±%	RMA Voltage Rating (All Capacitors)	Class
Black	0	1	20	–	1	20 (JAN 1948 RMA)	–	A
Brown	1	10	–	100	10	–	100	B
Red	2	100	–	200	100	2	200	C
Orange	3	1,000	30	300	1,000	3 (RMA)	300	D
Yellow	4	10,000	40	400	10,000	–	400	E
Green	5	10^5	5	500	–	5 (RMA)	500	F (JAN)
Blue	6	10^6	–	600	–	–	600	G (JAN)
Violet	7	–	–	700	–	–	700	–
Gray	8	–	–	800	–	–	800	I (RMA)
White	9	–	10 (RMA)	900	–	–	900	J (RMA)
Gold	–	0.1	–	–	0.1	5 (JAN)	1000	–
Silver	–	–	10 (JAN)	–	0.01	10	2000	–
No Color	–	–	–	–	–	20 Old (RMA)	500 Old (RMA)	–

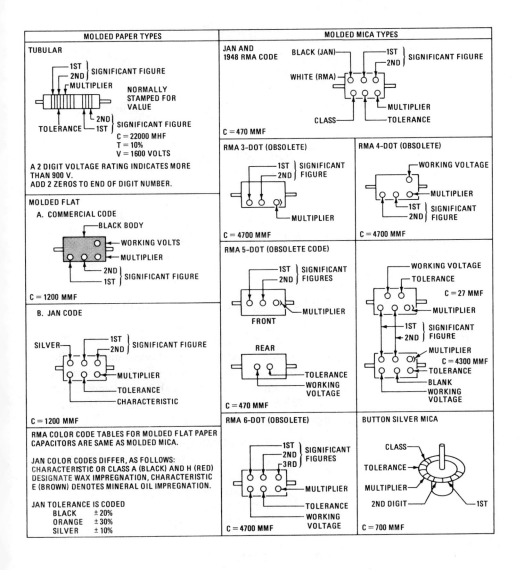

MOLDED PAPER TYPES	MOLDED MICA TYPES

MOLDED PAPER TYPES

TUBULAR

1ST } SIGNIFICANT FIGURE
2ND }
MULTIPLIER
NORMALLY STAMPED FOR VALUE
2ND } SIGNIFICANT FIGURE
TOLERANCE — 1ST }

C = 22000 MHF
T = 10%
V = 1600 VOLTS

A 2 DIGIT VOLTAGE RATING INDICATES MORE THAN 900 V.
ADD 2 ZEROS TO END OF DIGIT NUMBER.

MOLDED FLAT

A. COMMERCIAL CODE

BLACK BODY
WORKING VOLTS
MULTIPLIER
2ND } SIGNIFICANT FIGURE
1ST }

C = 1200 MMF

B. JAN CODE

SILVER
1ST } SIGNIFICANT FIGURE
2ND }
MULTIPLIER
TOLERANCE
CHARACTERISTIC

C = 1200 MMF

RMA COLOR CODE TABLES FOR MOLDED FLAT PAPER CAPACITORS ARE SAME AS MOLDED MICA.

JAN COLOR CODES DIFFER, AS FOLLOWS:
CHARACTERISTIC OR CLASS A (BLACK) AND H (RED) DESIGNATE WAX IMPREGNATION, CHARACTERISTIC E (BROWN) DENOTES MINERAL OIL IMPREGNATION.

JAN TOLERANCE IS CODED
BLACK   ± 20%
ORANGE ± 30%
SILVER ± 10%

MOLDED MICA TYPES

JAN AND 1948 RMA CODE
BLACK (JAN) — 1ST }
2ND } SIGNIFICANT FIGURE
WHITE (RMA) —
MULTIPLIER
CLASS —
TOLERANCE

C = 470 MMF

RMA 3-DOT (OBSOLETE)

1ST } SIGNIFICANT
2ND } FIGURE
MULTIPLIER

C = 4700 MMF

RMA 4-DOT (OBSOLETE)

WORKING VOLTAGE
MULTIPLIER
1ST } SIGNIFICANT
2ND } FIGURE

C = 4700 MMF

RMA 5-DOT (OBSOLETE CODE)

1ST } SIGNIFICANT
2ND } FIGURES
MULTIPLIER
FRONT

REAR
TOLERANCE
WORKING VOLTAGE

C = 470 MMF

WORKING VOLTAGE
TOLERANCE
C = 27 MMF
MULTIPLIER
1ST } SIGNIFICANT
2ND } FIGURE
MULTIPLIER
C = 4300 MMF
TOLERANCE
BLANK
WORKING VOLTAGE

RMA 6-DOT (OBSOLETE)

1ST }
2ND } SIGNIFICANT
3RD } FIGURES
MULTIPLIER
TOLERANCE
WORKING VOLTAGE

C = 4700 MMF

BUTTON SILVER MICA

CLASS
TOLERANCE
MULTIPLIER
2ND DIGIT
1ST

C = 700 MMF

RESISTOR DATA

Resistors, Fixed, Composition

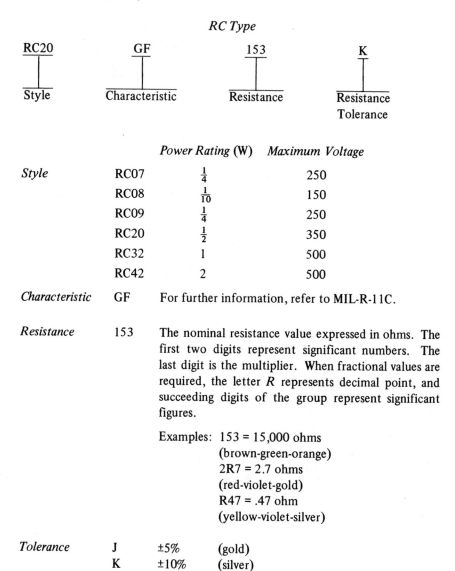

RC Type

RC20   GF   153   K

Style   Characteristic   Resistance   Resistance Tolerance

		Power Rating (W)	Maximum Voltage
Style	RC07	$\frac{1}{4}$	250
	RC08	$\frac{1}{10}$	150
	RC09	$\frac{1}{4}$	250
	RC20	$\frac{1}{2}$	350
	RC32	1	500
	RC42	2	500

Characteristic   GF   For further information, refer to MIL-R-11C.

Resistance   153   The nominal resistance value expressed in ohms. The first two digits represent significant numbers. The last digit is the multiplier. When fractional values are required, the letter R represents decimal point, and succeeding digits of the group represent significant figures.

Examples: 153 = 15,000 ohms
(brown-green-orange)
2R7 = 2.7 ohms
(red-violet-gold)
R47 = .47 ohm
(yellow-violet-silver)

Tolerance	J	±5%	(gold)
	K	±10%	(silver)

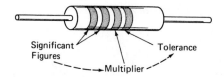

Resistors, Fixed, Film

RN Style

RN10	T	1003	F
Style	Characteristic	Resistance	Resistance Tolerance

		Power Rating (W)	*Maximum Voltage*
Style	RN10	$\frac{1}{4}$	300
	RN15	$\frac{1}{2}$	350
	RN20	$\frac{1}{2}$	350
	RN25	1	500
	RN30	2	750

Characteristic	R	Temperature coefficient
	X	For further information refer to MIL-Standard

Resistance 1003 First three digits represent significant numbers. The last digit is the multiplier. The letter *R* represents decimal point when values less than 100 ohms are required.

Examples: 1003 = 100,000 ohms
19R0 = 19.0 ohms
5R40 = 5.4 ohms

Tolerance	F	1%
	G	2%
	J	5%

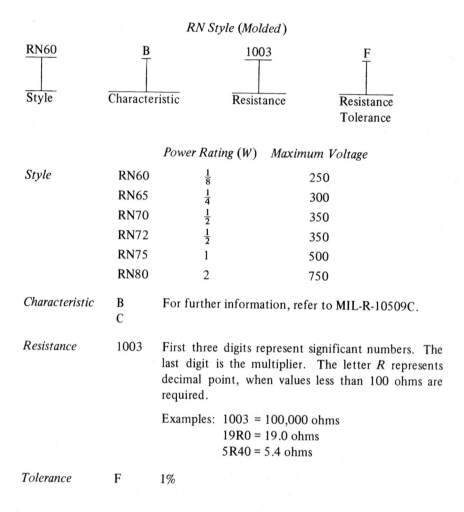

RN Style (Molded)

RN60	B	1003	F
Style	Characteristic	Resistance	Resistance Tolerance

Power Rating (W)   Maximum Voltage

Style		Power Rating (W)	Maximum Voltage
	RN60	$\frac{1}{8}$	250
	RN65	$\frac{1}{4}$	300
	RN70	$\frac{1}{2}$	350
	RN72	$\frac{1}{2}$	350
	RN75	1	500
	RN80	2	750

Characteristic	B C	For further information, refer to MIL-R-10509C.

Resistance   1003   First three digits represent significant numbers. The last digit is the multiplier. The letter *R* represents decimal point, when values less than 100 ohms are required.

Examples: 1003 = 100,000 ohms
19R0 = 19.0 ohms
5R40 = 5.4 ohms

Tolerance   F   1%

KEY PRECISION SILICONE OIL-FILLED CARBON-FILM RESISTORS

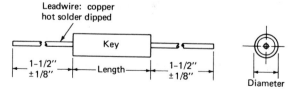

Fig. 14-1. Silicone oil-filled resistor.

Table 14-4. Resistor Physical and Electrical Characteristics.

| Type | | Rating | | Voltage | | Resistance | | Physical Dimensions | | | | |
| Key | MIL-R 10509B | Key Watts | Mil. Watts | Key Cont VDC | Mil. Cont VDC | OHMS | | Length | | Diameter | | Leads AWG |
						Minimum	Maximum	Minimum	Maximum	Minimum	Maximum	
KC50	–	0.1	–	150	–	50	100 K	.234	.266	.094	.125	26
KC55	–	0.125	–	150	–	5	500 K	.266	.297	.141	.172	22
KC60	RN60B	0.25	0.125	300	250	5	1 meg	.429	.437	.145	.165	22
KC65	RN65B	0.5	0.25	350	300	10	2 meg	.625	.655	.239	.249	20
KC70	RN70B	0.5	0.5	450	350	10	5 meg	.808	.842	.240	.260	20
KC75	RN75B	1.0	1.0	600	500	10	10 meg	1.073	.1.113	.375	.415	20
KC80	RN80B	2.0	2.0	1000	750	50	20 meg	2.225	.2.275	.375	.415	20

421

Table 14-5. Physical and Electrical Characteristics (Resin-Coated).

| Type | | Rating | | Voltage | | Resistance OHMS | | Physical Dimension | | |
Key	MIL-R 10509B	Key Watts	MIL Watts	Key VDC	MIL VDC	Minimum	Maximum	Maximum Length	Maximum Diameter	Leads AWG
E2 KV2	—	0.1	—	100	—	50	100 K	.250	.093	26
E5 KV5	—	0.125	—	150	—	5	500 K	.312	.109	22
E10 KV10	RN10	0.25	0.25	300	300	10	1 meg	.430	.110	22
E20 KV20	RN20	0.5	0.5	350	350	2	2 meg	.625	.170	20
E22 KV22	—	0.5	—	500	—	100	5 meg	.812	.170	20
— KV25	RN25	1.0	1.0	1000	500	10	10 meg	1.000	.295	20
— KV30	RN30	2.0	2.0	2000	750	50	20 meg	2.18	.295	20

Note: The E-Series has the same dimensions as the KV-series except that the E-series is epoxy-coated.

KEY PRECISION METAL-FILM RESISTORS

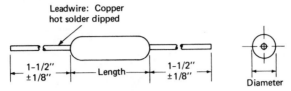

Fig. 14-2. Metal film resistor.

Table 14-6. Type-KM Characteristics.

Type Key	MIL-R 10509C	Watts 125°C	Volts	Minimum Ohms	Maximum Ohms	Maximum Diameter	Maximum Length	Lead AWG No.
KM2	–	0.1	100	100	50 K	.100	.250	24
KM5	–	0.125	150	100	100 K	.125	.312	22
KM60	RN60C	0.125	250	100	300 K	.165	.437	22
KM65	RN65C	0.250	300	100	500 K	.250	.656	22
KM70	RN70C	0.50	350	100	1 meg	.312	.875	20
KM75	RN75C	1.0	500	100	2 meg	.437	1.125	20

Table 14-7. Type-EM Characteristics.

Type Key	Watts 125°C	Volts	Minimum Ohms	Maximum Ohms	Maximum Diameter	Maximum Length	Lead AWG No.
EM2	0.1	100	100	50 K	.100	.250	24
EM5	0.125	150	100	100 K	.125	.312	22
EM60	0.125	250	100	300 K	.165	.437	22
EM65	0.250	300	100	500 K	.250	.656	22
EM70	0.50	350	100	1 megohm	.312	.875	20
EM75	1.0	500	100	2 megohm	.437	1.125	20

KEY PRECISION WIRE-WOUND RESISTORS

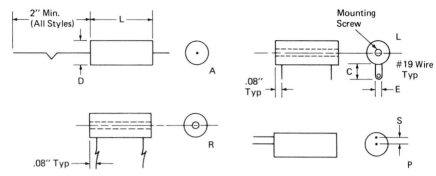

Fig. 14-3. Precision wire-wound resistor.

Table 14-8. Physical and Electrical Characteristics.

Key Type	Diameter (In.)		Length (In.)		Key Watts	Maximum Resistance[a]	
42	$\frac{1}{4}$		$\frac{1}{4}$	.250	0.1	50 K	
43			$\frac{3}{8}$	.375	0.15	130 K	
44	.250		$\frac{1}{2}$	.500	0.2	200 K	
46			$\frac{3}{4}$	.750	0.3	500 K	
48			1	1.0	0.4	700 K	
63			$\frac{3}{8}$	.375	0.125	200 K	
64	$\frac{3}{8}$		$\frac{1}{2}$	.500	0.35	400 K	
66			$\frac{3}{4}$	.750	0.5	750 K	
68	.3750		1	1.00	0.75	1	Meg
84			$\frac{1}{2}$	.500	0.5	800 K	
85			$\frac{5}{8}$	.625	0.625	1	Meg
86	$\frac{1}{2}$		$\frac{3}{4}$	.750	0.750	1.25	Meg
88			1	1.00	1.0	2	Meg
812	.500		$1\text{-}\frac{1}{2}$	1.50	1.5	3	Meg
814			$1\text{-}\frac{3}{4}$	1.75	1.75	3.5	Meg
816			2	2.00	2.00	5	Meg
1010	$\frac{5}{8}$	.6250	$1\text{-}\frac{1}{4}$	1.25	1.5	4	Meg
1016			2	2.00	2.00	8	Meg
1616	1	1.00	2	2.00	2.00	30	Meg

[a]Based on 0.001-in. resistance wire; 800/ft.

Table 14-9. Key and MIL Types Equivalent Resistors.

MIL-R-93	Key[a]	MIL-R-9444	Diameter	Length	Watts
RB09	L84	–	$\frac{7}{16}$	$\frac{1}{2}$	$\frac{1}{8}$
RB15	L84	–	$\frac{5}{8}$	$\frac{1}{2}$	$\frac{1}{4}$
RB16	L85	–	$\frac{5}{8}$	$\frac{5}{8}$	$\frac{1}{3}$
RB17	L88	–	$\frac{5}{8}$	1	$\frac{1}{2}$
RB18	L1010	–	$\frac{3}{4}$	$1\text{-}\frac{1}{4}$	$\frac{1}{2}$
RB19	L1616	–	$\frac{7}{8}$	$2\text{-}\frac{1}{8}$	1
RB52	L68	–	$\frac{3}{8}$	1	$\frac{1}{4}$
RB53	A66	–	$\frac{3}{8}$	$\frac{3}{4}$	$\frac{1}{3}$
RB54	A46	–	$\frac{1}{4}$	$\frac{3}{4}$	$\frac{1}{4}$
RB55	A44	AFRT-10	$\frac{1}{4}$	$\frac{1}{2}$	$\frac{1}{8}$
–	A46	AFRT-11	$\frac{1}{4}$	$\frac{3}{4}$	$\frac{1}{4}$
–	A66	AFRT-12	$\frac{3}{8}$	$\frac{3}{4}$	$\frac{1}{3}$
–	A68	AFRT-13	$\frac{3}{8}$	1	$\frac{3}{4}$
–	A88	AFRT-14	$\frac{1}{2}$	1	1
–	A812	AFRT-15	$\frac{1}{2}$	$1\text{-}\frac{1}{2}$	$1\text{-}\frac{1}{2}$
–	A816	AFRT-16	$\frac{1}{2}$	2	2
–	L812	AFRT-17	$\frac{1}{2}$	$1\text{-}\frac{1}{2}$	$\frac{1}{4}$
–	L85	AFRT-18	$\frac{1}{2}$	$\frac{5}{8}$	$\frac{1}{2}$
–	L88	AFRT-19	$\frac{1}{2}$	1	1
–	L812	AFRT-20	$\frac{1}{2}$	$1\text{-}\frac{1}{2}$	$1\text{-}\frac{1}{2}$
–	L816	AFRT-21	$\frac{1}{2}$	2	2

[a]A = axial, L = lug mounting, R = radial lead, P = printed-circuit type (not used here).

Table 14-10. Mounting Information.[a]

Key No.	C (In.)	E (In.)	S	Mounting Screw No.
40	0.125	0.125	0.10	2
60	0.375	0.188	0.10	4
80	0.375	0.188	0.20	6
100	0.375	0.188	0.20	6
160	0.375	0.188	0.40	6

[a]All lead wire No. 20 AWG bright gold Flash.

Military Resistors

MIL-R-11 (RC) RC07GF153K

RC07 Style/Power	GF Characteristic	153 Resistance Value	K Tolerance
(See Tables 1 and 2)	"G" = 70° C max. ambient temperature for full load operation. "F" = temperature coeffient which varies (with resistance) from ± 625 ppm/° C to ± 3100 ppm/° C.	First two digits are significant, 3rd digit "number of zeros." 153 = 15,000 Ω	(See Table 4) K = ± 10%

★MIL-R-39008 (RCR) RCR07G153JS

RCR07 Style/Power	G Characteristic	153 Resistance Value	J Tolerance	S Failure Rate
(See Tables 1 and 2)	"G" indicates a max. ambient temperature of 70° C for full load operation, and a TC which varies (with resistance) from ± 625 ppm/° C to ± 1900 ppm/° C.	First two digits are significant, 3rd digit "number of zeros." 153 = 15,000 Ω	(See Table 4) J = ± 5%	(See Table 5) S = .001%/1000 hours

MIL-R-10509 (RN) RN55D1003F

RN55 Style/Power	D Characteristic	1003 Resistance Value	F Tolerance
(See Tables 1 and 2)	(See Table 3) D = ± 100 ppm/° C	First three digits are significant, 4th digit "number of zeros." 1003 = 100,000 Ω	(See Table 4) F = ± 1%

★MIL-R-55182 (RNR) RNR55H1003FS

RNR55 Style, Terminal and Power	H Characteristic	1003 Resistance Value	F Tolerance	S Failure Rate
(See Tables 1 and 2) RNR = Solderable Leads RNC = Solderable/Weldable Leads RNN = Nickel Leads	(See Table 3) H = ± 50 ppm/° C	First three digits are significant, 4th digit "number of zeros." 1003 = 100,000 Ω	(See Table 4) F = ± 1%	(See Table 5) S = .001%/1000 hours

MIL-R-22684 (RL) RL07S153J

RL07 Style/Power	S Terminal (Lead)	153 Resistance Value	J Tolerance
(See Tables 1 and 2)	"S" = Solderable	First two digits are significant, 3rd digit "number of zeros." 153 = 15,000 Ω	(See Table 4) J = ± 5%

★MIL-R-39017 (RLR) RLR07C1502GR

RLR07 Style/Power	C Terminal (Lead)	1502 Resistance Value	G Tolerance	R Failure Rate
(See Tables 1 and 2)	"C" = Solderable/Weldable	First three digits are significant, 4th digit "number of zeros." 1502 = 15,000 Ω	(See Table 4) G = ± 2%	(See Table 5) R = .01%/1000 hours

★ NOTE: The Established Reliability specification (i.e., MIL-R-39008, MIL-R-39017, and MIL-R-55182) supersede MIL-R-11, MIL-R-22684, and MIL-R-10509 respectively, for all *new* design. Resistors qualified to the three Established Reliability specifications may be substituted, without limitation, wherever the older MIL devices are specified.

TABLE 1 – Resistor Style

RC	Fixed Composition Resistor (MIL-R-11)
RCR	Fixed Composition Resistor, Established Reliability (MIL-R-39008)
RL	Fixed Film Resistor (MIL-R-22684)
RLR	Fixed Film Resistor, Established Reliability (MIL-R-39017)
RN	Fixed Resistor, High Stability (MIL-R-10509)
RNR	Fixed Film Resistor, Established Reliability (MIL-R-55182)

TABLE 2 – Resistor Power

Nominal Body Length" x Dia."	Size	Power (@ 70° Unless Otherwise Stated)
RC · RCR · RL · RLR		
.145 x .062	05	'1/8
.250 x .090	07	1/4
.375 x .138	20	1/2
.562 x .225	32	1
.688 x .138	42	2
RN (Characteristic C) · RNR		
.150 x .065	50	1/20 @ 125° C
.250 x .109	55	1/10 @ 125° C
.375 x .125	60	1/8 @ 125° C
.625 x .188	65	1/4 @ 125° C
.750 x .250	70	1/2 @ 125° C
1.062 x .375	75	1 @ 125° C
RN (Characteristic D)		
.250 x .109	55	1/8
.375 x .125	60	1/4
.625 x .188	65	1/2
.750 x .250	70	3/4

TABLE 3 – Characteristics

RN	
B	± 500 ppm/°C
C	± 50 ppm/°C (T2)
D	± 100 ppm/°C (T0, T1)
E	± 25 ppm/°C (T9)
F	± 50 ppm/°C

RNR/ RNC	
H	± 50 ppm/°C (T2)
J	± 25 ppm/°C (T9)
K	± 100 ppm/°C (T0, T1)

NOTE: There is no temperature coefficient designation for the RL numbering system, all units are ± 100 ppm/°C.

TABLE 4 – Tolerance

K	= ± 10%
J	= ± 5%
G	= ± 2%
F	= ± 1%
D	= ± 0.5%
C	= ± 0.25%
B	= ± 0.10%

TABLE 5 – Failure Rate

1000 Hours (60% Confidence)	
M	= 1.0%
P	= 0.1%
R	= .01%
S	= .001%

426

FIXED RESISTORS

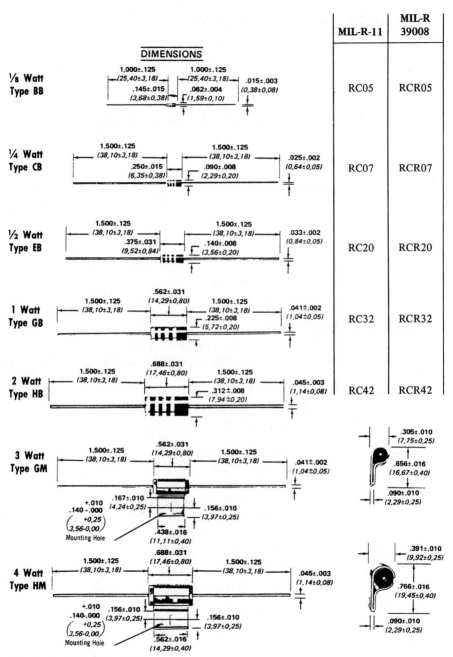

	MIL-R-11	MIL-R 39008
⅛ Watt Type BB	RC05	RCR05
¼ Watt Type CB	RC07	RCR07
½ Watt Type EB	RC20	RCR20
1 Watt Type GB	RC32	RCR32
2 Watt Type HB	RC42	RCR42
3 Watt Type GM		
4 Watt Type HM		

DIMENSIONS

⅛ Watt Type BB
1.000±.125 (25,40±3,18) 1.000±.125 (25,40±3,18) .015±.003 (0,38±0,08)
.145±.015 (3,68±0,38) .062±.004 (1,59±0,10)

¼ Watt Type CB
1.500±.125 (38,10±3,18) 1.500±.125 (38,10±3,18) .025±.002 (0,64±0,05)
.250±.015 (6,35±0,38) .090±.008 (2,29±0,20)

½ Watt Type EB
1.500±.125 (38,10±3,18) 1.500±.125 (38,10±3,18) .033±.002 (0,84±0,05)
.375±.031 (9,52±0,84) .140±.008 (3,56±0,20)

1 Watt Type GB
1.500±.125 (38,10±3,18) .562±.031 (14,29±0,80) 1.500±.125 (38,10±3,18) .041±.002 (1,04±0,05)
.225±.008 (5,72±0,20)

2 Watt Type HB
1.500±.125 (38,10±3,18) .688±.031 (17,46±0,80) 1.500±.125 (38,10±3,18) .045±.003 (1,14±0,08)
.312±.008 (7,94±0,20)

3 Watt Type GM
1.500±.125 (38,10±3,18) .562±.031 (14,29±0,80) 1.500±.125 (38,10±3,18) .041±.002 (1,04±0,05)
.167±.010 (4,24±0,25) .156±.010 (3,97±0,25)
.140 +.010 -.000 (3,56 +0,25 -0,00) Mounting Hole
.438±.016 (11,11±0,40)
.305±.010 (7,75±0,25) .656±.016 (16,67±0,40) .090±.010 (2,29±0,25)

4 Watt Type HM
1.500±.125 (38,10±3,18) .688±.031 (17,46±0,80) 1.500±.125 (38,10±3,18) .045±.003 (1,14±0,08)
.156±.010 (3,97±0,25) .156±.010 (3,97±0,25)
.140 +.010 -.000 (3,56 +0,25 -0,00) Mounting Hole
.562±.016 (14,29±0,40)
.391±.010 (9,92±0,25) .766±.016 (19,45±0,40) .090±.010 (2,29±0,25)

Dimensions shown in *ITALICS* are in Millimeters.

427

CAPACITORS, FIXED, MICA-DIELECTRIC

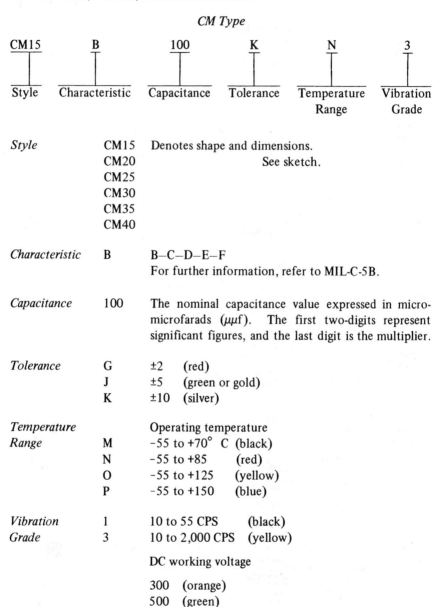

CM Type

CM15	B	100	K	N	3
Style	Characteristic	Capacitance	Tolerance	Temperature Range	Vibration Grade

Style	CM15	Denotes shape and dimensions.
	CM20	See sketch.
	CM25	
	CM30	
	CM35	
	CM40	

Characteristic	B	B–C–D–E–F
		For further information, refer to MIL-C-5B.

Capacitance 100 The nominal capacitance value expressed in micro-microfarads ($\mu\mu f$). The first two-digits represent significant figures, and the last digit is the multiplier.

Tolerance	G	±2	(red)
	J	±5	(green or gold)
	K	±10	(silver)

Temperature		Operating temperature	
Range	M	−55 to +70° C	(black)
	N	−55 to +85	(red)
	O	−55 to +125	(yellow)
	P	−55 to +150	(blue)

Vibration	1	10 to 55 CPS	(black)
Grade	3	10 to 2,000 CPS	(yellow)

DC working voltage

300	(orange)
500	(green)

CAPACITORS

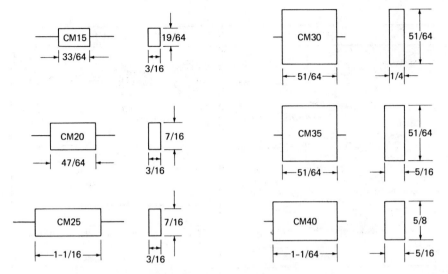

Fig. 14-4. Sizes of capacitor type.

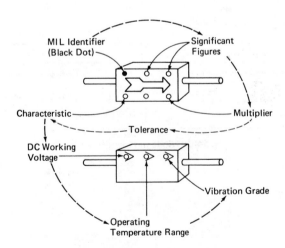

How to interpret color code dot system

Fig. 14-5. Capacitor code and size.

DIODE PIN IDENTIFICATION

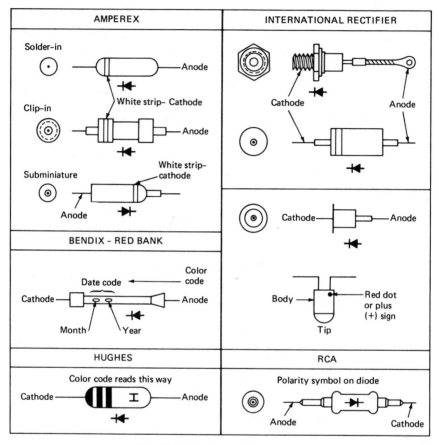

Fig. 14-6. Diode pin identification.

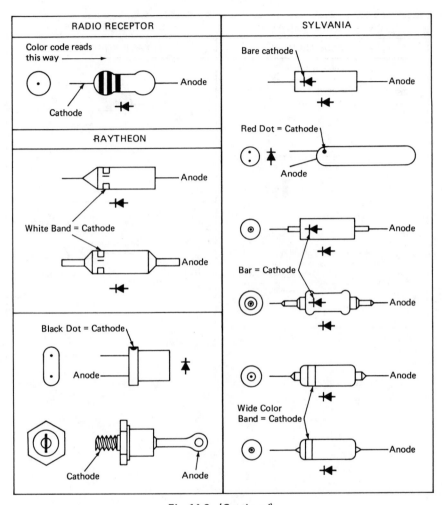

Fig. 14-6. (*Continued*)

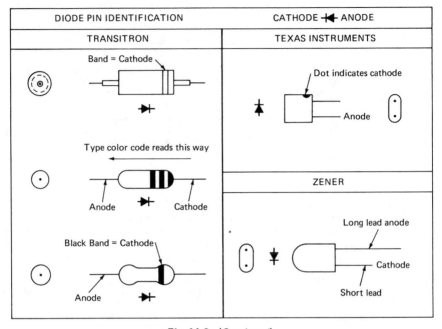

Fig. 14-6. (*Continued*)

TRANSISTOR PIN IDENTIFICATION

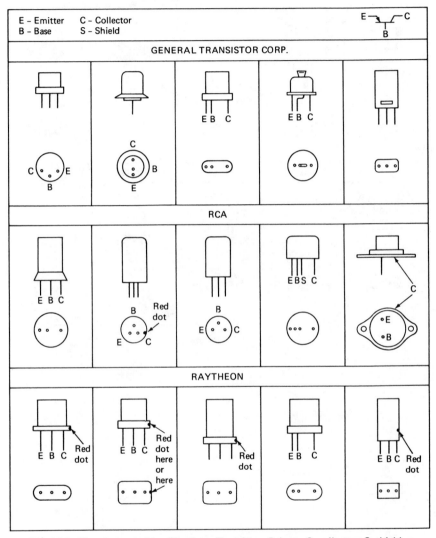

Fig. 14-7. Transistor pin identification. *E*, emitter; *B*, base; *C*, collector; *S*, shield.

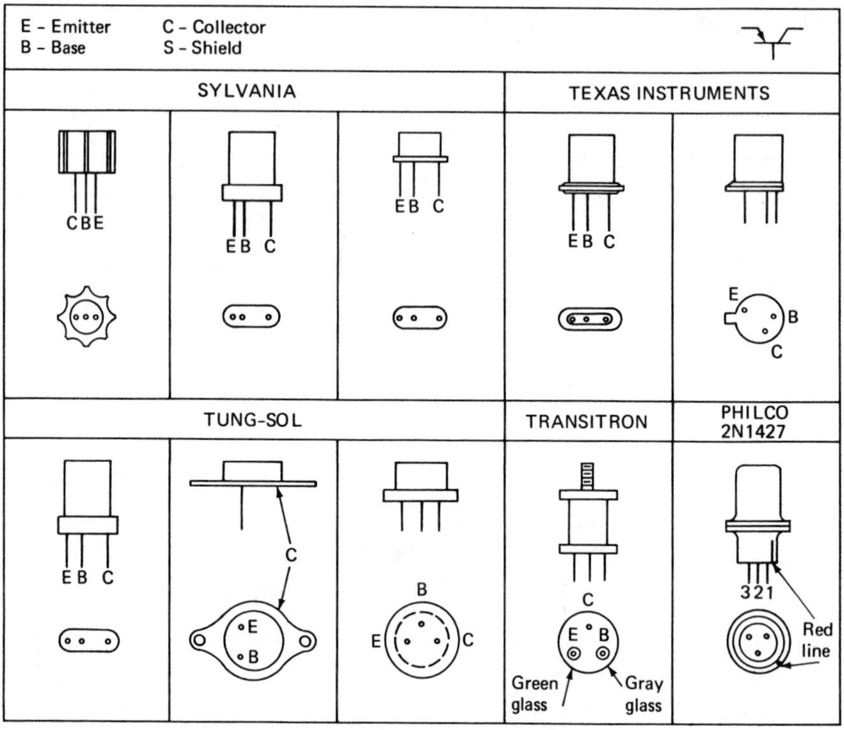

Fig. 14-7. (*Continued*)

SILICON POWER RECTIFIERS

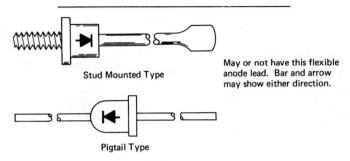

Stud Mounted Type

May or not have this flexible anode lead. Bar and arrow may show either direction.

Pigtail Type

Type and number will be stamped on side of body.

Where the arrow-and-bar symbol appears, the positive side is denoted by the arrow and the negative side by the bar.

In most cases, the coding of rectifiers or crystal diodes are as follows:

Red dot   Positive side
Black dot   Negative side
Arrow   Plus side
Bar   Negative side
Yellow dot   AC

The *IN* series of crystal diodes are usually marked for polarity as shown in the following sketch.

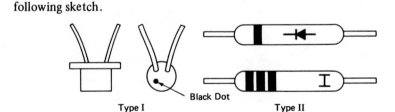

Type I   Black Dot   Type II

Type II: Silicon and germanium diodes are usually marked with one or more colored bands. The first band indicates the negative lead; it may or may not show the bar-and-arrow symbol.

Fig. 14-8. Silicon power rectifiers.

POTENTIOMETER PIN LOCATIONS

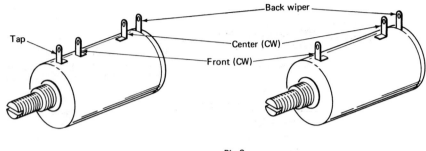

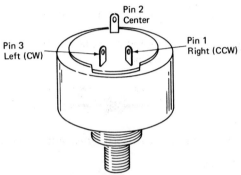

Miniature 3–Pin Potentiometer

Fig. 14-9. Potentiometers.

STANDARD RELAY-RACK PANEL DIMENSIONS

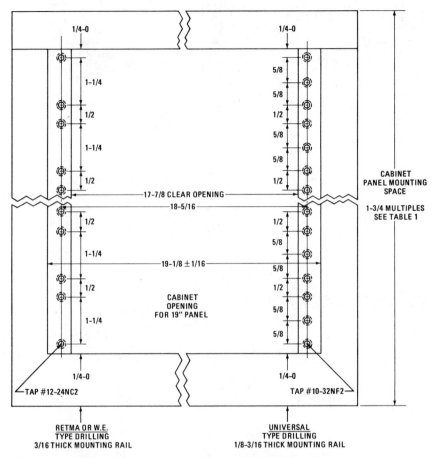

Fig. 14-10. Standard relay rack.

Panel Combinations

Cabinet Panel Mtg. Space	Quan.	Panel Size	Cabinet Panel Mtg. Space	Quan.	Panel Size
35	4	$8\frac{3}{4}$	$61\frac{1}{4}$	7	$8\frac{3}{4}$
$36\frac{3}{4}$	7	$5\frac{1}{4}$	$68\frac{1}{4}$	13	$5\frac{1}{4}$
42	6	7	70	8	$8\frac{3}{4}$
$43\frac{3}{4}$	5	$8\frac{3}{4}$	77	11	7

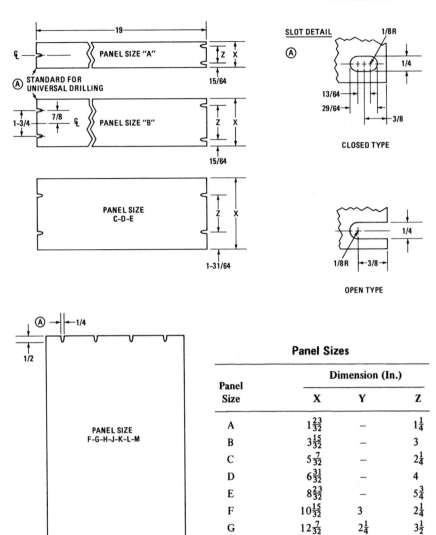

Fig. 14-10. (*Continued*)

Panel Sizes

Panel Size	Dimension (In.)		
	X	Y	Z
A	$1\frac{23}{32}$	—	$1\frac{1}{4}$
B	$3\frac{15}{32}$	—	3
C	$5\frac{7}{32}$	—	$2\frac{1}{4}$
D	$6\frac{31}{32}$	—	4
E	$8\frac{23}{32}$	—	$5\frac{3}{4}$
F	$10\frac{15}{32}$	3	$2\frac{1}{4}$
G	$12\frac{7}{32}$	$2\frac{1}{4}$	$3\frac{1}{2}$
H	$13\frac{31}{32}$	3	4
J	$15\frac{23}{32}$	$4\frac{3}{4}$	4
K	$17\frac{15}{32}$	$6\frac{1}{2}$	4
L	$19\frac{7}{32}$	$5\frac{3}{4}$	$5\frac{1}{4}$
M	$20\frac{31}{32}$	$7\frac{1}{2}$	$5\frac{1}{4}$

Table 14-11. Ultimate Strength in Pounds of Various Sizes and Grades of Bolts and Rivets.

(Reprinted through the courtesy of the Skidmore-Wilhelm Manufacturing Co., of Cleveland, Ohio. Manufacturers of Bolt Tension Calibrators.)

Material Strength (psi)	Stress Area[a] (In.^{2})	55,000 Low Carbon Steel Naval Bronze	60,000 Aluminum 24S-T4	65,000 Cold Rolled Low Carbon Steel	70,000 Yellow Brass	80,000 Medium Carbon Steel Low Silicon Bronze Monel	90,000 Medium Carbon Steel Type 304 Stainless	110,000 Medium Carbon Steel Heat-Treated	120,000 Medium Carbon Steel Heat-Treated	130,000 Alloy Steels Heat-Treated	150,000 Alloy Steels Heat-Treated	170,000 Alloy Steels Heat-Treated
Bolt Size						Bolts						
2-56	.0036	198	216	234	252	288	324	396	432	468	540	612
4-40	.0060	330	360	390	420	480	540	660	720	780	900	1,020
6-32	.0090	495	540	585	630	720	810	990	1,080	1,170	1,350	1,530
8-32	.0139	764	834	903	973	1,112	1,251	1,529	1,668	1,807	2,085	2,363
10-24	.0175	960	1,050	1,140	1,220	1,400	1,570	1,920	2,100	2,270	2,620	2,970
10-32	.0200	1,100	1,200	1,300	1,400	1,600	1,800	2,200	2,400	2,600	3,000	3,400
$\frac{1}{4}$-20	.0318	1,750	1,910	2,070	2,230	2,540	2,860	3,500	3,820	4,130	4,770	5,410
$\frac{1}{4}$-28	.0364	2,000	2,180	2,370	2,550	2,910	3,280	4,000	4,370	4,730	5,460	6,190
$\frac{5}{16}$-18	.0524	2,880	3,140	3,410	3,670	4,190	4,720	5,760	6,290	6,810	7,860	8,910
$\frac{5}{16}$-24	.0581	3,190	3,490	3,780	4,070	4,650	5,230	6,390	6,970	7,550	8,710	9,880
$\frac{3}{8}$-16	.0775	4,260	4,650	5,040	5,420	6,200	6,970	8,520	9,300	10,100	11,600	13,200
$\frac{3}{8}$-24	.0878	4,830	5,270	5,710	6,150	7,020	7,900	9,660	10,500	11,400	13,200	14,900
$\frac{7}{16}$-14	.1063	5,850	6,380	6,910	7,440	8,500	9,570	11,700	12,800	13,800	15,900	18,100

Table 14-11. (Continued)

Material Strength (psi)	Stress Area[a] (In.²)	55,000 Low Carbon Steel Naval Bronze	60,000 Aluminum 24S-T4	65,000 Cold Rolled Low Carbon Steel	70,000 Yellow Brass	80,000 Medium Carbon Steel Low Silicon Bronze Monel	90,000 Medium Carbon Steel Type 304 Stainless	110,000 Medium Carbon Steel Heat-Treated	120,000	130,000	150,000 Alloy Steels Heat-Treated	170,000
Bolt Size						Bolts						
7/16-20	.1187	6,530	7,120	7,710	8,310	9,500	10,700	13,100	14,200	15.400	17,800	20,200
1/2-13	.1419	7,800	8,510	9,220	9,930	11,300	12,800	15,600	17,000	18,400	21,300	24,100
1/2-20	.1599	8,790	9,590	10,400	11,200	12,800	14,400	17,600	19,200	20,800	24,000	27,200
9/16-12	.1819	10,000	10,900	11,800	12,700	14,500	16,400	20,000	21,800	23,600	27,300	30,900
9/16-18	.2029	11,200	12,200	13,200	14,200	16,200	18,300	22,300	24,300	26,400	30,400	34,500
5/8-11	.2260	12,400	13,600	14,700	15,800	18,100	20,300	24,900	27,100	29,400	33,900	38,400
5/8-18	.2559	14,100	15,300	16,600	17,000	20,500	23,000	28,100	30,700	33,300	38,400	43,500
3/4-10	.3344	18,400	20,100	21,700	23,400	26,700	30,100	36,800	40,100	43,500	50,200	56,800
3/4-16	.3729	20,500	22,400	24,200	26,100	29,800	33,600	41,000	44,700	48,500	55,900	63,400
7/8-9	.4617	25,400	27,700	30,000	32,300	36,900	41,600	50,800	55,400	60,000	69,300	78,500

Size											
$\frac{7}{8}$-14	.5095	28,000	30,600	33,100	35,700	40,800	45,800	56,000	61,100	66,200	76,400	86,600
1- 8	.6057	33,300	36,300	39,400	42,400	48,500	54,500	66,600	72,700	78,700	90,800	103,000
1-14	.6799	37,400	40,800	44,200	47,600	54,400	61,200	74,800	81,600	88,400	102,000	115,600
$1\frac{1}{8}$- 7	.7632	42,000	45,800	49,600	53,400	61,100	68,700	84,000	91,600	99,200	114,500	129,700
$1\frac{1}{8}$-12	.8557	47,100	51,300	55,600	59,900	68,500	77,000	94,100	102,700	111,200	128,400	145,500
$1\frac{1}{4}$- 7	.9691	53,300	58,100	63,000	67,800	77,500	87,200	106,600	116,300	126,000	145,400	174,700
$1\frac{1}{4}$-12	1.0729	59,000	64,400	69,700	75,100	85,800	96,600	118,000	128,700	139,500	160,900	172,400

Rivets

		23K Cu.	24IC 1100	43IC 2117-74	60K 24S-74
$\frac{1}{16}$	.0031	71	74	133	186
$\frac{3}{32}$	.0069	158	165	296	414
$\frac{1}{8}$	.0123	282	295	528	738
$\frac{5}{32}$	.0192	441	460	825	1,152
$\frac{3}{16}$	.0276	634	662	1,186	1,656
$\frac{1}{4}$	.0491	1,129	1,178	2,111	2,946
$\frac{5}{16}$	.0767	1,764	1,840	3,298	4,602

[a]Stress area is calculated as the area of the circle whose diameter is the mean between the root and pitch diameters. This closely approximates the actual stress condition.

OUTLINE DRAWINGS

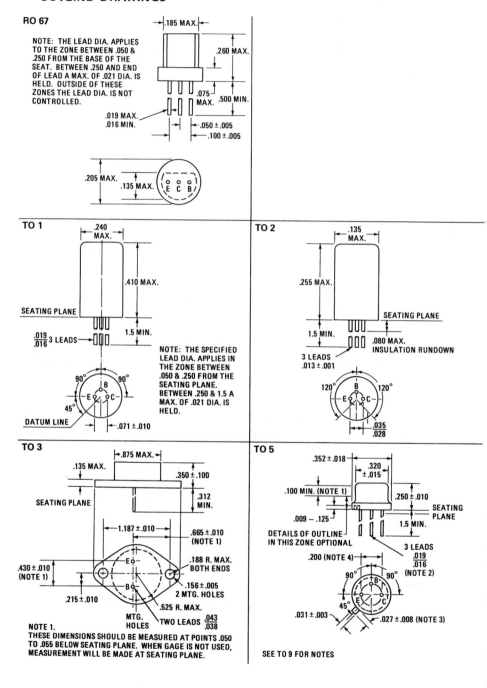

RO 67

NOTE: THE LEAD DIA. APPLIES TO THE ZONE BETWEEN .050 & .250 FROM THE BASE OF THE SEAT. BETWEEN .250 AND END OF LEAD A MAX. OF .021 DIA. IS HELD. OUTSIDE OF THESE ZONES THE LEAD DIA. IS NOT CONTROLLED.

.185 MAX.
.260 MAX.
.075 MAX.
.500 MIN.
.019 MAX.
.016 MIN.
.050 ±.005
.100 ±.005
.205 MAX.
.135 MAX.
E C B

TO 1

.240 MAX.
.410 MAX.
SEATING PLANE
.019/.016 3 LEADS
1.5 MIN.
90° 90°
B
E C
45°
DATUM LINE
.071 ±.010

NOTE: THE SPECIFIED LEAD DIA. APPLIES IN THE ZONE BETWEEN .050 & .250 FROM THE SEATING PLANE. BETWEEN .250 & 1.5 A MAX. OF .021 DIA. IS HELD.

TO 2

.135 MAX.
.255 MAX.
SEATING PLANE
1.5 MIN.
.080 MAX. INSULATION RUNDOWN
3 LEADS .013 ±.001
120° 120°
B
E C
.035/.028

TO 3

.875 MAX.
.135 MAX.
.350 ±.100
.312 MIN.
SEATING PLANE
1.187 ±.010
.665 ±.010 (NOTE 1)
.430 ±.010 (NOTE 1)
E
.188 R. MAX. BOTH ENDS
B
.156 ±.005 2 MTG. HOLES
.215 ±.010
.525 R. MAX.
MTG. HOLES
TWO LEADS .043/.038

NOTE 1.
THESE DIMENSIONS SHOULD BE MEASURED AT POINTS .050 TO .055 BELOW SEATING PLANE. WHEN GAGE IS NOT USED, MEASUREMENT WILL BE MADE AT SEATING PLANE.

TO 5

.352 ±.018
.320 ±.015
.100 MIN. (NOTE 1)
.250 ±.010
.009 – .125
SEATING PLANE
1.5 MIN.
DETAILS OF OUTLINE IN THIS ZONE OPTIONAL
3 LEADS .019/.016
.200 (NOTE 4)
(NOTE 2)
90° 90°
B
E C
45°
.031 ±.003
.027 ±.008 (NOTE 3)

SEE TO 9 FOR NOTES

TO 6

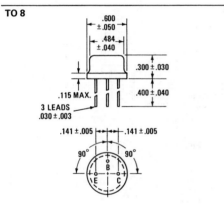

1.188 MAX.

SEATING PLANE

.625 MAX.

.120 MAX.

.438

1.625 MIN.

INSULATED LOCATER PIN

10-32 UNF-24

.345 R.

E

CO

B

TO 7

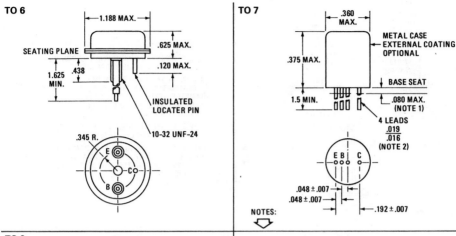

.360 MAX.

METAL CASE EXTERNAL COATING OPTIONAL

.375 MAX.

BASE SEAT

1.5 MIN.

.080 MAX. (NOTE 1)

4 LEADS .019 .016 (NOTE 2)

E B C

.048 ± .007

.048 ± .007

.192 ± .007

NOTES:

TO 8

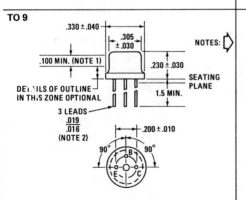

.600 ± .050

.484 ± .040

.300 ± .030

.115 MAX.

.400 ± .040

3 LEADS .030 ± .003

.141 ± .005

.141 ± .005

90° 90°

B

E C

NOTES FOR TO 7:

NOTE 1. EXTERNALLY COATED DEVICES SHALL NOT HAVE COATING ON THE LEADS BEYOND THIS ZONE.

NOTE 2. THE SPECIFIED LEAD DIAMETER APPLIES IN THE ZONE BETWEEN 0.050 AND 0.250 FROM THE SEATING PLANE BETWEEN 0.250 AND 1.50 A MAXIMUM OF 0.021 DIAMETER IS HELD. OUTSIDE OF THESE ZONES, THE LEAD DIAMETER IS NOT CONTROLLED.

TO 9

.330 ± .040

.305 ± .030

NOTES:

.100 MIN. (NOTE 1)

.230 ± .030

SEATING PLANE

DETAILS OF OUTLINE IN THIS ZONE OPTIONAL

1.5 MIN.

3 LEADS .019 .016 (NOTE 2)

.200 ± .010

90° 90°

B

E C

NOTES FOR TO 5, 9, 11, 12, 16, 28, 33, 39, 42, 43.

THIS DEVICE IS FOR SOCKETED, SINGLE-SIDED CIRCUIT-BOARD, WIRE IN & SIMILAR APPLICATIONS WHERE SOLDER BRIDGING MAY OCCUR. A DIELECTRIC WASHER OR OTHER STANDOFF DEVICE MAY BE NECESSARY.
NOTE 1: THIS ZONE IS CONTROLLED FOR AUTOMATIC HANDLING. THE VARIATION IN ACTUAL DIA. WITHIN THIS ZONE SHALL NOT EXCEED .010.
NOTE 2: THE SPECIFIED LEAD DIA. APPLIES IN THE ZONE BETWEEN .050 & .250 FROM THE SEATING PLANE.
BETWEEN .250 & 1.5 A MAX. OF .021 DIA. IS HELD. OUTSIDE OF THESE ZONES THE LEAD DIA. IS NOT CONTROLLED.
NOTE 3: MEASURED FROM MAX. DIA. OF THE ACTUAL DEVICE.
NOTE 4: LEADS HAVING MAX. DIA. .019 MEASURED IN GAGING PLANE .054 + .001 BELOW THE SEATING PLANE SHALL BE WITHIN .007 OF THEIR TRUE LOCATIONS RELATIVE TO A MAX-WIDTH TAB.

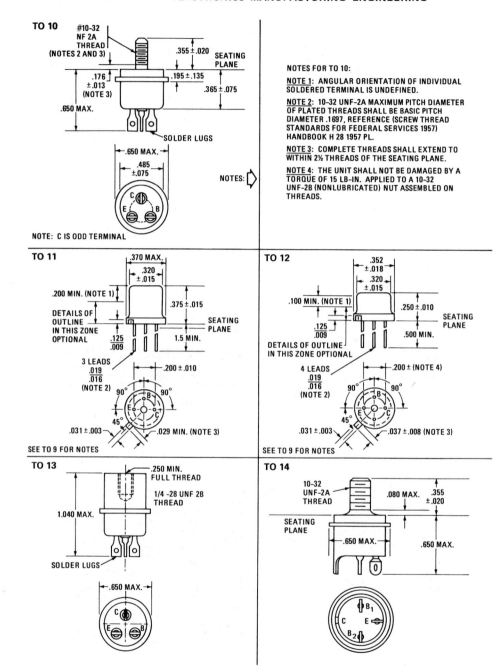

TO 10

#10-32 NF 2A THREAD (NOTES 2 AND 3)

.355 ± .020

SEATING PLANE

.176 ± .013 (NOTE 3)

.195 ± .135

.365 ± .075

.650 MAX.

SOLDER LUGS

.650 MAX.

.485 ± .075

NOTES: ▷

C
E   B

NOTE: C IS ODD TERMINAL

NOTES FOR TO 10:

NOTE 1: ANGULAR ORIENTATION OF INDIVIDUAL SOLDERED TERMINAL IS UNDEFINED.

NOTE 2: 10-32 UNF-2A MAXIMUM PITCH DIAMETER OF PLATED THREADS SHALL BE BASIC PITCH DIAMETER .1697, REFERENCE (SCREW THREAD STANDARDS FOR FEDERAL SERVICES 1957) HANDBOOK H 28 1957 PL.

NOTE 3: COMPLETE THREADS SHALL EXTEND TO WITHIN 2½ THREADS OF THE SEATING PLANE.

NOTE 4: THE UNIT SHALL NOT BE DAMAGED BY A TORQUE OF 15 LB-IN. APPLIED TO A 10-32 UNF-2B (NONLUBRICATED) NUT ASSEMBLED ON THREADS.

TO 11

.370 MAX.

.320 ± .015

.200 MIN. (NOTE 1)

DETAILS OF OUTLINE IN THIS ZONE OPTIONAL

.375 ± .015

SEATING PLANE

.125 / .009

1.5 MIN.

3 LEADS .019 / .016 (NOTE 2)

.200 ± .010

90°   90°

45°

.031 ± .003

.029 MIN. (NOTE 3)

SEE TO 9 FOR NOTES

TO 12

.352 ± .018

.320 ± .015

.100 MIN. (NOTE 1)

.250 ± .010

SEATING PLANE

.125 / .009

.500 MIN.

DETAILS OF OUTLINE IN THIS ZONE OPTIONAL

4 LEADS .019 / .016 (NOTE 2)

.200 ± (NOTE 4)

90°   90°

45°

.031 ± .003

.037 ± .008 (NOTE 3)

SEE TO 9 FOR NOTES

TO 13

.250 MIN. FULL THREAD

1/4 –28 UNF 2B THREAD

1.040 MAX.

SOLDER LUGS

.650 MAX.

C
E   B

TO 14

10-32 UNF-2A THREAD

.080 MAX.

.355 ± .020

SEATING PLANE

.650 MAX.

.650 MAX.

B₁
C   E
B₂

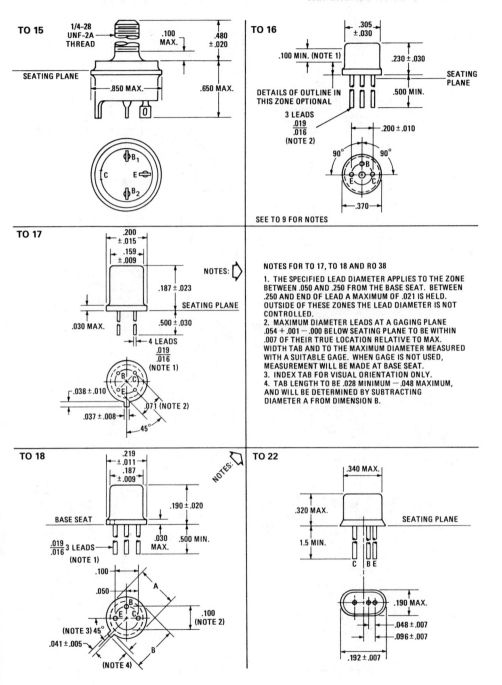

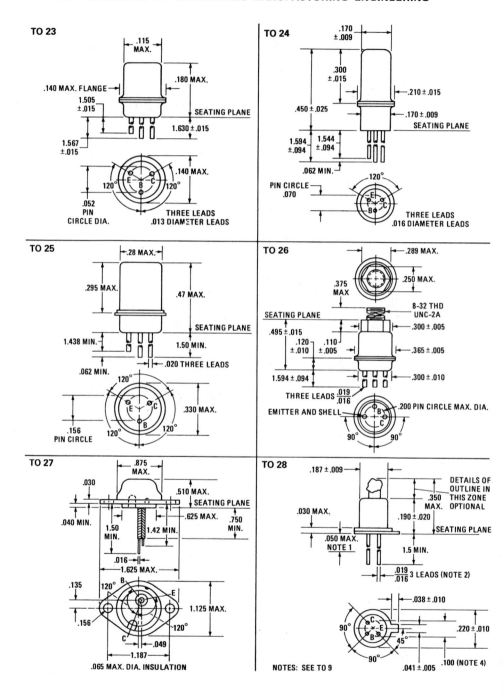

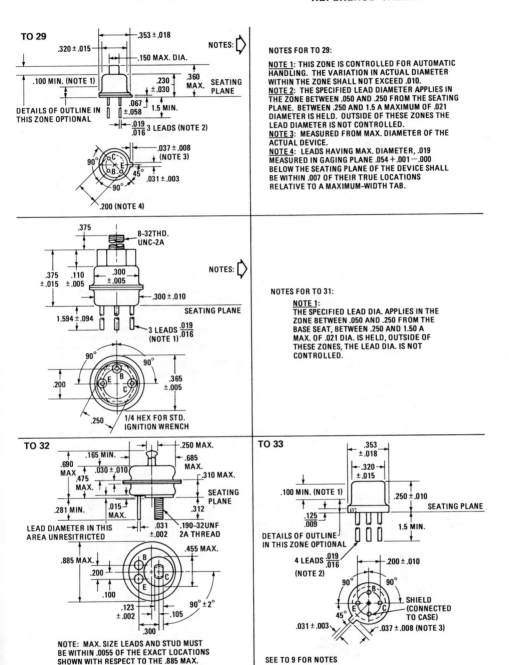

TO 29

.353 ± .018
.320 ± .015
.150 MAX. DIA.
NOTES:

.100 MIN. (NOTE 1)
.230 ± .030
.360 MAX. SEATING PLANE
DETAILS OF OUTLINE IN THIS ZONE OPTIONAL
.067 ± .058
1.5 MIN.
.019/.016 3 LEADS (NOTE 2)

.037 ± .008 (NOTE 3)
90°
C E
B
45°
90°
.031 ± .003
.200 (NOTE 4)

NOTES FOR TO 29:

NOTE 1: THIS ZONE IS CONTROLLED FOR AUTOMATIC HANDLING. THE VARIATION IN ACTUAL DIAMETER WITHIN THE ZONE SHALL NOT EXCEED .010.
NOTE 2: THE SPECIFIED LEAD DIAMETER APPLIES IN THE ZONE BETWEEN .050 AND .250 FROM THE SEATING PLANE. BETWEEN .250 AND 1.5 A MAXIMUM OF .021 DIAMETER IS HELD. OUTSIDE OF THESE ZONES THE LEAD DIAMETER IS NOT CONTROLLED.
NOTE 3: MEASURED FROM MAX. DIAMETER OF THE ACTUAL DEVICE.
NOTE 4: LEADS HAVING MAX. DIAMETER, .019 MEASURED IN GAGING PLANE .054 + .001 − .000 BELOW THE SEATING PLANE OF THE DEVICE SHALL BE WITHIN .007 OF THEIR TRUE LOCATIONS RELATIVE TO A MAXIMUM-WIDTH TAB.

.375
8-32THD. UNC-2A
NOTES:

.375 ± .015
.110 ± .005
.300 ± .005
.300 ± .010
SEATING PLANE
1.594 ± .094
3 LEADS .019/.016 (NOTE 1)

90°
90°
.200
E B
C
.365 ± .005
.250
1/4 HEX FOR STD. IGNITION WRENCH

NOTES FOR TO 31:

NOTE 1:
THE SPECIFIED LEAD DIA. APPLIES IN THE ZONE BETWEEN .050 AND .250 FROM THE BASE SEAT, BETWEEN .250 AND 1.50 A MAX. OF .021 DIA. IS HELD, OUTSIDE OF THESE ZONES, THE LEAD DIA. IS NOT CONTROLLED.

TO 32

.250 MAX.
.165 MIN.
.685 MAX.
.690 MAX.
.030 ± .010
.310 MAX.
.475 MAX.
SEATING PLANE
.281 MIN.
.015 MAX.
.312
LEAD DIAMETER IN THIS AREA UNRESTRICTED
.031 ± .002
.190-32UNF 2A THREAD

.455 MAX.
.885 MAX.
B
.200
C
E
.100
.123 ± .002
.105
.300
90° ± 2°

NOTE: MAX. SIZE LEADS AND STUD MUST BE WITHIN .0055 OF THE EXACT LOCATIONS SHOWN WITH RESPECT TO THE .885 MAX. DIA. MEASURED POINTS .015 MAX. BELOW.

TO 33

.353 ± .018
.320 ± .015
.100 MIN. (NOTE 1)
.250 ± .010
SEATING PLANE
.125/.009
1.5 MIN.
DETAILS OF OUTLINE IN THIS ZONE OPTIONAL
4 LEADS .019/.016 (NOTE 2)
.200 ± .010

90°
90°
B
45°
E
C
.031 ± .003
SHIELD (CONNECTED TO CASE)
.037 ± .008 (NOTE 3)

SEE TO 9 FOR NOTES

TO 36

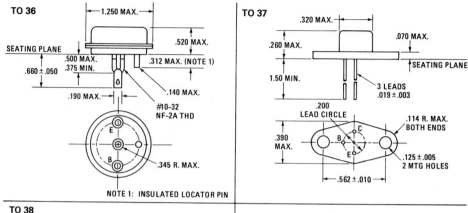

SEATING PLANE

1.250 MAX.

.520 MAX.

.500 MAX.
.375 MIN.

.660 ± .050

.312 MAX. (NOTE 1)

.190 MAX.

.140 MAX.

#10-32 NF-2A THD

E B

.345 R. MAX.

NOTE 1: INSULATED LOCATOR PIN

TO 37

.320 MAX.

.260 MAX.

.070 MAX.

SEATING PLANE

1.50 MIN.

3 LEADS
.019 ± .003

.200 LEAD CIRCLE

.390 MAX.

C
B
E

.114 R. MAX. BOTH ENDS

.125 ± .005
2 MTG HOLES

.562 ± .010

TO 38

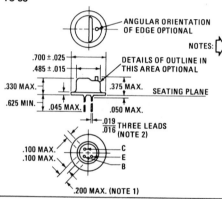

ANGULAR ORIENTATION OF EDGE OPTIONAL

NOTES:

.700 ± .025
.485 ± .015

DETAILS OF OUTLINE IN THIS AREA OPTIONAL

.330 MAX.
.625 MIN.
.045 MAX.

.375 MAX.
SEATING PLANE

.050 MAX.

.019
.016
THREE LEADS
(NOTE 2)

.100 MAX.
.100 MAX.

C
E
B

.200 MAX. (NOTE 1)

NOTES FOR TO 38:

NOTE 1: MAXIMUM DIAMETER LEADS MEASURED AT A GAGING PLANE .054 ± .001-.000 BELOW THE SEATING PLANE SHALL BE WITHIN .010 OF THEIR TRUE LOCATIONS WITH RESPECT TO THE .725 DIAMETER.

NOTE 2: THE SPECIFIED LEAD DIAMETER APPLIES IN THE ZONE BETWEEN .050 AND .250 FROM THE SEATING PLANE. IN THE ZONE BETWEEN .050 AND .625 FROM THE SEATING PLANE DIAMETER OF LEADS SHALL NOT EXCEED .021. DIAMETER IS UNCONTROLLED BEYOND .625 FROM SEATING PLANE.

TO 39

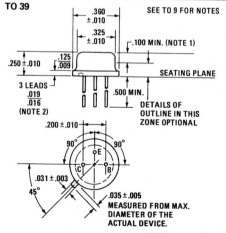

.360 ± .010

SEE TO 9 FOR NOTES

.325 ± .010

.100 MIN. (NOTE 1)

.125
.009

.250 ± .010

SEATING PLANE

3 LEADS
.019
.016
(NOTE 2)

.500 MIN.

DETAILS OF OUTLINE IN THIS ZONE OPTIONAL

.200 ± .010

90°   90°

E

C   B

.031 ± .003

45°

.035 ± .005
MEASURED FROM MAX. DIAMETER OF THE ACTUAL DEVICE.

TO 40

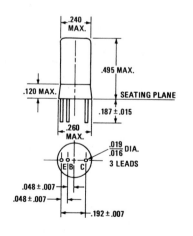

.240 MAX.

.495 MAX.

.120 MAX.

SEATING PLANE

.187 ± .015

.260 MAX.

.019
.016
DIA.
3 LEADS

E B C

.048 ± .007
.048 ± .007

.192 ± .007

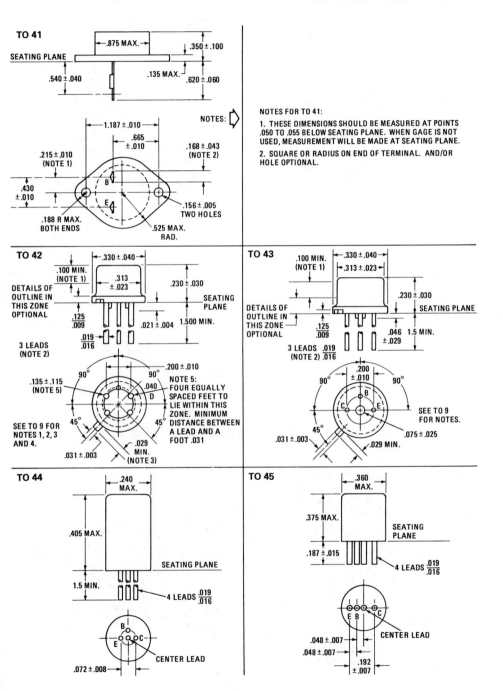

TO 41

SEATING PLANE

.875 MAX.
.350 ± .100
.540 ± .040
.135 MAX.
.620 ± .060

NOTES:

1.187 ± .010
.665 ± .010
.168 ± .043 (NOTE 2)
.215 ± .010 (NOTE 1)
.430 ± .010
.188 R MAX. BOTH ENDS
.156 ± .005 TWO HOLES
.525 MAX. RAD.

NOTES FOR TO 41:

1. THESE DIMENSIONS SHOULD BE MEASURED AT POINTS .050 TO .055 BELOW SEATING PLANE. WHEN GAGE IS NOT USED, MEASUREMENT WILL BE MADE AT SEATING PLANE.

2. SQUARE OR RADIUS ON END OF TERMINAL. AND/OR HOLE OPTIONAL.

TO 42

.100 MIN. (NOTE 1)
.330 ± .040
.313 ± .023
.230 ± .030
DETAILS OF OUTLINE IN THIS ZONE OPTIONAL
SEATING PLANE
.125 / .009
.021 ± .004
1.500 MIN.
.019 / .016
3 LEADS (NOTE 2)

.200 ± .010
.135 ± .115 (NOTE 5)
90°
90°
.040
D
NOTE 5: FOUR EQUALLY SPACED FEET TO LIE WITHIN THIS ZONE. MINIMUM DISTANCE BETWEEN A LEAD AND A FOOT .031
45°
45°
SEE TO 9 FOR NOTES 1, 2, 3 AND 4.
.029 MIN. (NOTE 3)
.031 ± .003

TO 43

.100 MIN. (NOTE 1)
.330 ± .040
.313 ± .023
.230 ± .030
DETAILS OF OUTLINE IN THIS ZONE OPTIONAL
SEATING PLANE
.125 / .009
.046 ± .029
1.5 MIN.
3 LEADS (NOTE 2)
.019 / .016

.200 ± .010
90°
90°
B
C
E
SEE TO 9 FOR NOTES.
45°
.075 ± .025
.031 ± .003
.029 MIN.

TO 44

.240 MAX.
.405 MAX.
SEATING PLANE
1.5 MIN.
4 LEADS .019 / .016

B
E
C
CENTER LEAD
.072 ± .008

TO 45

.360 MAX.
.375 MAX.
SEATING PLANE
.187 ± .015
4 LEADS .019 / .016

E B C
CENTER LEAD
.048 ± .007
.048 ± .007
.192 ± .007

TO 46

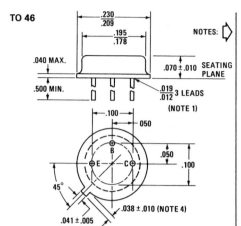

NOTES: ▷

NOTES FOR TO 46, TO 47

1. THE SPECIFIED LEAD DIAMETER APPLIES TO THE ZONE BETWEEN .050 AND .250 FROM THE BASE SEAT. BETWEEN .250 AND THE END OF LEAD A MAXIMUM OF .021 IS HELD. OUTSIDE OF THESE ZONES THE LEAD DIAMETER IS NOT CONTROLLED.
2. MAXIMUM DIAMETER LEADS AT A GAGING PLAN $.054\,^{+.001}_{-.000}$ BELOW BASE SEAT TO BE WITHIN .007 OF THEIR TRUE LOCATION RELATIVE TO MAX. WIDTH TAB AND TO THE MAXIMUM DIAMETER MEASURED WITH A SUITABLE GAGE. WHEN GAGE IS NOT USED, MEASUREMENT WILL BE MADE AT SEATING PLANE.
3. INDEX TAB FOR VISUAL ORIENTATION ONLY.
4. MEASURED FROM MAX. DIAMETER OF THE ACTUAL DEVICE.

TO 47

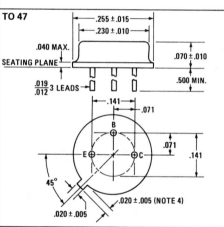

TO 48

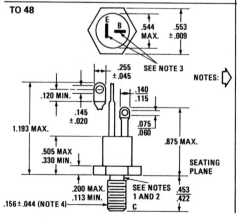

NOTES: ▷

NOTES FOR TO 48

1. COMPLETE THREADS TO EXTEND TO WITHIN 2½ THREADS OF HEAD.
2. DIA. OF UNTHREADED PORTION .249 MAX. .220 MIN.
3. ANGULAR ORIENTATION OF THESE TERMINALS IS UNDEFINED.
4. MAX. PITCH DIA. OF PLATED THREADS SHALL BE BASIC PITCH DIA. (.2268) REF. (SCREW THREAD STANDARDS FOR FEDERAL SERVICES 1957) HANDBOOK H28 1957 PI.
5. A CHAMFER (OR UNDERCUT) ON ONE OR BOTH ENDS OF HEXAGONAL PORTIONS IS OPTIONAL.
6. SQUARE OR RADIUS ON END OF TERMINAL IS OPTIONAL.

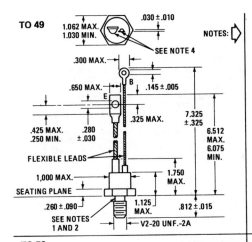

TO 49

NOTES FOR TO 49

1. COMPLETE THREADS TO EXTEND TO WITHIN 2½ THREADS OF HEAD.
2. DIA. OF UNTHREADED PORTION .435 MAX., .425 MIN.
3. SCREW THD. STANDARDS FOR FEDERAL SERVICES (1957 HANDBOOK H28 PI) APPLY TO UNF 2A THD.
4. ANGULAR ORIENTATION OF THESE TERMINALS IS UNDEFINED.
5. A CHAMFER (OR UNDERCUT) ON ONE OR BOTH ENDS OF HEXAGONAL PORTIONS IS OPTIONAL.
6. SQUARE OR RADIUS ON END OF TERMINAL IS OPTIONAL.

TO 50
TO 51

	D-MIN.	D-MAX.
TO 50	.170	.190
TO 51	.140	.165

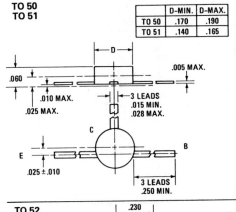

TO 52

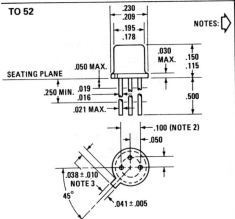

NOTES FOR TO 52:

1. LEAD DIAMETER .016 MIN., .019 MAX. APPLIES BETWEEN .05 MAX. AND .250 MIN. FROM SEATING PLANE. LEAD DIAMETER .021 MAX. APPLIES BETWEEN .250 MIN. AND .50 FROM SEATING PLANE. DIAMETER IS UNCONTROLLED IN .05 AND .50 FROM SEATING PLANE.
2. LEADS HAVING MAX. DIAMETER MEASURED IN GAGING PLANE .054 +.001-.000 BELOW THE SEATING PLANE OF THE DEVICE SHALL·BE WITHIN .007 OF THEIR TRUE LOCATIONS RELATIVE TO A MAX. WIDTH TAB.
3. MEASURED FROM MAX. DIAMETER OF THE ACTUAL DEVICE.

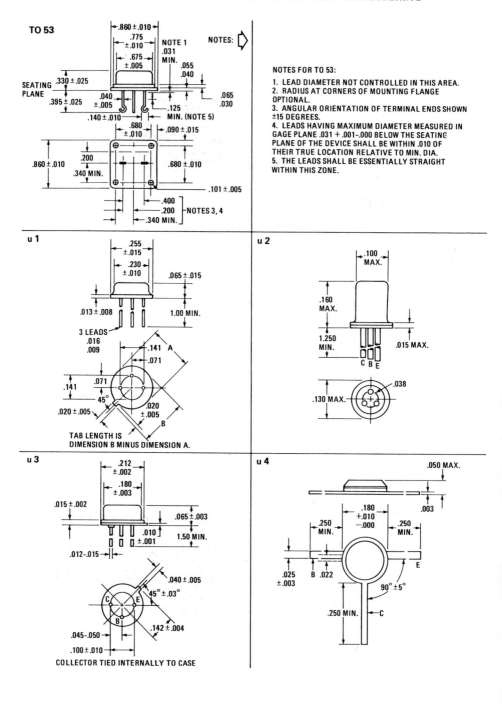

TO 53

NOTES: ⇨

NOTES FOR TO 53:

1. LEAD DIAMETER NOT CONTROLLED IN THIS AREA.
2. RADIUS AT CORNERS OF MOUNTING FLANGE OPTIONAL.
3. ANGULAR ORIENTATION OF TERMINAL ENDS SHOWN ±15 DEGREES.
4. LEADS HAVING MAXIMUM DIAMETER MEASURED IN GAGE PLANE .031 + .001-.000 BELOW THE SEATING PLANE OF THE DEVICE SHALL BE WITHIN .010 OF THEIR TRUE LOCATION RELATIVE TO MIN. DIA.
5. THE LEADS SHALL BE ESSENTIALLY STRAIGHT WITHIN THIS ZONE.

u 1

TAB LENGTH IS DIMENSION B MINUS DIMENSION A.

u 2

u 3

COLLECTOR TIED INTERNALLY TO CASE

u 4

Package Information

INTEGRATED CIRCUIT PACKAGE OUTLINES

2-PIN PACKAGES

2-Lead Metal Can Package

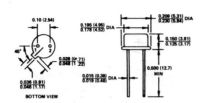

2-Lead Flat Package

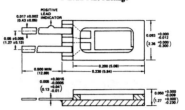

3-PIN PACKAGES

TO-5 Package

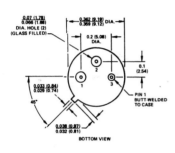

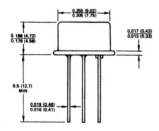

TO-52 Package

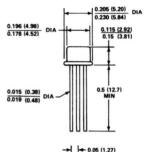

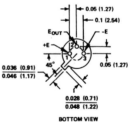

Dimensions shown in inches and (mm).
Lead No. 1 Identified by Dot or Notch.

8-PIN PACKAGES

TO-99 Package

(TO-99 STYLE)
(JEDEC REF MO-002AB)

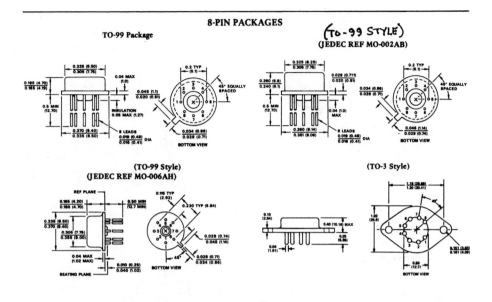

(TO-99 Style)
(JEDEC REF MO-006AH)

(TO-3 Style)

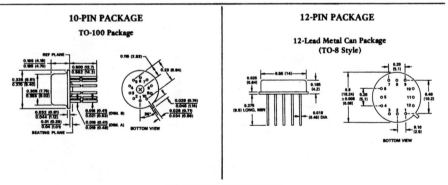

10-PIN PACKAGE
TO-100 Package

12-PIN PACKAGE

12-Lead Metal Can Package
(TO-8 Style)

Dimensions shown in inches and (mm).
Lead No. 1 Identified by Dot or Notch.

14-PIN PACKAGES

14-Lead Ceramic Package

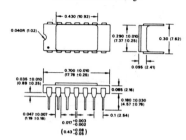

14-Lead Ceramic DIP Package

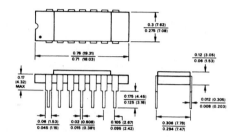

14-Pin Plastic DIP Package

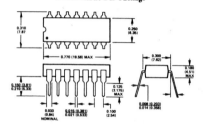

14-Pin Plastic Package

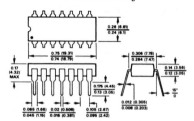

14-Pin CERDIP Package

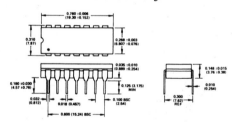

Dimensions shown in inches and (mm).
Lead No. 1 Identified by Dot or Notch.

14-PIN PACKAGES
(Continued)

14-Pin Hybrid Package

14-Pin Hybrid Package

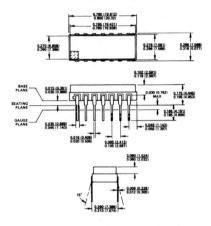

14-Pin Hybrid Package

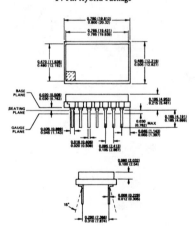

14-Pin Metal DIP Package

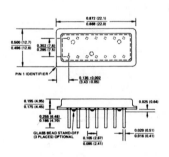

Dimensions shown in inches and (mm).
Lead No. 1 Identified by Dot or Notch.

16-PIN PACKAGES

16-Pin Ceramic DIP Package

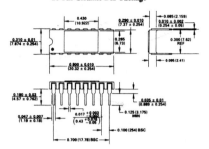

16-Pin Ceramic DIP Package

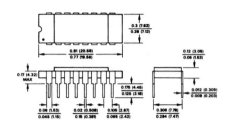

16-Pin Plastic DIP Package

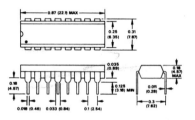

16-Pin Plastic DIP Package

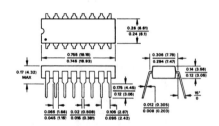

16-Pin CERDIP Package

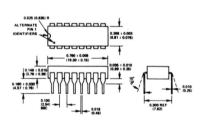

16-Pin CERDIP Package

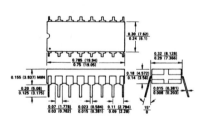

Dimensions shown in inches and (mm).
Lead No. 1 Identified by Dot or Notch.

18-PIN PACKAGES

18-Pin Ceramic DIP Package

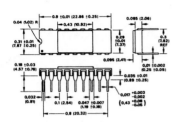

18-Pin Ceramic DIP Package

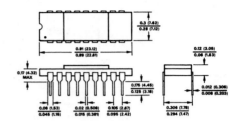

18-Pin Plastic DIP Package

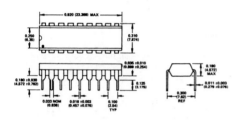

18-Pin Plastic Package

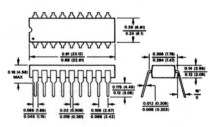

18-Pin CERDIP Package

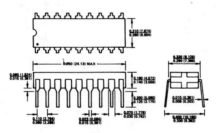

Dimensions shown in inches and (mm).
Lead No. 1 Identified by Dot or Notch.

18-PIN PACKAGES
(Continued)

18-Pin Hybrid Package

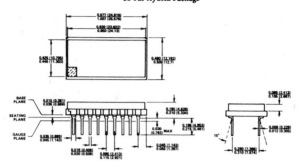

20-PIN PACKAGES

20-Pin Ceramic DIP Package

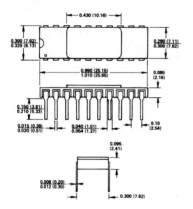

20-Pin Ceramic DIP Package

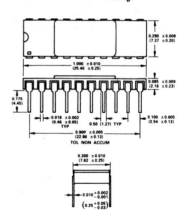

Dimensions shown in inches and (mm).
Lead No. 1 Identified by Dot or Notch.

20-PIN PACKAGES
(Continued)

20-Pin Plastic DIP Package

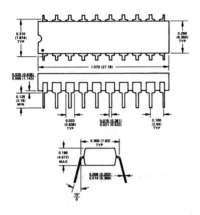

20-Pin Plastic DIP Package

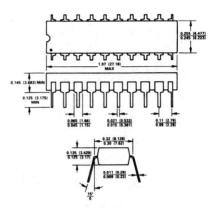

20-Pin CERDIP Package

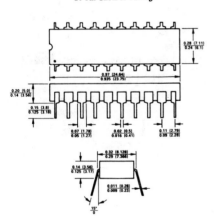

20-Pin Hybrid Package

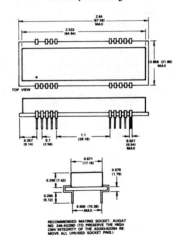

RECOMMENDED MATING SOCKET; AUGAT NO. 240-AG39D (TO PRESERVE THE HIGH CMV INTEGRITY OF THE AD293/AD294 REMOVE ALL UNUSED SOCKET PINS.)

Dimensions shown in inches and (mm).
Lead No. 1 Identified by Dot or Notch.

24-PIN PACKAGES

24-Lead Ceramic DIP Package

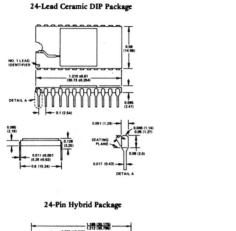

24-Lead Plastic Package

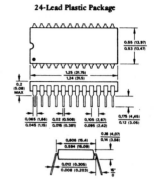

24-Pin Hybrid Package

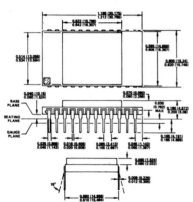

24-Pin Hybrid Package

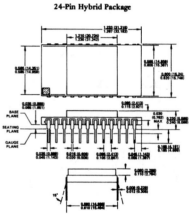

Dimensions shown in inches and (mm).
Lead No. 1 Identified by Dot or Notch.

24-PIN PACKAGES
(Continued)

24-Pin Hybrid Package

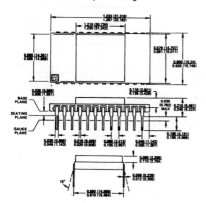

24-Pin Hybrid Package

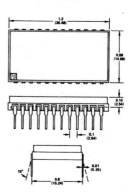

24-Pin Hybrid Package

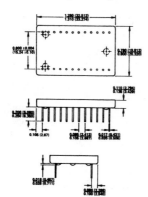

24-Pin Hybrid Package

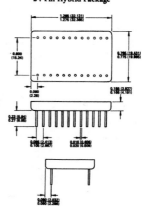

Dimensions shown in inches and (mm).
Lead No. 1 Identified by Dot or Notch.

28-PIN PACKAGES

28-Pin Ceramic DIP Package

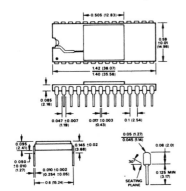

28-Pin Ceramic Package

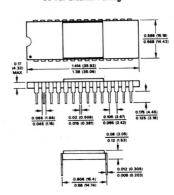

28-Lead Plastic DIP Package

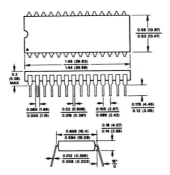

Dimensions shown in inches and (mm).
Lead No. 1 Identified by Dot or Notch.

28-PIN PACKAGES
(Continued)

28-Pin Hybrid Package   28-Pin Hybrid Package

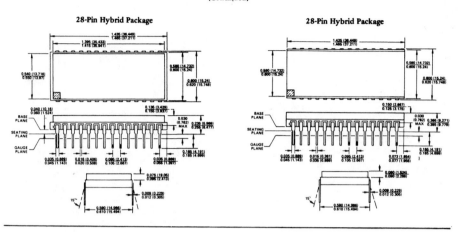

32-PIN PACKAGES

32-Pin Hybrid Package   32-Pin Hybrid Package

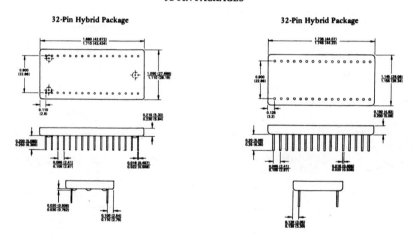

Dimensions shown in inches and (mm).
Lead No. 1 Identified by Dot or Notch.

32-PIN PACKAGES
(Continued)

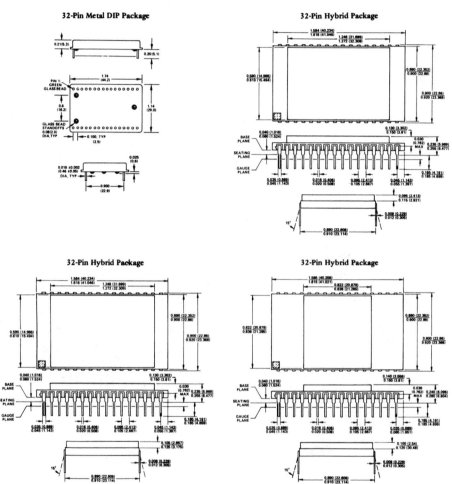

Dimensions shown in inches and (mm).
Lead No. 1 Identified by Dot or Notch.

32-PIN PACKAGES
(Continued)

32-Pin Hybrid Package

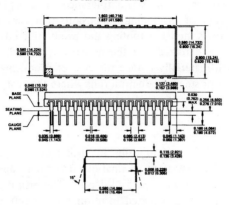

40-PIN PACKAGES

40-Pin Ceramic DIP Package

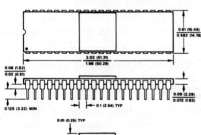

40-Pin Plastic DIP Package

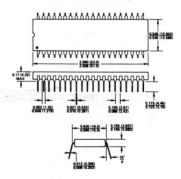

Dimensions shown in inches and (mm).
Lead No. 1 Identified by Dot or Notch.

TWIST DRILL DATA

Scope

This subsection covers the sizes and tolerances obtainable with standard twist drills. Specification of machining dimensions that cannot be produced with readily available tools increases tooling and production costs. The sizes of twist drills have been standardized and the design of parts which can accommodate standard drills will materially reduce the cost of producing a part. This is especially true in pilot or small production runs where the cost cannot be distributed over a large number of parts.

Drill Specifications

Table 14-12 lists the standard drills by drill number, letter, or fractional or metric diameter, the equivalent decimal diameter, the upper and lower limits of drill diameter, and the maximum drill depth. The drill diameter tolerances are the closest that can be anticipated in normal manufacture. Where design permits, additional tolerances should be allowed. The following notes apply to the data presented in Table 14-12.

1. Dimensions are in inches except where millimeters (mm) are specified.
2. All fractional drill sizes and those letter and number sizes marked with an asterisk (*) are recommended for use by the United States of America Standards Institute and the Society of Automotive Engineers.
3. Drill sizes above $1\frac{1}{2}$ inches are available in increments of $\frac{1}{64}$ inch to $1\frac{3}{4}$ inches, $\frac{1}{32}$ inch to $2\frac{1}{4}$ inches, and $\frac{1}{16}$ inch to $3\frac{1}{2}$ inches.
4. Drill depths are approximations based on new, standard drills most frequently used in industry. Longer drills are available in some of the larger sizes. However, excessive depth does not allow for runout and grinding. Where extra depth holes are unavoidable, the limits of Table GVI-1 must be increased by one-half the specified tolerance and the new limits specified on the drawing.

Table 14-12. Twist Drills and Tolerances.

Drill Size		Drill Diameter Limits		
Number, Letter, Etc.	Decimal Equivalent	Lower	Upper	Maximum Drill Depth
80	0.0135	0.0125	0.0175	
79	0.0145	0.0135	0.0185	
$\frac{1}{64}$	0.0156	0.0146	0.0196	0.150

Table 14-12. (Continued)

| Drill Size | | Drill Diameter Limits | | Maximum Drill Depth |
Number, Letter, Etc.	Decimal Equivalent	Lower	Upper	
78	0.0160	0.0150	0.0200	
*77	0.0180	0.0170	0.0220	
*76	0.0200	0.0190	0.0240	
75	0.0210	0.0200	0.0250	
74	0.0225	0.0215	0.0265	
*73	0.0240	0.0230	0.0280	0.250
72	0.0250	0.0240	0.0290	
*71	0.0260	0.0250	0.0300	
*70	0.0280	0.0270	0.0310	
69	0.0292	0.0282	0.0332	
0.75 mm	0.0295	0.0285	0.0335	0.350
68	0.0310	0.0300	0.0350	
$\frac{1}{32}$	0.0312	0.0302	0.0352	
67	0.0320	0.0310	0.0360	
*66	0.0330	0.0320	0.0370	
*65	0.0350	0.0340	0.0390	0.450
64	0.0360	0.0350	0.0400	
*63	0.0370	0.0360	0.0410	
62	0.0380	0.0370	0.0420	
*61	0.0390	0.0380	0.0430	0.550
60	0.0400	0.0390	0.0440	
59	0.0410	0.0400	0.0450	
58	0.0420	0.0410	0.0460	
*57	0.0430	0.0420	0.0470	
1.15 mm	0.0453	0.0443	0.0493	
56	0.0465	0.0455	0.0505	
$\frac{3}{64}$	0.0469	0.0459	0.0509	
1.25 mm	0.0492	0.0482	0.0532	0.650
1.30 mm	0.0512	0.0502	0.0552	
55	0.0520	0.0510	0.0560	
1.35 mm	0.0531	0.0521	0.0571	
*54	0.0550	0.0540	0.0549	

Table 14-12. (Continued)

| Drill Size | | Drill Diameter Limits | | Maximum Drill Depth |
Number, Letter, Etc.	Decimal Equivalent	Lower	Upper	
1.45 mm	0.0571	0.0561	0.0611	
1.50 mm	0.0591	0.0581	0.0631	
53	0.0595	0.0585	0.0635	
1.55 mm	0.0610	0.0600	0.0650	
$\frac{1}{16}$	0.0625	0.0615	0.0665	0.750
1.60 mm	0.0630	0.0620	0.0670	
52	0.0635	0.0625	0.0675	
1.65 mm	0.0649	0.0639	0.0689	
50	0.0670	0.0660	0.0710	
49	0.0730	0.0720	0.0770	
48	0.0760	0.0750	0.0890	
$\frac{5}{64}$	0.0781	0.0771	0.0821	0.850
47	0.0785	0.0775	0.0825	
46	0.0810	0.0800	0.0850	
45	0.0820	0.0810	0.0860	
2.10 mm	0.0827	0.0817	0.0867	
*44	0.0860	0.0850	0.0900	0.950
*43	0.0890	0.0880	0.0930	
2.30 mm	0.0906	0.0896	0.0946	
42	0.0935	0.0925	0.0975	
$\frac{3}{32}$	0.0938	0.0928	0.0978	
*41	0.0960	0.0950	0.1000	
40	0.0980	0.0970	0.0102	1.100
*39	0.0995	0.0985	0.1035	
38	0.1015	0.1005	0.1055	
2.60 mm	0.1024	0.1014	0.1064	
*37	0.1040	0.1030	0.1080	
*36	0.1065	0.1055	0.1105	
$\frac{7}{64}$	0.1094	0.1084	0.1134	0.1250
35	0.1100	0.1090	0.1140	
34	0.1110	0.1100	0.1150	
*33	0.1130	0.1120	0.1170	
*32	0.1160	0.1150	0.1200	

Table 14-12. (Continued)

Drill Size		Drill Diameter Limits		
Number, Letter, Etc.	Decimal Equivalent	Lower	Upper	Maximum Drill Depth
*31	0.1200	0.1190	0.1240	
3.10 mm	0.1220	0.1210	0.1260	
$\frac{1}{8}$	0.1250	0.1240	0.1290	1.350
*30	0.1285	0.1275	0.1335	
3.30 mm	0.1200	0.1289	0.1349	
3.30 mm	0.1339	0.1329	0.1389	
*29	0.1360	0.1350	0.1410	
3.50 mm	0.1378	0.1368	0.1428	1.450
28	0.1405	0.1395	0.1455	
$\frac{9}{64}$	0.1406	0.1396	0.1456	
*27	0.1440	0.1430	0.1490	
*26	0.1470	0.1460	0.1520	
25	0.1495	0.1485	0.1545	1.550
*24	0.1520	0.1510	0.1570	
23	0.1540	0.1530	0.1590	
$\frac{5}{32}$	0.1562	0.1552	0.1612	
22	0.1570	0.1560	0.1620	
21	0.1590	0.1580	0.1640	
*20	0.1610	0.1600	0.1660	1.700
*19	0.1660	0.1650	0.1710	
*18	0.1695	0.1685	0.1745	
$\frac{11}{64}$	0.1719	0.1709	0.1769	
*17	0.1730	0.1720	0.1780	
16	0.1770	0.1760	0.1820	
15	0.1800	0.1790	0.1850	1.800
14	0.1820	0.1810	0.1870	
13	0.1850	0.1840	0.1900	
$\frac{3}{16}$	0.1875	0.1865	0.1925	
12	0.1890	0.1880	0.1940	
*11	0.1910	0.1900	0.1960	1.900
*10	0.1935	0.1925	0.1985	
*9	0.1960	0.1950	0.2010	
*8	0.1990	0.1980	0.2040	
7	0.2010	0.2000	0.2060	

Table 14-12. (Continued)

Drill Size		Drill Diameter Limits		Maximum Drill Depth
Number, Letter, Etc.	Decimal Equivalent	Lower	Upper	
$\frac{13}{64}$	0.2031	0.2021	0.2081	
6	0.2040	0.2030	0.2090	
5	0.2055	0.2045	0.2105	2.000
*4	0.2090	0.2080	0.2140	
*3	0.2130	0.2120	0.2180	
$\frac{7}{32}$	0.2188	0.2178	0.2238	
2	0.2210	0.2200	0.2260	
5.70 mm	0.2244	0.2234	0.2294	
*1	0.2280	0.2270	0.2330	
A	0.2340	0.2330	0.2400	
$\frac{15}{64}$	0.2344	0.2334	0.2394	
B	0.2380	0.2370	0.2430	
6.10 mm	0.2402	0.2392	0.2452	2.2000
C	0.2420	0.2410	0.2470	
*D	0.2460	0.2450	0.2510	
$\frac{1}{4}$(E)	0.2500	0.2490	0.2550	
$\frac{1}{4}$(E)	0.2500	0.2490	0.2550	
6.4 mm	0.2520	0.2510	0.2580	
*F	0.2570	0.2560	0.2630	
*G	0.2610	0.2600	0.2670	
$\frac{17}{64}$	0.2656	0.2646	0.2716	2.300
H	0.2660	0.2650	0.2720	
*I	0.2720	0.2710	0.2780	
*J	0.2770	0.2760	0.2830	
K	0.2810	0.2800	0.2870	
$\frac{9}{32}$	0.2812	0.2802	0.2872	
7.25 mm	0.2854	0.2844	0.2914	
L	0.2900	0.2890	0.2960	
7.4 mm	0.2913	0.2903	0.2973	2.400
M	0.2950	0.2940	0.3010	
$\frac{19}{64}$	0.2969	0.2959	0.3029	
*N	0.3020	0.3010	0.3080	
7.80 mm	0.3071	0.3061	0.3131	
$\frac{5}{16}$	0.3125	0.3115	0.3185	
*O	0.3160	0.3150	0.3220	2.500

Table 14-12. (*Continued*)

Drill Size		Drill Diameter Limits		Maximum
Number, Letter, Etc.	Decimal Equivalent	Lower	Upper	Drill Depth
P	0.3230	0.3220	0.3290	
$\frac{21}{64}$	0.3281	0.3271	0.3341	
Q	0.3320	0.3310	0.3380	
R	0.3390	0.3380	0.3450	
$\frac{11}{32}$	0.3438	0.3428	0.3498	
S	0.3480	0.3470	0.3540	2.700
9.00 mm	0.3543	0.3533	0.3603	
T	0.3580	0.3570	0.3640	
$\frac{23}{64}$	0.3594	0.3584	0.3654	
*U	0.3680	0.3670	0.3740	
$\frac{3}{8}$	0.3750	0.3740	0.3810	
V	0.3770	0.3760	0.3830	2.800
9.60 mm	0.3780	0.3770	0.3840	
*W	0.3860	0.3850	0.3920	
$\frac{25}{64}$	0.3906	0.3896	0.3966	
*X	0.3970	0.3960	0.4030	2.900
Y	0.4040	0.4030	0.4100	
$\frac{13}{32}$	0.4062	0.4052	0.4122	
Z	0.4130	0.4120	0.4190	
$\frac{27}{64}$	0.4219	0.4209	0.4179	3.250
$\frac{7}{16}$	0.4375	0.4365	0.4435	
$\frac{29}{64}$	0.4531	0.4521	0.4591	
$\frac{15}{32}$	0.4688	0.4678	0.4748	3.500
$\frac{31}{64}$	0.4844	0.4834	0.4904	
$\frac{1}{2}$	0.5000	0.4990	0.5060	
$\frac{33}{64}$	0.5156	0.5146	0.5226	
$\frac{17}{32}$	0.5312	0.5302	0.5382	
$\frac{35}{64}$	0.5469	0.5459	0.5539	
$\frac{9}{16}$	0.5625	0.5615	0.5695	3.750
$\frac{37}{64}$	0.5781	0.5771	0.5861	
$\frac{19}{32}$	0.5937	0.5927	0.6017	
$\frac{39}{64}$	0.6094	0.6084	0.6174	
$\frac{5}{8}$	0.6250	0.6240	0.6330	
$\frac{41}{64}$	0.6406	0.6396	0.6486	

TAP DRILL DATA

Table 14-13. Tap Drill Sizes.

Probable Percentage of Full Thread Produced in Tapped Hole Using Stock Sizes of Drill

Tap	Tap Drill	Decimal Equivalent of Tap Drill	Theoretical % of Thread	Probable Oversize (Mean)	Probable Hole Size	% of Thread
1/4-28	3	.2130	80	.0038	.2168	72
	7/32	.2188	67	.0038	.2226	59
	2	.2210	63	.0038	.2248	55
5/16-18	F	.2570	77	.0038	.2608	72
	G	.2610	71	.0041	.2651	66
	17/64	.2656	65	.0041	.2697	59
	H	.2660	64	.0041	.2701	59
5/16-24	H	.2660	86	.0041	.2701	78
	I	.2720	75	.0041	.2761	67
	J	.2770	66	.0041	.2811	58
3/8-16	5/16	.3125	77	.0044	.3169	72
	O	.3160	73	.0044	.3204	68
	P	.3230	64	.0044	.3274	59
3/8-24	21/64	.3281	87	.0044	.3325	79
	Q	.3320	79	.0044	.3364	71
	R	.3390	67	.0044	.3434	58
7/16-14	T	.3580	86	.0046	.3626	81
	23/64	.3594	84	.0046	.3640	79
	U	.3680	75	.0046	.3726	70
	3/8	.3750	67	.0046	.3796	62
	V	.3770	65	.0046	.3816	60
7/16-20	W	.3860	79	.0046	.3906	72
	25/64	.3906	72	.0046	.3952	65
	X	.3970	62	.0046	.4016	55
1/2-13	27/64	.4219	78	.0047	.4266	73
	7/16	.4375	63	.0047	.4422	58
1/2-20	29/64	.4531	72	.0047	.4578	65
9/16-12	15/32	.4688	87	.0048	.4736	82
	31/64	.4844	72	.0048	.4892	68
9/16-18	1/2	.500	87	.0048	.5048	80
	33/64	.5156	65	.0048	.5204	58
5/8-11	17/32	.5313	79	.0049	.5362	75
	35/64	.5469	66	.0049	.5518	62
5/8-18	9/16	.5625	87	.0049	.5674	80
	37/64	.5781	65	.0049	.5831	58

Tap	Tap Drill	Decimal Equivalent of Tap Drill	Theoretical % of Thread	Probable Oversize (Mean)	Probable Hole Size	% of Thread
0-80	56	.0465	83	.0015	.0480	74
	3/64	.0469	81	.0015	.0484	71
1-64	54	.0550	89	.0015	.0565	81
	53	.0595	67	.0015	.0610	59
1-72	53	.0595	75	.0015	.0610	67
	1/16	.0625	58	.0015	.0640	50
2-56	51	.0670	82	.0017	.0687	74
	50	.0700	69	.0017	.0717	62
	49	.0730	56	.0017	.0747	49
2-64	50	.0700	79	.0017	.0717	70
	49	.0730	64	.0017	.0747	56
3-48	48	.0760	85	.0019	.0779	78
	5/64	.0781	77	.0019	.0800	70
	47	.0785	76	.0019	.0804	69
	46	.0810	67	.0019	.0829	60
	45	.0820	63	.0019	.0839	56
3-56	46	.0810	78	.0019	.0829	69
	45	.0820	73	.0019	.0839	65
	44	.0860	56	.0019	.0879	48
4-40	44	.0860	80	.0020	.0880	74
	43	.0890	71	.0020	.0910	65
	42	.0935	57	.0020	.0955	51
	3/32	.0938	56	.0020	.0958	50
4-48	42	.0935	68	.0020	.0955	61
	3/32	.0938	68	.0020	.0958	60
	41	.0960	59	.0020	.0980	52
5-40	40	.0980	83	.0023	.1003	76
	39	.0995	79	.0023	.1018	71
	38	.1015	72	.0023	.1038	65
	37	.1040	65	.0023	.1063	58
5-44	38	.1015	79	.0023	.1038	72
	37	.1040	71	.0023	.1063	63
	36	.1065	63	.0023	.1088	55

Table 14-13. (Continued)

Tap Size	Drill	Drill Decimal	%	Oversize	Probable Hole	%
6–32	37	.1040	84	.0023	.1063	78
	36	.1065	78	.0026	.1091	71
	7/64	.1094	70	.0026	.1120	64
	35	.1100	69	.0026	.1126	63
	34	.1110	67	.0026	.1136	60
	33	.1130	62	.0026	.1156	55
6–40	34	.1110	83	.0026	.1136	75
	33	.1130	77	.0026	.1156	69
	32	.1160	68	.0026	.1186	60
8–32	29	.1360	69	.0029	.1389	62
	28	.1405	58	.0029	.1434	51
8–36	29	.1360	78	.0029	.1389	70
	28	.1405	68	.0029	.1434	57
	9/64	.1406	68	.0029	.1435	57
10–24	27	.1440	85	.0032	.1472	79
	26	.1470	79	.0032	.1502	74
	25	.1495	75	.0032	.1527	69
	24	.1520	70	.0032	.1552	64
	23	.1540	67	.0032	.1572	61
	5/32	.1563	62	.0032	.1595	56
	22	.1570	61	.0032	.1602	55
10–32	5/32	.1563	83	.0032	.1595	75
	22	.1570	81	.0032	.1602	73
	21	.1590	76	.0032	.1622	68
	20	.1610	71	.0032	.1642	64
	19	.1660	59	.0032	.1692	51
12–24	11/64	.1719	82	.0035	.1754	75
	17	.1730	79	.0035	.1765	73
	16	.1770	72	.0035	.1805	66
	15	.1800	67	.0035	.1835	60
	14	.1820	63	.0035	.1855	56
12–28	16	.1770	84	.0035	.1805	77
	15	.1800	78	.0035	.1835	70
	14	.1820	73	.0035	.1855	66
	13	.1850	67	.0035	.1885	59
	3/16	.1875	61	.0035	.1910	54
1/4–20	9	.1960	83	.0038	.1998	77
	8	.1990	79	.0038	.2028	73
	7	.2010	75	.0038	.2048	70
	13/64	.2031	72	.0038	.2069	66
	6	.2040	71	.0038	.2078	65
	5	.2055	69	.0038	.2093	63
	4	.2090	63	.0038	.2128	57

Tap Size	Drill	Drill Decimal	%	Oversize	Probable Hole	%
3/4–10	41/64	.6406	84	.0050	.6456	80
	21/32	.6563	72	.0050	.6613	68
3/4–16	11/16	.6875	77	.0050	.6925	71
7/8–9	49/64	.7656	76	.0052	.7708	72
	25/32	.7812	65	.0052	.7864	61
7/8–14	51/64	.7969	84	.0052	.8021	79
	13/16	.8125	67	.0052	.8177	62
1″–8	55/64	.8594	87	.0059	.8653	83
	7/8	.875	77	.0059	.8809	73
	57/64	.8906	67	.0059	.8965	64
	29/32	.9063	58	.0059	.9122	54
1″–12	29/32	.9063	87	.0060	.9123	81
	59/64	.9219	72	.0060	.9279	67
	15/16	.9375	58	.0060	.9435	52
1″–14	59/64	.9219	84	.0060	.9279	78
	15/16	.9375	67	.0060	.9435	61

FASTENER HEAD STYLES

†SCREW HEADS (MACHINE, WOOD, TAPPING, STOVE BOLTS)

Binding Head:

Formerly referred to as "Straight Side Binding Heads" this style is most generally used in the electrical trades with an undercut under the head for the binding of wire. Sometimes specified without undercut where appearance is the principal factor.

Truss Head:

Sometimes called "Oval Binding," "Oven Head" or "Stove Head" this large diameter, low clearance head is used both for appearance and low height and sometimes to span clearance holes or slots where wide tolerances are necessary.

Pan Head:

Principally a developed design for tapping screws where a low, attractive head style is desirable. The semi-squared outer periphery of the head offers full edges for high torque driving.

Jackson Head:

This style which is nothing more than an under-sized oval (countersunk) head screw is used extensively in the architectural hardware trades for assembling moldings, etc., because of its neat, diminutive and finished style.

Knurled Head:

A high, circular head with ridges on the outer rim for thumb adjustment or manual fastening.

RIVET HEADS

Round Head:

Like the typical screw head, this is the most common rivet head and is specified for most general applications.

Flat Head:

Unlike the same nomenclature in screw heads the rivet "flat" head is a low, round, flat type of head not countersunk. Specified where a low head will solve a clearance or other design problem.

Countersunk Head:

The sister head to the screw standard "Flat Head." For countersunk applications where flush surfaces are indicated. Angle of countersink on small diameter rivets is approximately 90 degrees; on rivets one-half inch and over, 78 degrees.

†SCREW HEADS (MACHINE, WOOD, TAPPING, STOVE BOLTS (Continued)

Welding Screw Head:

Characterized by multiple welding lugs on the top or underside of the head which are fused to the assembly by means of projection welding to provide a permanent threaded member.

Tinners Rivet Head:

Same general contour as a "Flat Head" but to different standards. The head has a larger bearing surface to provide a greater distribution of load on light gauge metal. Tinners rivets are used principally in sheet metal work.

†Most Screw Products can be supplied in Phillips Recessed Heads as well as slotted.

Head Dimensions
Decimal Inches
Finished Hexagon Head Machine Bolts and Cap Screws

Nominal Size or Basic Major Diameter of Thread	Body Dia. Minimum (Maximum Equal to Nominal Size)	F Width Across Flats			G Width Across Corners		H Height			R Radius of Fillet	
		Max.	(Basic)	Min.	Max.	Min.	Nom.	Max.	Min.	Max.	Min.
1/4 .2500	.2450	7/16	.4375	.428	.505	.488	5/32	.163	.150	.023	.009
5/16 .3125	.3065	1/2	.5000	.489	.577	.557	13/64	.211	.195	.023	.009
3/8 .3750	.3690	9/16	.5625	.551	.650	.628	15/64	.243	.226	.023	.009
7/16 .4375	.4305	5/8	.6250	.612	.722	.698	9/32	.291	.272	.023	.009
1/2 .5000	.4930	3/4	.7500	.736	.866	.840	5/16	.323	.302	.023	.009
9/16 .5625	.5545	13/16	.8125	.798	.938	.910	23/64	.371	.348	.041	.021
5/8 .6250	.6170	15/16	.9375	.922	1.083	1.051	25/64	.403	.378	.041	.021
3/4 .7500	.7410	1 1/8	1.1250	1.100	1.299	1.254	15/32	.483	.455	.041	.021
7/8 .8750	.8660	1 5/16	1.3125	1.285	1.516	1.465	35/64	.563	.531	.062	.041
1 1.0000	.9900	1 1/2	1.5000	1.469	1.732	1.675	39/64	.627	.591	.093	.062
1 1/8 1.1250	1.1140	1 11/16	1.6875	1.631	1.949	1.859	11/16	.718	.658	.093	.062
1 1/4 1.2500	1.2390	1 7/8	1.8750	1.812	2.165	2.066	25/32	.813	.749	.093	.062
1 3/8 1.3750	1.3630	2 1/16	2.0625	1.994	2.382	2.273	27/32	.878	.810	.093	.062
1 1/2 1.5000	1.4880	2 1/4	2.2500	2.175	2.598	2.480	15/16	.974	.902	.093	.062
1 3/4 1.7500	1.7350	2 5/8	2.6250	2.538	3.031	2.893	1 3/32	1.134	1.054	.093	.062
2 2.0000	1.9580	3	3.0000	2.900	3.464	3.306	1 7/32	1.263	1.175	.093	.062

Round Head, Square Neck Carriage Bolts

Size	Body Diam. Max.	A Diameter of Head Basic	A Min.	A Max.	H Height of Head	H Min.	H Max.	P Depth of Square Min.	P Max.	B Width of Square Min.	B Max.
#10–24	.199	7/16	.438	.469	3/32	.094	.114	.094	.125	.185	.199
1/4–20	.260	9/16	.563	.594	1/8	.125	.145	.125	.156	.245	.260
5/16–18	.324	11/16	.688	.719	5/32	.156	.176	.156	.187	.307	.324
3/8–16	.388	13/16	.782	.844	3/16	.188	.208	.188	.219	.368	.388
7/16–14	.452	15/16	.907	.969	7/32	.219	.239	.219	.250	.431	.452
1/2–13	.515	1 1/16	1.032	1.094	1/4	.250	.270	.250	.281	.492	.515
5/8–11	.642	1 5/16	1.219	1.344	5/16	.313	.344	.313	.344	.616	.642
3/4–10	.768	1 9/16	1.469	1.594	3/8	.375	.406	.375	.406	.741	.768
7/8–9	.895	1 13/16	1.719	1.844	7/16	.438	.469	.438	.469	.865	.895
1–8	1.022	2 1/16	1.969	2.094	1/2	.500	.531	.500	.531	.990	1.022

Lag Bolts (Screws)

Diameter of Bolt		Threads per Inch	Thread Dimensions P Pitch	B Flat at Root	T Depth of Thread	R Root Dia.	F Width Across Flats Max.	F (Basic)	F Min.	H Height Nom.	H Max.	H Min.
#10	.1900	11	.091	.039	.035	.120	9/32	.2810	.271	1/8	.140	.110
1/4	.2500	10	.100	.043	.039	.173	3/8	.3750	.362	11/64	.188	.156
5/16	.3125	9	.111	.048	.043	.227	1/2	.5000	.484	13/64	.220	.186
3/8	.3750	7	.143	.062	.055	.265	9/16	.5625	.544	1/4	.268	.232
7/16	.4375	7	.143	.062	.055	.328	5/8	.6250	.603	19/64	.316	.278
1/2	.5000	6	.167	.072	.064	.371	3/4	.7500	.725	21/32	.348	.308
5/8	.6250	5	.200	.086	.077	.471	15/16	.9375	.906	27/64	.444	.400
3/4	.7500	4 1/2	.222	.096	.085	.579	1 1/8	1.1250	1.088	1/2	.524	.476
7/8	.8750	4	.250	.108	.096	.683	1 5/16	1.3125	1.269	19/32	.620	.568
1	1.0000	3 1/2	.286	.123	.110	.780	1 1/2	1.5000	1.450	21/32	.684	.628
1 1/8	1.1250	3 1/4	.308	.133	.119	.887	1 11/16	1.6875	1.631	3/4	.780	.720
1 1/4	1.2500	3 1/4	.308	.133	.119	1.012	1 7/8	1.8750	1.812	27/32	.876	.812

Most lag bolts are supplied with A cut thread and full diameter unthreaded shank.

Head Dimensions—(Continued)
Decimal Inches

Hexagonal Socket Head Cap Screws

Nom.	D Body Diameter[1] Max.	Min.	A Head Diameter Max.	Min.	H Head Height Max.	Min.	S Head Side-Height Nom.	Max.	Min.	J Socket Width Across Flats Max.	Min.
2	0.0860	0.0840	0.140	0.136	0.086	0.083	0.079	0.081	0.078	0.0635	1/16
3	0.0990	0.0968	0.161	0.157	0.099	0.096	0.091	0.093	0.089	0.0791	5/64
4	0.1120	0.1096	0.183	0.178	0.112	0.109	0.103	0.105	0.101	0.0791	5/64
5	0.1250	0.1226	0.205	0.200	0.125	0.122	0.115	0.117	0.113	0.0947	3/32
6	0.1380	0.1353	0.226	0.221	0.138	0.134	0.127	0.129	0.125	0.0947	3/32
8	0.1640	0.1613	0.270	0.265	0.164	0.160	0.150	0.152	0.148	0.1270	1/8
10	0.1900	0.1867	5/16	0.306	0.190	0.185	0.174	0.176	0.172	0.1582	5/32
12	0.2160	0.2127	11/32	0.337	0.216	0.211	0.198	0.200	0.196	0.1582	5/32
1/4	0.2500	0.2464	3/8	0.367	1/4	0.244	0.229	0.232	0.226	0.1895	3/16
5/16	0.3125	0.3084	7/16	0.429	5/16	0.306	0.286	0.289	0.283	0.2207	7/32
3/8	0.3750	0.3705	9/16	0.553	3/8	0.368	0.344	0.347	0.341	0.3155	5/16
7/16	0.4375	0.4326	5/8	0.615	7/16	0.430	0.401	0.405	0.397	0.3155	5/16
1/2	0.5000	0.4948	3/4	0.739	1/2	0.492	0.458	0.462	0.454	0.3780	3/8
9/16	0.5625	0.5569	13/16	0.801	9/16	0.554	0.516	0.520	0.512	0.3780	3/8
5/8	0.6250	0.6191	7/8	0.863	5/8	0.616	0.573	0.577	0.569	0.5030	1/2
3/4	0.7500	0.7436	1	0.987	3/4	0.741	0.688	0.693	0.684	0.5655	9/16
7/8	0.8750	0.8680	1 1/8	1.111	7/8	0.865	0.802	0.807	0.797	0.5655	9/16
1	1.0000	0.9924	1 9/16	1.297	1	0.989	0.917	0.922	0.912	0.6290	5/8

[1]BODY DIAMETER (D) refers to the unthreaded portion of the screw and the maximum diameter conforms to the basic diameter or size of the screw.

Square Head Set Screws

Nominal Size	F Width Across Flats		G Width Across Corners Min.	H Height of Head			X Radius of Head Nom.
	Max.	Min.		Nom.	Max.	Min.	
#10	.1875	.180	.247	9/64	.148	.134	15/32
#12	.2160	.208	.292	5/32	.163	.147	35/64
1/4	.2500	.241	.331	3/16	.196	.178	5/8
5/16	.3125	.302	.415	15/64	.245	.224	25/32
3/8	.3750	.362	.497	9/32	.293	.270	15/16
7/16	.4375	.423	.581	21/64	.341	.315	1 3/32
1/2	.5000	.484	.665	3/8	.389	.361	1 1/4
9/16	.5625	.545	.748	27/64	.437	.407	1 13/32
5/8	.6250	.606	.833	15/32	.485	.452	1 9/16
3/4	.7500	.729	1.001	9/16	.582	.544	1 7/8
7/8	.8750	.852	1.170	21/32	.678	.635	2 3/16
1	1.0000	.974	1.337	3/4	.774	.726	2 1/2
1 1/8	1.1250	1.096	1.505	27/32	.870	.817	2 13/16
1 1/4	1.2500	1.219	1.674	15/16	.966	.908	3 1/8
1 3/8	1.3750	1.342	1.843	1 3/32	1.063	1.000	3 7/16
1 1/2	1.5000	1.464	2.010	1 1/8	1.159	1.091	3 3/4

Cup points are standard with stock sizes of all set screws. Other points are available on special order.

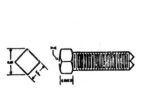

Head Dimensions—(Continued)
Decimal Inches

Hexagonal Socket Type Set Screws

D Nominal Diameter	C Cup and Flat Point Dia. Max.	C Min.	R Oval Point Radius	Y Cone Point Angle 118°±2° for these Lengths and Under	Y 90°±2° for these Lengths and Over	P Full Dog Point and Half Dog Point Diameter Max.	P Min.	Q Full Dog Point and Half Dog Point Full	q Half	J Socket Width Across Flats Max.	J Min.
5	0.067	0.057	3/32	1/8	3/16	0.083	0.078	0.06	0.03	0.0635	1/16
6	0.074	0.064	7/64	1/8	3/16	0.092	0.087	0.07	0.03	0.0635	1/16
8	0.087	0.076	1/8	3/16	1/4	0.109	0.103	0.08	0.04	0.0791	5/64
10	0.102	0.088	9/64	3/16	1/4	0.127	0.120	0.09	0.04	0.0947	3/32
12	0.115	0.101	5/32	3/16	1/4	0.144	0.137	0.11	0.06	0.0947	3/32
1/4	0.132	0.118	3/16	1/4	5/16	5/32	0.149	1/8	1/16	0.1270	1/8
5/16	0.172	0.156	13/64	5/16	3/8	13/64	0.195	5/32	5/64	0.1582	5/32
3/8	0.212	0.194	9/32	3/8	7/16	1/4	0.241	3/16	3/32	0.1895	3/16
7/16	0.252	0.232	21/64	7/16	1/2	19/64	0.287	7/32	7/64	0.2207	7/32
1/2	0.291	0.270	3/8	1/2	9/16	11/32	0.334	1/4	1/8	0.2520	1/4
9/16	0.332	0.309	27/64	9/16	5/8	25/64	0.379	9/32	9/64	0.2520	1/4
5/8	0.371	0.347	15/32	5/8	3/4	15/32	0.456	5/16	5/32	0.3155	5/16
3/4	0.450	0.425	9/16	3/4	7/8	9/16	0.549	3/8	3/16	0.3780	3/8
7/8	0.530	0.502	21/32	7/8	1	21/32	0.642	7/16	7/32	0.5030	1/2
1	0.609	0.579	3/4	1	1 1/8	3/4	0.734	1/2	1/4	0.5655	9/16

Maximum socket depth does not exceed three-fourths of minimum head height.
Cup points are standard with stock sizes of all set screws. Other points are available.

Headless Slotted Set Screws

Nominal Size	D Body Diameter of Screw	R Radius of Oval Point Screw	I Radius of Headless Crown	J Width of Slot	T Depth of Slot	C Diameter of Cup and Flat Points		P Diameter of Dog Point		Q Height of Dog	
						Max.	Min.	Max.	Min.	Full	Half
5	0.125	0.094	0.125	0.023	0.031	0.067	0.057	0.083	0.078	0.060	0.030
6	0.138	0.109	0.138	0.025	0.035	0.074	0.064	0.092	0.087	0.070	0.035
8	0.164	0.125	0.164	0.029	0.041	0.087	0.076	0.109	0.103	0.080	0.040
10	0.190	0.141	0.190	0.032	0.048	0.102	0.088	0.127	0.120	0.090	0.045
12	0.216	0.156	0.216	0.036	0.054	0.115	0.101	0.144	0.137	0.110	0.055
1/4	0.250	0.188	0.250	0.045	0.063	0.132	0.118	0.156	0.119	0.125	0.063
5/16	0.312	0.234	0.313	0.051	0.078	0.172	0.156	0.203	0.195	0.156	0.078
3/8	0.375	0.281	0.375	0.064	0.094	0.212	0.194	0.250	0.241	0.188	0.094
7/16	0.438	0.328	0.438	0.072	0.109	0.252	0.232	0.297	0.287	0.219	0.109
1/2	0.500	0.375	0.500	0.081	0.125	0.291	0.270	0.344	0.344	0.250	0.125
9/16	0.562	0.422	0.563	0.091	0.141	0.332	0.309	0.391	0.379	0.281	0.140
5/8	0.625	0.469	0.625	0.102	0.156	0.371	0.347	0.469	0.456	0.313	0.156
3/4	0.750	0.563	0.750	0.129	0.188	0.450	0.425	0.563	0.549	0.375	0.188

Cup points are standard with stock sizes of all set screws. Other points available on special order.

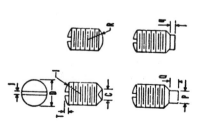

Head Dimensions—(Continued)
Decimal Inches
Flat Head (F) Machine Screws, Tapping Screws, Wood Screws, Stove Bolts

Nominal Size	Diameter of Screw Basic	A Head Diameter			E Height of Head		J Width of Slot		T Depth of Slot		Phillips Recessed D Diameter of Recess		C Depth of Recess
		Max.	Min.	Absol. Min.	Max.	Min.	Max.	Min.	Max.	Min.	Max.	Min.	Max.
2	.086	.172	.156	.147	.051	.040	.031	.023	.023	.015	.102	.089	.063
3	.099	.199	.181	.171	.059	.048	.035	.027	.027	.017	.107	.094	.068
4	.112	.225	.207	.195	.067	.055	.039	.031	.030	.020	.128	.115	.089
5	.125	.252	.232	.220	.075	.062	.043	.035	.034	.022	.154	.141	.086
6	.138	.279	.257	.244	.083	.069	.048	.039	.038	.024	.174	.161	.106
7	.151	.305	.283	.268	.091	.076	.048	.039	.041	.027	.182	.169	.114
8	.164	.332	.308	.292	.100	.084	.054	.045	.045	.029	.189	.176	.121
8											**.204**	**.191**	**.136**
9	.177	.358	.334	.316	.108	.091	.054	.045	.049	.032	.214	.201	.146
10	.190	.385	.359	.340	.116	.098	.060	.049	.053	.034	.204	.191	.136
10											**.258**	**.245**	**.146**
12	.216	.438	.410	.389	.132	.112	.067	.056	.060	.039	.268	.255	.156
12											**.283**	**.270**	**.171**
14	.242	.491	.461	.437	.148	.127	.075	.064	.068	.044	.283	.270	.171
1/4	.250	.507	.477	.452	.153	.131	.075	.064	.070	.046	.283	.270	.171
16	.268	.544	.512	.485	.164	.141	.075	.064	.075	.049	.303	.290	.191
18	.294	.597	.563	.534	.180	.155	.084	.072	.083	.054	.365	.352	.216
5/16	.312	.635	.600	.568	.191	.165	.094	.081	.088	.058	.365	.352	.216
3/8	.375	.762	.722	.685	.230	.200	.094	.081	.106	.070	.393	.380	.245
7/16	.438	.812	.771	.723	.223	.190	.094	.081	.103	.066	.409	.396	.261
1/2	.500	.875	.831	.775	.223	.186	.106	.091	.103	.065	.424	.411	.276

Edges of head may be rounded.
Dimensions shown in bold type apply to Wood Screws only.

Round Head (R) Machine Screws, Tapping Screws, Wood Screws, Stove Bolts

Nominal Size	Diameter of Screw Basic	A Head Diameter		Standard Slotted						Phillips Recessed		
				E Height of Head		J Width of Slot		T Depth of Slot		D Diameter of Recess		C Depth of Recess
		Max.	Min.	Max.	Min.	Max.	Min.	Max.	Min.	Max.	Min.	Max.
2	.036	.162	.146	.069	.059	.031	.023	.048	.037	.100	.087	.053
2										**.114**	**.101**	**.064**
3	.099	.187	.169	.078	.067	.035	.027	.053	.040	.109	.096	.062
3										**.122**	**.109**	**.073**
4	.112	.211	.193	.086	.075	.039	.031	.058	.044	.118	.105	.072
4										**.130**	**.117**	**.083**
5	.125	.236	.217	.095	.083	.043	.035	.063	.047	.154	.141	.074
6	.138	.260	.240	.103	.091	.048	.039	.068	.051	.162	.149	.084
7	.151	.285	.264	.111	.099	.048	.039	.072	.055	.170	.157	.092
8	.164	.309	.287	.120	.107	.054	.045	.077	.058	.178	.165	.101
9	.177	.334	.311	.128	.115	.054	.045	.082	.062			
10	.190	.359	.334	.137	.123	.060	.050	.087	.065	.195	.182	.119
12	.216	.408	.382	.153	.139	.067	.056	.096	.073	.249	.236	.125
14	.242	.457	.429	.170	.155	.075	.064	.106	.080	.265	.252	.142
1/4	.250	.472	.443	.175	.160	.075	.064	.109	.082	.268	.255	.147
16	.268	.506	.476	.187	.171	.075	.064	.115	.087	.281	.268	.159
18	.294	.555	.523	.204	.187	.084	.072	.125	.094	.329	.316	.167
5/16	.312	.590	.557	.216	.198	.084	.072	.132	.099	.308	.295	.187
3/8	.375	.708	.670	.256	.237	.094	.081	.155	.117	.387	.374	.228
7/16	.438	.750	.707	.328	.307	.094	.081	.196	.148	.402	.389	.241
1/2	.500	.813	.766	.355	.332	.106	.091	.211	.159	.416	.403	.256

Dimensions shown in bold type apply to Wood Screws only.

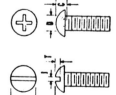

Head Dimensions—(Continued)
Decimal Inches

Fillister Head (P) Machine Screws, Tapping Screws

Nominal Size	Diameter of Screw Basic	A Head Diameter Max.	A Head Diameter Min.	Standard Slotted H Height of Head Max.	H Height of Head Min.	J Width of Slot Max.	J Width of Slot Min.	T Depth of Slot Max.	T Depth of Slot Min.	E Total Height of Head Max.	E Total Height of Head Min.	Phillips Recessed D Diameter of Recess Max.	D Diameter of Recess Min.	C Depth of Recess Max.
2	.086	.140	.124	.062	.053	.031	.023	.037	.025	.083	.066	.104	.091	.059
3	.099	.161	.145	.070	.061	.035	.027	.043	.030	.095	.077	.112	.099	.068
4	.112	.183	.166	.079	.069	.039	.031	.048	.035	.107	.088	.122	.109	.078
5	.125	.205	.187	.088	.078	.043	.035	.054	.040	.120	.100	.148	.135	.067
6	.138	.226	.208	.096	.086	.048	.039	.060	.045	.132	.111	.166	.153	.091
8	.164	.270	.250	.113	.102	.054	.045	.071	.054	.156	.133	.182	.169	.108
10	.190	.313	.292	.130	.118	.060	.050	.083	.064	.180	.156	.199	.186	.124
12	.216	.357	.334	.148	.134	.067	.056	.094	.074	.205	.178	.259	.246	.141
1/4	.250	.414	.389	.170	.155	.075	.064	.109	.087	.237	.207	.281	.268	.161
5/16	.312	.518	.490	.211	.194	.084	.072	.137	.110	.295	.262	.322	.309	.235
3/8	.375	.622	.590	.253	.233	.094	.081	.164	.133	.355	.315	.393	.380	.233
7/16	.438	.625	.589	.265	.242	.094	.081	.170	.135	.368	.321	.413	.400	.259
1/2	.500	.750	.710	.297	.273	.106	.091	.190	.151	.412	.362	.435	.422	.280

Oval Head (O) Machine Screws, Tapping Screws, Wood Screws

Nominal Size of Screw	Diameter of Screw Basic	A Head Diameter Max	A Head Diameter Min	A Head Diameter Ab-solute Min	H Height of Head Max (Standard Slotted)	H Height of Head Min	J Width of Slot Max	J Width of Slot Min	T Depth of Slot Max	T Depth of Slot Min	E Total Height of Head Max	E Total Height of Head Min	D Diameter of Recess Max (Phillips Recessed)	D Diameter of Recess Min	C Depth of Recess Max
2	.086	.172	.156	.147	.051	.040	.031	.023	.045	.037	.080	.063	.112	.099	.069
3	.099	.199	.181	.171	.059	.048	.035	.027	.052	.043	.092	.073	.124	.111	.081
4	.112	.225	.207	.195	.067	.055	.039	.031	.059	.049	.104	.084	.136	.123	.094
5	.125	.252	.232	.220	.075	.062	.043	.035	.067	.055	.116	.095	.158	.145	.085
6	.138	.279	.257	.244	.083	.069	.048	.039	.074	.060	.128	.105	.178	.165	.105
7	.151	.305	.283	.268	.091	.076	.048	.039	.081	.066	.140	.116	.183	.170	.111
8	.164	.332	.308	.292	.100	.084	.054	.045	.088	.072	.152	.126	.192	.179	.119
8													.205	.192	.131
9	**.177**	**.358**	**.334**	**.316**	**.108**	**.091**	**.054**	**.045**	**.095**	**.078**	**.164**	**.137**	**.216**	**.203**	**.144**
10	.190	.385	.359	.340	.116	.098	.060	.050	.103	.084	.176	.148	.209	.196	.137
10													.261	.248	.142
12	.216	.438	.410	.389	.132	.112	.067	.056	.117	.096	.200	.169	.270	.257	.152
12													.283	.270	.165
14	.242	.491	.461	.437	.148	.127	.075	.064	.132	.108	.224	.190	.288	.275	.165
1/4	.250	.507	.477	.452	.153	.131	.075	.064	.136	.112	.232	.197	.290	.277	.173
16	.268	.544	.512	.485	.164	.141	.075	.064	.146	.120	.248	.212	.332	.319	.214
18	.294	.597	.563	.534	.180	.155	.084	.072	.160	.132	.272	.233	.381	.368	.226
5/16	.312	.635	.600	.568	.191	.165	.084	.072	.171	.141	.290	.249	.390	.377	.238
3/8	.375	.762	.722	.685	.230	.200	.094	.081	.206	.170	.347	.300	.410	.397	.257
7/16	.438	.812	.771	.723	.223	.190	.094	.081	.210	.174	.345	.295	.422	.409	.269
1/2	.500	.875	.831	.775	.223	.186	.106	.091	.216	.176	.354	.299	.437	.424	.283

Edges of head may be rounded.
Dimensions shown in bold type apply to Wood Screws only.

Head Dimensions—(Continued)
Decimal Inches
Binding Head (B) Machine Screws
Standard Slotted

Nominal Size	Diameter of Screw Basic	A Head Diameter		E Total Height of Head		F Height of Oval		J Width of Slot		T Depth of Slot		U Diameter of Undercut		X Depth of Undercut	
		Max.	Min.	Max.	Min.	Max.	Min.	Max.	Min.	Max.	Min.	Max.	Min.	Max.	Min.
2	.086	.181	.171	.050	.041	.018	.013	.031	.023	.030	.024	.141	.124	.010	.005
3	.099	.208	.197	.059	.048	.022	.016	.035	.027	.036	.029	.162	.143	.011	.006
4	.112	.235	.223	.068	.056	.025	.018	.039	.031	.042	.034	.184	.161	.012	.007
5	.125	.263	.249	.078	.064	.029	.021	.043	.035	.048	.039	.205	.180	.014	.009
6	.138	.290	.275	.087	.071	.032	.024	.048	.039	.053	.044	.226	.199	.015	.010
8	.164	.344	.326	.105	.087	.039	.029	.054	.045	.065	.054	.269	.236	.017	.012
10	.190	.399	.378	.123	.102	.045	.034	.060	.050	.077	.064	.312	.274	.020	.015
12	.216	.454	.430	.141	.117	.052	.039	.067	.056	.089	.074	.354	.311	.023	.018
1/4	.250	.513	.488	.165	.138	.061	.046	.075	.064	.105	.088	.410	.360	.026	.021
5/16	.312	.641	.609	.209	.174	.077	.059	.084	.072	.134	.112	.513	.450	.032	.027
3/8	.375	.769	.731	.253	.211	.094	.071	.094	.081	.163	.136	.615	.540	.039	.034

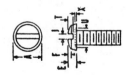

Truss Head (T) Machine Screws, Tapping Screws

Nominal Size	Diameter of Screw Basic	A Head Diameter		Standard Slotted E Height of Head		J Width of Slot		T Depth of Slot		Phillips Recessed D Diameter of Recess		C Depth of Recess
		Max.	Min.	Max.	Min.	Max.	Min.	Max.	Min.	Max.	Min.	Max.
2	.086	.194	.180	.053	.044	.031	.023	.031	.022	.104	.091	.059
3	.099	.226	.211	.061	.051	.035	.027	.036	.026	.110	.097	.066
4	.112	.257	.241	.069	.059	.039	.031	.040	.030	.112	.099	.069
5	.125	.289	.272	.078	.066	.043	.035	.045	.034	.128	.115	.085
6	.138	.321	.303	.086	.074	.048	.039	.050	.037	.158	.145	.084
7	.151	.352	.333	.094	.081	.048	.039	.054	.041	.165	.152	.091
8	.164	.384	.364	.102	.088	.054	.045	.058	.045	.173	.160	.099
10	.190	.448	.425	.118	.103	.060	.050	.068	.053	.188	.175	.115
12	.216	.511	.487	.134	.118	.067	.056	.077	.061	.248	.235	.128
1/4	.250	.573	.546	.150	.133	.075	.064	.087	.070	.263	.250	.143
5/16	.312	.598	.666	.183	.162	.084	.072	.106	.085	.352	.330	.193
3/8	.375	.823	.787	.215	.191	.094	.081	.124	.100	.388	.375	.226

Head Dimensions—(Continued)

Decimal Inches

Pan Head (D) Machine Screws, Tapping Screws

Nominal Size	Diameter of Screw Basic	A Head Diameter Max.	A Head Diameter Min.	Standard Slotted E Height of Slotted Head Max.	E Min.	J Width of Slot Max.	J Min.	T Depth of Slot Max.	T Min.	Phillips Recessed R Height of Recessed Head Max.	R Min.	D Diameter of Recess Max.	D Min.	C Depth of Recess Max.
2	.086	.167	.155	.053	.045	.031	.023	.031	.022	.062	.053	.104	.091	.059
3	.099	.193	.180	.060	.051	.035	.027	.036	.026	.071	.062	.112	.099	.068
4	.112	.219	.205	.068	.058	.039	.031	.040	.030	.080	.070	.122	.109	.078
5	.125	.245	.231	.075	.065	.043	.035	.045	.034	.089	.079	.158	.145	.083
6	.138	.270	.256	.082	.072	.048	.039	.050	.037	.097	.087	.166	.153	.091
7	.151	.296	.281	.089	.079	.048	.039	.054	.041	.106	.096	.176	.163	.100
8	.164	.322	.306	.096	.085	.054	.045	.058	.045	.115	.105	.182	.169	.108
10	.190	.373	.357	.110	.099	.063	.050	.068	.053	.133	.122	.199	.186	.124
12	.216	.425	.407	.125	.112	.067	.056	.077	.061	.151	.139	.259	.246	.141
1/4	.250	.492	.473	.144	.130	.075	.064	.087	.070	.175	.162	.281	.268	.161
5/16	.312	.615	.594	.178	.162	.084	.072	.106	.085	.218	.203	.350	.337	.193
3/8	.375	.740	.716	.212	.195	.094	.081	.130	.113	.261	.244	.393	.380	.233

Round (Button) Head Small Solid Rivets

Nominal	D Diameter of Body Max.	D Min.	A Diameter of Head Max.	A Min.	H Height of Head Max.	H Min.	r Radius of Head Approx.	Length Tolerance Plus	Length Tolerance Minus
3/32	.096	.090	.182	.162	.077	.065	.084	.016	.016
1/8	.127	.121	.235	.215	.100	.088	.111	.016	.016
5/32	.158	.152	.290	.268	.124	.110	.138	.016	.016
3/16	.191	.182	.348	.322	.147	.133	.166	.016	.016
7/32	.222	.213	.405	.379	.172	.158	.195	.016	.016
1/4	.253	.244	.460	.430	.196	.180	.221	.016	.016
9/32	.285	.273	.518	.484	.220	.202	.249	.016	.016
5/16	.317	.305	.572	.538	.243	.225	.276	.016	.016
11/32	.348	.336	.630	.592	.267	.247	.304	.016	.016
3/8	.380	.365	.684	.646	.291	.271	.332	.016	.016
7/16	.443	.428	.798	.754	.339	.317	.387	.016	.016

Countersunk Head Small Solid Rivets

Nominal	D Diameter of Body Max.	D Min.	A Diameter of Head Max.	A Min.	H Height of Head	Length Tolerance Plus	Length Tolerance Minus
3/32	.096	.090	.176	.171	.040	.016	.016
1/8	.127	.121	.235	.227	.053	.016	.016
5/32	.158	.152	.293	.284	.066	.016	.016
3/16	.191	.182	.351	.340	.079	.016	.016
7/32	.222	.213	.413	.400	.094	.016	.016
1/4	.253	.244	.469	.455	.106	.016	.016
9/32	.285	.273	.528	.511	.119	.016	.016
5/16	.317	.305	.588	.569	.133	.016	.016
11/32	.348	.336	.646	.626	.146	.016	.016
3/8	.380	.365	.704	.682	.159	.016	.016
7/16	.443	.428	.823	.797	.186	.016	.016

Flat Head Small Solid Rivets

Nominal	D Diameter of Body Max.	D Min.	A Diameter of Head Max.	A Min.	H Height of Head Max.	H Min.	Length Tolerance Plus	Length Tolerance Minus
3/32	.096	.090	.200	.180	.038	.026	.016	.016
1/8	.127	.121	.260	.240	.048	.036	.016	.016
5/32	.158	.152	.323	.301	.059	.045	.016	.016
3/16	.191	.182	.387	.361	.069	.055	.016	.016
7/32	.222	.213	.453	.427	.080	.065	.016	.016
1/4	.253	.244	.515	.485	.091	.075	.016	.016
9/32	.285	.273	.579	.545	.103	.085	.016	.016
5/16	.317	.305	.641	.607	.113	.095	.016	.016
11/32	.348	.336	.705	.667	.124	.104	.016	.016
3/8	.380	.365	.769	.731	.135	.115	.016	.016
7/16	.443	.428	.896	.852	.157	.135	.016	.016

Head Dimensions—(Continued)
Decimal Inches
Tinners Rivets

Size No.†	D Diameter of Body		A Diameter of Head		H Height of Head			Length	
	Max.	Min.	Max.	Min.	Max.	Min.	Nom.	Max.	Min.
6 oz.	.081	.075	.213	.193	.028	.016	1/8	.135	.115
8 oz.	.091	.085	.225	.205	.036	.024	5/32	.166	.146
10 oz.	.097	.091	.250	.230	.037	.025	11/64	.182	.162
12 oz.	.107	.101	.265	.245	.037	.025	3/16	.198	.178
14 oz.	.111	.105	.275	.255	.038	.026	3/16	.198	.178
1 lb.	.113	.107	.285	.265	.040	.028	13/64	.213	.193
1¼ lb.	.122	.116	.295	.275	.045	.033	7/32	.229	.209
1½ lb.	.132	.126	.316	.294	.046	.034	15/64	.244	.224
1¾ lb.	.136	.130	.331	.309	.049	.035	1/4	.260	.240
2 lb.	.146	.140	.341	.319	.050	.036	17/64	.276	.256
2½ lb.	.150	.144	.311	.289	.069	.055	9/32	.291	.271
3 lb.	.163	.154	.329	.303	.073	.059	5/16	.323	.303
3½ lb.	.168	.159	.348	.322	.074	.060	21/64	.338	.318
4 lb.	.179	.170	.368	.342	.076	.062	11/32	.354	.334
5 lb.	.190	.181	.388	.362	.084	.070	3/8	.385	.365
6 lb.	.206	.197	.419	.393	.090	.076	25/64	.401	.381
7 lb.	.223	.214	.431	.405	.094	.080	13/32	.416	.396
8 lb.	.227	.218	.475	.445	.101	.085	7/16	.448	.428
9 lb.	.241	.232	.490	.460	.103	.087	29/64	.463	.443
10 lb.	.241	.232	.505	.475	.104	.088	15/32	.479	.459
12 lb.	.263	.251	.532	.498	.108	.090	1/2	.510	.490
14 lb.	.288	.276	.577	.543	.113	.095	33/64	.525	.505
16 lb.	.304	.292	.597	.563	.128	.110	17/32	.541	.521
18 lb.	.347	.335	.706	.668	.156	.136	19/32	.603	.583

†Size numbers refer to the approximate weight of 1,000 rivets.

Hexagon and Square Machine Screw and Stove Bolt Nuts

Nominal Size or Basic Major Diameter of Thread		F Width Across Flats		G Width Across Corners				H Thickness		
		Max. (Basic)	Min.	Square Max.	Square Min.	Hex. Max.	Hex. Min.	Nom.	Max.	Min.
0	.0600	5/32 .1562	.150	.221	.206	.180	.171	3/64	.050	.043
1	.0730	5/32 .1562	.150	.221	.206	.180	.171	3/64	.050	.043
2	.0860	3/16 .1875	.180	.265	.247	.217	.205	1/16	.066	.057
3	.0990	3/16 .1875	.180	.265	.247	.217	.205	1/16	.066	.057
4	.1120	1/4 .2500	.241	.354	.331	.289	.275	3/32	.098	.087
5	.1250	5/16 .3125	.302	.442	.415	.361	.344	7/64	.114	.102
6	.1380	5/16 .3125	.302	.442	.415	.361	.344	7/64	.114	.102
8	.1640	11/32 .3438	.332	.486	.456	.397	.378	1/8	.130	.117
10	.1900	3/8 .3750	.362	.530	.497	.433	.413	1/8	.130	.117
12	.2160	7/16 .4375	.423	.619	.581	.505	.482	5/32	.161	.148
1/4	.2500	7/16 .4375	.423	.619	.581	.505	.482	3/16	.193	.178
5/16	.3125	9/16 .5625	.545	.795	.748	.650	.621	7/32	.225	.208
3/8	.3750	5/8 .6250	.607	.884	.833	.722	.692	1/4	.257	.239

Finished Hexagon Castellated (Castle) Nuts

Nominal Size or Basic Major Diameter of Thread		F Width Across Flats		G Width Across Corners		H Thickness			Nominal Height of Flats	S Slot Width	T Slot Depth	Radius of Fillet (±.010)	Dia. of Cylindrical Part
		Max. (Basic)	Min.	Max.	Min.	Nom.	Max.	Min.					
1/4	.2500	7/16 .4375	.428	.505	.488	9/32	.288	.274	3/16	.078	.094	3/32	.371
5/16	.3125	1/2 .5000	.489	.577	.557	21/64	.336	.320	15/64	.094	.094	3/32	.425
3/8	.3750	9/16 .5625	.551	.650	.628	13/32	.415	.398	9/32	.125	.125	3/32	.478
7/16	.4375	11/16 .6875	.675	.794	.768	29/64	.463	.444	19/64	.125	.156	3/32	.582
1/2	.5000	3/4 .7500	.736	.866	.840	9/16	.573	.552	13/32	.156	.156	1/8	.637
9/16	.5625	7/8 .8750	.861	1.010	.982	39/64	.621	.598	27/64	.156	.188	5/32	.744
5/8	.6250	15/16 .9375	.922	1.083	1.051	23/32	.731	.706	1/2	.188	.219	5/32	.797
3/4	.7500	1 1/8 1.1250	1.088	1.299	1.240	13/16	.827	.798	9/16	.188	.250	3/16	.941
7/8	.8750	1 5/16 1.3125	1.269	1.516	1.447	29/32	.922	.890	21/32	.188	.250	3/16	1.097
1	1.0000	1 1/2 1.5000	1.450	1.732	1.653	1	1.018	.982	23/32	.250	.281	3/16	1.254
1 1/8	1.1250	1 11/16 1.6875	1.631	1.949	1.859	1 5/32	1.176	1.136	13/16	.250	.344	1/4	1.411
1 1/4	1.2500	1 7/8 1.8750	1.812	2.165	2.066	1 1/4	1.272	1.228	7/8	.312	.375	1/4	1.570
1 3/8	1.3750	2 1/16 2.0625	1.994	2.382	2.273	1 13/32	1.399	1.351	1	.312	.375	1/4	1.726
1 1/2	1.5000	2 1/4 2.2500	2.175	2.598	2.480	1 1/2	1.526	1.474	1 1/16	.375	.438	1/4	1.881

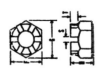

Head Dimensions—*(Continued)*
Decimal Inches

Finished Hexagon Nuts (Full and Jam)

Nominal Size or Basic Major Diameter of Thread		F Width Across Flats			G Width Across Corners		H Thickness Full Nuts			h Thickness Jam Nuts		
		Max.	(Basic)	Min.	Max.	Min.	Nom.	Max.	Min.	Nom.	Max.	Min.
1/4	.2500	7/16	.4375	.428	.505	.488	7/32	.226	.212	5/32	.163	.150
5/16	.3125	1/2	.5000	.489	.577	.557	17/64	.273	.258	3/16	.195	.180
3/8	.3750	9/16	.5625	.551	.650	.628	21/64	.337	.320	7/32	.227	.210
7/16	.4375	11/16	.6875	.675	.794	.768	3/8	.385	.365	1/4	.260	.240
1/2	.5000	3/4	.7500	.736	.866	.840	7/16	.448	.427	5/16	.323	.302
9/16	.5625	7/8	.8750	.861	1.010	.982	31/64	.496	.473	5/16	.324	.301
5/8	.6250	15/16	.9375	.922	1.083	1.051	35/64	.559	.535	3/8	.387	.363
3/4	.7500	1 1/8	1.1250	1.088	1.299	1.240	41/64	.665	.617	27/64	.446	.398
7/8	.8750	1 5/16	1.3125	1.269	1.516	1.447	3/4	.776	.724	31/64	.510	.458
1	1.0000	1 1/2	1.5000	1.450	1.732	1.653	55/64	.887	.831	35/64	.575	.519
1 1/8	1.1250	1 11/16	1.6875	1.631	1.949	1.859	31/32	.999	.939	39/64	.639	.579
1 1/4	1.2500	1 7/8	1.8750	1.812	2.165	2.066	1 1/16	1.094	1.030	23/32	.751	.687
1 3/8	1.3750	2 1/16	2.0625	1.994	2.382	2.273	1 11/64	1.206	1.138	25/32	.815	.747
1 1/2	1.5000	2 1/4	2.2500	2.175	2.598	2.480	1 9/32	1.317	1.245	27/32	.880	.808

Heavy Semi-Finished Hexagon Nuts (Full and Jam)

Nominal Size or Basic Major Diameter of Thread		F Width Across Flats			G Width Across Corners		H Thickness (Heavy Nuts)			h Thickness (Heavy Jam Nuts)		
		Max.	(Basic)	Min.	Max.	Min.	Nom.	Max.	Min.	Nom.	Max.	Min.
1/4	.2500	1/2	.5000	.488	.577	.556	15/64	.250	.218	11/64	.188	.156
5/16	.3125	9/16	.5625	.546	.650	.622	19/64	.314	.280	13/64	.220	.186
3/8	.3750	11/16	.6875	.669	.794	.763	23/64	.377	.341	15/64	.252	.216
7/16	.4375	3/4	.7500	.728	.866	.830	27/64	.441	.403	17/64	.285	.247
1/2	.5000	7/8	.8750	.850	1.010	.969	31/64	.504	.464	19/64	.317	.277
9/16	.5625	15/16	.9375	.909	1.083	1.037	35/64	.568	.526	21/64	.349	.307
5/8	.6250	1 1/16	1.0625	1.031	1.227	1.175	39/64	.631	.587	23/64	.381	.337
3/4	.7500	1 1/4	1.2500	1.212	1.443	1.382	47/64	.758	.710	27/64	.446	.398
7/8	.8750	1 7/16	1.4375	1.394	1.660	1.589	55/64	.885	.833	31/64	.510	.458
1	1.0000	1 5/8	1.6250	1.575	1.876	1.796	63/64	1.012	.956	35/64	.575	.519
1 1/8	1.1250	1 13/16	1.8125	1.756	2.093	2.002	1 7/64	1.139	1.079	39/64	.639	.579
1 1/4	1.2500	2	2.0000	1.938	2.309	2.209	1 7/32	1.251	1.187	23/32	.751	.687
1 3/8	1.3750	2 3/16	2.1875	2.119	2.526	2.416	1 11/32	1.378	1.310	25/32	.815	.747
1 1/2	1.5000	2 3/8	2.3750	2.300	2.742	2.622	1 15/32	1.505	1.433	27/32	.879	.808
1 5/8	1.6250	2 9/16	2.5625	2.481	2.959	2.828	1 19/32	1.632	1.556	29/32	.944	.868
1 3/4	1.7500	2 3/4	2.7500	2.662	3.175	3.035	1 23/32	1.759	1.679	31/32	1.009	.929
1 7/8	1.8750	2 15/16	2.9375	2.844	3.392	3.242	1 27/32	1.886	1.802	1 1/32	1.073	.989
2	2.0000	3 1/8	3.1250	3.025	3.608	3.449	1 31/32	2.013	1.925	1 3/32	1.138	1.050
2 1/4	2.2500	3 1/2	3.5000	3.388	4.041	3.862	2 13/64	2.251	2.155	1 13/64	1.251	1.155
2 1/2	2.5000	3 7/8	3.8750	3.750	4.474	4.275	2 29/64	2.505	2.401	1 29/64	1.505	1.401

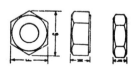

Set Screw Dimensions—Square Head Cup Point

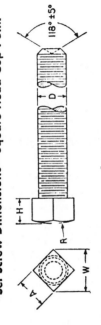

118° ±5°

Nominal Size or Basic Major Diameter of Thread	D Minimum Body Diameter	A Width Across Flats Maximum	A Width Across Flats Minimum	W Width Across Corners Minimum	H Height of Head Nominal	H Height of Head Maximum	H Height of Head Minimum	R Radius of Head
1/4	0.2500	0.2428	0.2500	0.241	0.331	3/16	0.196	0.178	5/8
5/16	0.3125	0.3043	0.3125	0.302	0.415	15/64	0.245	0.224	25/32
3/8	0.3750	0.3660	0.3750	0.362	0.497	9/32	0.293	0.270	15/16
7/16	0.4375	0.4277	0.4375	0.423	0.581	21/64	0.341	0.315	1 3/32
1/2	0.5000	0.4896	0.5000	0.484	0.665	3/8	0.389	0.361	1 1/4
9/16	0.5625	0.5513	0.5625	0.545	0.748	27/64	0.437	0.407	1 13/32
5/8	0.6250	0.6132	0.6250	0.606	0.833	15/32	0.485	0.452	1 9/16
3/4	0.7500	0.7372	0.7500	0.729	1.001	9/16	0.582	0.544	1 7/8
7/8	0.8750	0.8610	0.8750	0.852	1.170	21/32	0.678	0.635	2 3/16
1	1.0000	0.9848	1.0000	0.974	1.337	3/4	0.774	0.726	2 1/2
1 1/8	1.1250	1.1080	1.1250	1.096	1.505	27/32	0.870	0.817	2 13/16
1 1/4	1.2500	1.2330	1.2500	1.219	1.674	15/16	0.966	0.908	3 1/8
1 3/8	1.3750	1.3548	1.3750	1.342	1.843	1 1/32	1.063	1.000	3 7/16
1 1/2	1.5000	1.4798	1.5000	1.464	2.010	1 1/8	1.159	1.091	3 3/4

Inches

15
TERMINOLOGY

GENERAL TERMS

The alphabetical listing below includes most of the terms and phrases in general use in metallurgy and in fabrication of materials.

Age hardening. Hardening by aging, usually after rapid cooling or cold working. *See* Aging.

Aging. In a metal or alloy, a change in properties that generally occurs slowly at room temperature and more rapidly at higher temperatures. *See* Age hardening, Artificial aging, Natural aging, Precipitation hardening, Precipitation heat treatment, Quench aging, and Strain aging.

Air-hardening steel. A steel containing sufficient carbon and other alloying elements to harden fully during cooling in air or other gaseous media, from a temperature above its transformation range. The term should be restricted to steels that are capable of being hardened by cooling in air in fairly large sections, about 2 inches or more in diameter. *See* Self-hardening steel.

Alclad. Composite sheet produced by bonding either corrosion-resistant aluminum alloy or aluminum of high purity to base metal of structurally stronger aluminum alloy.

Alkali metal. A metal in group IA of the periodic system (namely, lithium, sodium, potassium, rubidium, cesium, francium). These metals form strong alkaline hydroxides; hence, the name.

Alkaline earth metal. A metal in group IIA of the periodic system (including beryllium, magnesium, calcium, strontium, barium, and radium), so named because the oxides of calcium, strontium, and barium were found by early chemists to be alkaline in reaction.

Allotropy. The reversible phenomenon by which certain metals may exist in more than one crystal structure. If not reversible, the phenomenon is termed *polymorphism.*

Alloy. A substance having metallic properties and being composed of two or more chemical elements, of which at least one is an elemental metal.

Alpha ferrite. See Ferrite.

Alpha iron. The body-centered cubic form of pure iron, stable below $1670°$F.

Alumel. A nickel-base alloy containing about 2.5% Mn, 2% Al, and 1% Si, used chiefly as a component of pyrometric thermocouples.

Aluminizing. Forming an aluminum or aluminum alloy coating on a metal by hot dipping, hot spraying, or diffusion.

Amorphous. Not having a crystal structure; noncrystalline.

Annealing. Heating to and holding at a suitable temperature and then cooling at a suitable rate, for such purposes as reducing hardness, improving machinability, facilitating cold working, producing a desired microstructure, or obtaining desired mechanical, physical, or other properties. When applicable, the following more specific terms should be used: black annealing, blue annealing, box annealing, bright annealing, flame annealing, full annealing, graphitizing, intermediate annealing, isothermal annealing, malleableizing, process annealing, quench annealing, recrystallization annealing, and spheroidizing.

When applied to ferrous alloys, the term *annealing*, without qualification, implies full annealing.

When applied to nonferrous alloys, the term implies a heat treatment designed to soften a cold-worked structure by recrystallization or subsequent grain growth or to soften an age hardened alloy by causing a nearly complete precipitation of the second phase in relatively coarse form.

Any process of annealing will usually reduce stresses, but if the treatment is applied for the sole purpose of such relief, it should be designed *stress relieving.*

Anode. The electrode by which electrons leave (current enters) an operating system such as a battery, an electrolytic cell, an X-ray tube, or a vacuum tube. In the case of the battery, the anode is negative; in the other three, positive. In a battery or electrolytic cell, the anode is the electrode at which oxidation occurs. *Contrast with* Cathode.

Anode effect. The effect produced by polarization of the anode in the electrolysis of fused salts. It is characterized by a sudden increase in voltage and a corresponding decrease in current due to the virtual separation of the anode from the electrolyte by a gas film.

Anodizing. Forming a conversion coating on a metal surface by anodic oxidation; most frequently applied to aluminum.

Arc brazing. Brazing with an electric arc, usually with two nonconsumable electrodes.

Arc cutting. Metal cutting with an arc between an electrode and the metal itself. The terms *carbon-arc cutting* and *metal-arc cutting* refer, respectively, to the use of a carbon or metal electrode.

Arc welding. Welding with an electric arc-welding electrode. *See* Electrode.

Artificial aging. Aging above room temperature. *See* Aging; Precipitation heat treatment. *Compare with* Natural aging.

Atomic-hydrogen welding. Arc welding with heat from an arc between two tungsten or other suitable electrodes in a hydrogen atmosphere. The use of pressure and filler metal is optional.

Austempering. Quenching a ferrous alloy from a temperature above the transformation range, in a medium having a rate of heat abstraction high enough to prevent the formation of high-temperature transformation products, and then holding the alloy, until transformation is complete, at a temperature below that of pearlite formation and above that of martensite formation.

Austenite. A solid solution of one or more elements in face-centered cubic iron. Unless otherwise designated (such as nickel austenite), the solute is generally assumed to be carbon.

Austenitic steel. A alloy steel whose structure is normally austenitic at room temperature.

Austenitizing. Forming austenite by heating a ferrous alloy into the transformation range (partial austenitizing) or above the transformation range (complete austenitizing).

Backhand welding. Welding in which the back of the principal hand (torch or electrode hand) of the welder faces the direction of travel. It has special significane in gas welding because it provides postheating.

Backstep sequence. A longitudinal welding sequence in which the direction of general progress is opposite to that of welding the individual increments.

Bainite. A decomposition product of austenite consisting of an aggregate of ferrite and carbide. In general, it forms at temperatures lower than those where very fine pearlite forms and higher than that where martensite begins to form on cooling. Its appearance is feathery if formed in the upper part of the temperature range; acicular, resembling tempered martensite, if formed in the lower part.

Bead weld. A weld composed of one or more string or weave beads deposited on an unbroken surface.

Bearing load. A compressive load supported by a member, usually a tube or collar along a line where contact is made with a pin, rivet, axle, or shaft.

Bearing strength. The maximum bearing load at failure divided by the effective bearing area. In a pinned or riveted joint, the effective area is calculated as the product of the diameter of the hole and the thickness of the bearing member.

Bending moment. The algebraic sum of the couples or the moments of the external forces, or both, to the left or right of any section on a member subjected to bending by couples or transverse forces, or both.

Beveling. See Chamfering.

Biaxial stress. A state of stress in which only one of the principal stresses is zero, the other two usually being in tension.

Billet. A solid, semifinished round or square product that has been hot worked by forging, rolling, or extrusion. An iron or steel billet has a minimum width or thickness of $1\frac{1}{2}$ inches and the cross-sectional area varies from $2\frac{1}{4}$ to 36 square inches. For nonferrous metals, it may also be a casting suitable for finished or semifinished rolling or for extrusion. *See* Extrusion billet.

Black light. Electronmagnetic radiation not visible to the human eye. The portion of the spectrum generally used in fluorescent inspection falls in the ultraviolet region between 3300 to 4000 angstrom units, with the peak at 3650 angstrom units.

Blank nitriding. Simulating the nitriding operation without introducing nitrogen. This is usually accomplished by using an inert material in place of the nitriding agent or by applying a suitable protective coating to the ferrous alloy.

Block brazing. Brazing with heat from hot blocks.

Bottom drill. A flat-ended twist drill used to convert a cone at the bottom of a drilled hole into a cylinder.

Bottoming tap. A tap with a chamfer of 1 to $1\frac{1}{2}$ threads in length.

Brass. An alloy consisting mainly of copper (over 50%) and zinc, to which smaller amounts of other elements may be added.

Braze welding. Welding in which a groove, fillet, plug, or slot weld is made, using a nonferrous filler metal having a melting point lower than that of the base metal but higher than $800°F$. The filler metal is not distributed by capillarity.

Brazing. Joining metals by following a thin layer, capillary thickness, of nonferrous filler metal into the space between them. Bonding results from the intimate contact produced by the dissolution of a small amount of base metal in the molten filler metal, without fusion of the base metal. Sometimes the filler metal is put in place as a thin solid sheet or as a clad layer, and the composite is heated as in furnace brazing. The term *brazing* is used where the temperature exceeds some arbitrary value, such as $800°F$; the term *soldering* is used for temperatures lower than the arbitrary value.

Brazing alloy. See Brazing filler metal.

Brazing filler metal. A nonferrous filler metal used in brazing and braze welding.

Brazing sheet. Brazing filler metal in sheet form a flat-rolled metal clad with brazing filler metal on one or both sides.

Breaking stress. See Fracture stress (1).

Brinell hardness test. A test for determining the hardness of a material by forcing a hard steel or carbide ball of specified diameter into it under a specified load. The result is expressed as the Brinell hardness number, which is the value obtained by dividing the applied load in kilograms by the surface area of the resulting impression in square millimeters.

Brittle crack propagation. A very sudden propagation of a crack with the absorption of no energy except that stored elastically in the body. Microscopic examination may reveal some deformation even though it is not noticeable to the unaided eye.

Brittle fracture. Fracture with little or no plastic deformation.

Brittleness. The quality of a material that leads to crack propagation without appreciable plastic deformation.

Broach. A bar-shaped cutting tool provided with a series of cutting edges or teeth that increase in size or change in shape from the starting to finishing end. The tool cuts in the axial direction when pushed or pulled and is used to shape either holes or outside surfaces.

Bronze. A cooper-rich copper-tin alloy with or without small proportions of elements such as zinc and phosphorus. By extension, certain copper-base alloys containing considerably less tin than other alloying elements, such as manganese bronze (copper-zinc plus manganese, tin, and iron) and leaded tin bronze (copper-lead plus tin and sometimes zinc). Also, certain other essentially binary copper-base alloys containing no tin, such as aluminum bronze (copper-aluminum), silicon bronze (copper-silicon), and beryllium bronze (copper-beryllium). Also, trade designations for certain specific copper-base alloys that are actually brasses, such as architectural bronze (57% Cu, 40% Zn, 3% Ph) and commercial bronze (90% Cu, 10% Zn).

Brush plating. Plating with a concentrated solution or gel held in or fed to an absorbing medium, pad, or brush carrying the anode (usually insoluble). The brush is moved back and forth over the area of the cathode to be plated.

Buckling. Producing a bulge, bend, bow, kink, or other wavy condition in sheets or plates by compressive stresses.

Butt seam welding. See Seam welding.

Butt welding. Welding a butt joint.

Carat. A unit weight of diamond (abbreviated *c*); the international metric carat (abbreviated *mc*) is 200 mg.

Carbide. A compound of carbon with one or more metallic elements.

Carbide tools. Cutting tools, made of tungsten carbide, titanium carbide, tantalum carbide, or combinations of these, in a matrix of cobalt or nickel, having sufficient wear resistance and heat resistance to permit high machine speeds.

Carbon-arc cutting. Metal cutting by melting with the heat of an arc between a carbon electrode and the base metal.

Carbon-arc welding. Welding in which an arc is maintained between a nonconsumable carbon electrode and the work.

Carbonitriding. Introducing carbon and nitrogen into a solid ferrous alloy by holding above Ac_1 in an atmosphere that contains suitable gases such as hydrocarbons, carbon monoxide, and ammonia. The carbonitrided alloy is usually quench-hardened.

Carbonization. Conversion of a substance into elemental carbon. (Should not be confused with *carburization.*)

Carbon steel. Steel containing carbon up to about 2% and only residual quantities of other elements except those added for deoxidation, with silicon usually limited to 0.60% and manganese to about 1.65%. Also termed *plain carbon steel, ordinary steel,* and *straight carbon steel.*

Carburizing. Introducing carbon into a solid ferrous alloy by holding above Ac_1 in contact with a suitable carbonaceous material, which may be a solid, liquid, or gas. The carburized alloy is usually quench-hardened.

Case hardening. Hardening a ferrous alloy so that the outer portion, or case, is made substantially harder than the inner portion, or core. Typical processes used for case hardening are carburizing, cyaniding, carbonitriding, nitriding, induction hardening, and flame hardening.

Cast-alloy tool. A cutting tool made by casting a cobalt-base alloy and used at machining speeds between those for high-speed steels and sintered carbides.

Casting. (1) An object at or near finished shape, obtained by solidification of a substance in a mold. (2) Pouring molten metal into a mold to produce an object of desired shape.

Casting copper. Fire-refined, tough pitch copper usually cast from melted secondary metal into ingot and ingot bars only, and used for making foundry castings but not wrought products.

Casting shrinkage. (1) *Liquid shrinkage:* the reduction in volume of liquid metal as it cools to the liquidus. (2) *Solidification shrinkage:* the reduction in volume of metal from the beginning to ending of solidification. (3) *Solid shrinkage:* the reduction in volume of metal from the solidus to room temperature. (4) *Total shrinkage:* the sum of the shrinkage in parts (1), (2), and (3).

Casting strains. Strains in a casting caused by casting stresses that develop as the casting cools.

Casting stresses. Stresses set up in a casting because of geometry and casting shrinkage.

Cast iron. An iron containing carbon in excess of the solubility in the austenite that exists in the alloy at the eutectic temperature. For the various forms (gray cast iron, white cast iron, malleable cast iron, and nodular cast iron), the word *cast* is often left out, resulting in "gray iron," "white iron," "malleable iron," and "nodular iron," respectively.

Cast steel. Steel in the form of castings.

Cast structure. The internal physical structure of a casting evidence by shape and orientation of crystals and segregation of impurities.

Catalyst. A substance capable of changing the rate of a reaction without itself undergoing any net change.

Cathode. The electrode where electrons enter (current leaves) an operating system such as a battery, an electrolytic cell, an X-ray tube, or a vacuum tube. In the first of these, it is positive; in the other three, negative. In a battery or electrolytic cell, it is the electrode where reduction occurs. *Contrast with* Anode.

Cathode compartment. In an electrolytic cell, the enclosure formed by a diaphragm around the cathode.

Cathode copper. Copper deposited at the cathode in electrolytic refining.

Cathode-ray tube. A special form of vacuum tube in which a focused beam of electrons is caused to strike a surface coated with a phosphor. This beam is deflected so that it traces an orthogonal presentation of two separate signals; a third independent signal may be presented as a variation of the intensity of the electron beam, and in turn, the fluorescent intensity.

Caustic dip. A strongly alkaline solution into which metal is immersed for etching, neutralizing acid, or removing organic materials such as grease or paints.

Cavitation. The formation and instantaneous collapse of innumerable tiny voids or cavities within a liquid subjected to rapid and intense pressure changes. Cavitation produced by ultrasonic radiation is sometimes used to give violent localized agitation. That caused by severe turbulent flow often leads to cavitation damage.

Center drilling. Drilling a conical hole (pit) in one end of a workpiece.

Centerless grinding. Grinding the outside or inside of a workpiece mounted on rollers rather than on centers. The workpiece may be in the form of a cylinder or the frustum of a cone.

Centrifugal casting. A casting made by pouring metal into a mold that is rotated or revolved.

Cermet (ceramal). A body consisting of ceramic particles bonded with a metal.

Chafting fatigue. Fatigue initiated in a surface damaged by rubbing against another body. *See* Fretting.

Chamfer. (1) A beveled surface to elimiate an otherwise sharp corner. (2) A relieved angular cutting edge at a tooth corner.

Chamfer angle. (1) The angle between a referenced surface and the bevel. (2) On a milling cutter, the angle between a beveled surface and the axis of the cutter.

Chamfering. Making a sloping surface on the edge of a member. Also called *beveling*.

Charpy test. A pendulum-type single-blow impact test in which the specimen, usually notched, is supported at both ends as a simple beam and broken by a falling pendulum. The energy absorbed, as determined by the subsequent rise of the pendulum, is a measure of impact strength or notch toughness.

Chromate treatment. A treatment of metal in a solution of a hexavalent chromium compound to produce a conversion coating consisting of trivalent and hexavalent chromium compounds.

Chromating. Performing a chromate treatment.

Chromel. (1) A 90 Ni-10 Cr alloy used in thermocouples. (2) A series of nickel-chromium alloys, some with iron, used for heat-resistant applications.

Clay. An earthy or stony mineral aggregate consisting essentially of hydrous silicates of alumina, plastic when sufficiently pulverized and wetted, rigid when dry, and vitreous when fired at a sufficiently high temperature. Clay minerals most commonly used in the foundry are montmorillonites and kaolinites.

Cleavage. The splitting (fracture) of a crystal on a crystallographic plane of low index.

Cleavage fracture. A fracture, usually of a polycrystalline metal, in which most of the grains have failed by cleavage, resulting in bright reflecting facets. It is one type of crystalline fracture. *Contrast with* Shear fracture.

Coalescence. The union of particles of a dispersed phase into larger units, usually effected at temperatures below the fusion point.

Coarsening. *See* Grain growth.

Coating abrasive. An abrasive product. Sandpaper is an example in which a layer of abrasive particles is firmly attached to a paper, cloth, or fiber backing by means of glue or synthetic-resin adhesive.

Cobalt-60. A radioisotope with a half-life of 5.2 years and dominant-characteristic gamma-radiation energies of 1.17 and 1.33 MeV. It is used as a gamma-radiation source in industrial radiography and in therapy.

Coefficient of elasticity. *See* Modulus of elasticity.

Cohesion. Force of attraction between the molecules (or atoms) within a single phase. *Contrast with* Adhesion.

Cohesive strength. (1) The hypothetical stress in an unnotched bar causing tensile fracture without plastic deformation. (2) The stress corresponding to the forces between atoms. (3) Same as technical cohesive strength or disruptive strength.

Coil weld. A butt weld joining the ends of two metal sheets to make a continuous strip for coiling.

Coin silver. An alloy containing 90% silver, with copper being the usual alloying element.

Cold treatment. Cooling to a low temperature, often near $-100°F$, for the purpose of obtaining desired conditions or properties, such as dimensional or structural stability.

Cold welding. Solid-phase welding in which pressure, without pressure, without added heat, is used to cause interface movements that bring the atoms of the faying surfaces close enough together that a weld ensues.

Cold work. Permanent strain produced by an external force in a metal below its recrystallization temperature.

Cold working. Deforming metal plastically at a temperature lower than the recrystallization temperature.

Compound die. Any die so designed that it performs more than one operation on a part with one stroke of the press, such as blanking and piercing where all functions are performed simultaneously within the confines of the particular blank size being worked.

Compressive strength. The maximum compressive stress that a material is capable of developing, based on original area of cross section. In the case of a material that fails in compression by a shattering fracture, the compressive strength has a very definite value. In the case of materials that do not fail in compression by a shattering fracture, the value obtained for compressive strength is an arbitrary value depending upon the degree of distortion that is regarded as indicating complete failure of the material.

Continuous casting. A casting technique in which an ingot, billet, tube, or other shape is continuously solidified while it is being poured, so that its length is not determined by mold dimensions.

Continuous precipitation. Precipitation from a supersaturated solid solution accompanied by a gradual change of lattice parameter of the matrix with aging time. It is characteristic of the alloys that produce uniform precipitate throughout the grains. *See* Discontinuous precipitation.

Continuous weld. A weld extending continuously from one end of a joint to the other; where the joint is essentially circular, completely around the joint. *Contrast with* Intermittent weld.

Contour forming. See Stretch forming, Tangent bending, Wiper forming.

Contour machining. Machining of irregular surfaces, such as those generated in tracer turning, tracer boring, and tracer milling.

Contour milling. Milling of irregular surfaces. *See* Tracer milling.

Conversion coating. A coating consisting of a compound of the surface metal, produced by chemical or electromechanical treatments of the metal. (Examples are chromate coatings on zinc, cadmium, magesium, and aluminum; oxides or phosphate coatings on steel.)

Copper brazing. Brazing with copper as the filler metal.

Corona. In spot welding, an area sometimes surrounding the nugget at the faying surfaces, contributing slightly to overall bond strength.

Corrosion. The deterioration of a metal by chemical or electromechemical reaction with its environment.

Corrosion embrittlement. The severe loss of ductility of a metal resulting from corrosive attack, usually intergranular and often not visually apparent.

Corrosion fatigue. Effect of the application of repeated or fluctuating stresses in a corrosive environment characterized by shorter life than would be encountered as a result of either the repeated or fluctuating stresses alone or the corrosive environment alone.

Corundum. Natural abrasive of the aluminum oxide type that has higher purity than emery.

Coupon. A piece of metal from which a test specimen is to be prepared—often an extra piece as on a casting or forging.

Covalent bond. A bond between two or more atoms resulting from the completion of shells by the sharing of electrons.

Creep. Time-dependent strain occurring under stress. The creep strain occurring at a diminishing rate is called *primary creep*; that occurring at a minimum and almost constant rate, *secondary creep*; that occurring at an accelerating rate, *tertiary creep*.

Creep limit. (1) The maximum stress that will cause less than a specified quantity of creep in a given time. (2) The maximum nominal stress under which the creep strain rage decreases continuously with time under constant load and at constant temperature. Sometimes used synonymously with Creep strength.

Creep recovery. Time-dependent strain after release of load in a creep test.

Creep strength. (1) The constant nominal stress that will cause a specified quantity of creep in a given time at constant temperature. (2) The constant nominal stress that will cause a specified creep rate at constant temperature.

Crevice corrosion. A type of concentration-cell corrosion; corrosion of a metal that is caused by the concentration of dissolved salts, metal ions, oxygen or other gases, and such, in crevices or pockets remote from the principal fluid stream, with a resultant building-up of differential cells, which ultimately causes deep pitting.

Critical point. (1) The temperature or pressure at which a change in crystal structure, phase, or physical properties occurs. *See* Transformation temperature. (2) In an equilibrium diagram, that specific value of composition, temperature, and pressure, or combinations thereof, at which the phases of a heterogeneous system are in equilibrium.

Critical temperature. (1) Synonymous with critical point if the pressure is constant. (2) The temperature above which the vapor phase cannot be condensed to liquid by an increase in pressure.

Crystal. A solid composed of atoms, ions, or molecules arranged in a pattern which is repetitive in three dimensions.

Crystalline fracture. A fracture of a polycrystalline metal characterized by a grainy appearance.

Crystallization. The separation, usually from a liquid phase on cooling, of a solid crystalline phase.

Cyanide copper. Copper electrodeposited from an alkali-cyanide solution containing a complex ion made up of univalent copper and the cyanide radical; also, the solution itself.

Cyaniding. Introducing carbon and nitrogen into a solid ferrous alloy by holding above Ac_1 in contact with molten cyanide of suitable composition. The cyanided alloy is usually quench-hardened.

dc casting. *See* Direct-chill casting.

Decalescence. A phenomenon, associated with the transformation of alpha iron to gamma iron on the heating (superheating) or iron or steel, revealed by the darkening of the metal surface owing to the sudden decrease in temperature caused by the fast absorption of the latent heat of transformation.

Decarburization. The loss of carbon from the surface of a ferrous alloy as a result of heating in a medium that reacts with the carbon at the surface.

Decay curve. A graphic presentation of the manner in which a quantity decays with time or, rarely, with distance through matter; usually refers to radioactive decay or decay of electrical and acoustical signals.

Degreasing. Removing oil or grease from a surface. *See* Vapor degreasing.

Deionization. Removal of ions from solution by chemical means.

Dendrite. A crystal that has a treelike branching pattern, being most evident in cast metals slowly cooled through the solidification range.

Density ratio. Powered metal. The ratio of the determined density of a compact to the absolute density of metal of the same composition, usually expressed as a percentage.

Deoxidized copper. Copper from which cuprous oxide has been removed by adding a deoxider, such as phosphorus, to the molten bath.

Deoxidizer. A substance that can be added to molten metal to remove either free or combined oxygen.

Deoxidizing. (1) The removal of oxygen from molten metals by use of suitable deoxidizers. (2) Sometimes refers to the removal of undesirable elements other than oxygen by the introduction of elements or compounds that readily react with them. (3) In metal finishing, the removal of oxide films from metal surfaces by chemical or electronchemical reaction.

Depolarization. Reduction of polarization by changing the electrode film.

Depolarizer. A substance that produces depolarization.

Dezincification. Corrosion of some copper-zinc alloys involving loss of zinc and the formation of a spongy porous copper.

Diamond pyramid hardness test. An indentation hardness test employing a $136°$ diamond pyramid indenter and variable loads enabling the use of one hardness scale for all ranges of hardness from very soft lead to tungsten carbide.

Diamond tool. A diamond, shaped or formed to the contour of a single-pointed cutting tool, for use in the precision machining of nonferrous or nonmetallic materials.

Dichromate treatment. A chromate conversion coating produced on magnesium alloys in a boiling solution of sodium dichromate.

Die. (1) Various tools used to impart shape to material primarily because of the shape of the tool itself. Examples are blanking dies, cutting dies, drawing dies, forging dies, punching dies, and threading dies. (2) Powdered metal. The part or parts making up the confining form into which a powder is pressed. The parts of the die may include some or all of the following: die body, punches, core rods. Synonym: "mold."

Die block. The tool steel block into which the desired impressions are machined and from which forgings are produced.

Die body. Powdered metal. The stationary or fixed part of a die.

Die casting. (1) A casting made in a die. (2) A casting process where molten metal is forced under high pressure into the cavity of a metal mold.

Die welding. Forge welding between dies.

Dilatometer. An instrument for measuring the expansion or contraction in a metal resulting from changes in such factors as temperature or allotropy.

Dip brazing. Brazing by immersion in a molten salt or metal bath. Where a metal bath is employed, it may provide the filler metal.

Direct-chill (dc) casting. A continuous method of making ingots or billets for sheet or extrusion by pouring the metal into a short mold. The base of the mold is a platform that is gradually lowered while the metal solidifies, the frozen shell or metal acting as a retainer for the liquid metal below the wall of the mold. The ingot is usually cooled by the impingement of water directly on the mold or on the walls of the solid metal as it is lowered. The length of the ingot is limited by the depth to which the plateform can be lowered; therefore, it is often called *semicontinuous casting.*

Directional solidification. The solidification of molten metal in a casting in such a manner that feed metal is always available for that portion that is just solidifying.

Direct quenching. Quenching carburized parts directly from the carburizing operation.

Dow process. A process for the production of magnesium electrolysis of molten magnesium chloride.

Draft. (1) The angle or taper on the surface of a punch or die, or the parts made with them, which facilitates the removal of the work. (2) The change in cross section in rolling or wiredrawing. (3) Taper put on the surfaces of a pattern so that it can be withdrawn successfully from the mold.

Drift. (1) A flat piece or steel of tapering width used to remove taper shank drills and other tools from their holders. (2) A tapered rod used to force mismated holes in line for riveting or bolting. Sometimes called a *drift pin.*

Dross. The scum that forms on the surface of molten metals largely because of oxidation but sometimes because of the rising of impurities to the surface.

Dry cyaniding. See Carbonitriding.

Dry method. In magnetic-particle inspection, a method in which a dry powder is used to detect magnetic-leakage fields.

Dry sand mold. A mold of sand and then dried.

Ductile crack propagation. Slow crack propagation that is accompanied by noticeable plastic deformation and requires energy to be supplied from outside the body.

Dye penetrant. Penetrant with dye added to render it more readily visible under normal conditions.

Eddy-current testing. Nondestructive testing method in which eddy-current flow is induced in the test object. Changes in the flow caused by variations in the object are reflected into a nearby coil or coils for subsequent analysis by suitable instrumentation and techniques.

Elastic deformation. Change of dimensions accompanying stress in the elastic range, original dimensions being restored upon release of stress.

Elastic hysteresis. Erroneously used for mechanical hysteresis. The effect is inelastic.

Elasticity. That property of a material by virtue of which it tends to recover its original size and shape after deformation.

Elastic limit. The maximum stress to which a material may be subjected without any permanent strain remaining upon complete release of stress.

Elastic modulus. See Modulus of elasticity.

Electrochemical corrosion. Corrosion that occurs when current flows between cathodic and anodic areas metallic surfaces.

Electrochemical series. See Electromotive series.

Electrode. (1) In arc welding, a current-carrying rod which supports the arc between the rod and work, or between two rods as in twin carbon-arc welding. It may or may not furnish filler metal. (2) In resistance welding, a part of a resistance welding machine through which current and, in most cases, pressure are applied directly to the work. The electrode may be in the form of a rotating wheel, rotating roll, bar, cylinder, plate, clamp, chuck, or modification thereof. (3) An electrical conductor for leading current into or out of a medium.

Electroforming. Making parts by electrodeposition on a removable form.

Electrogalvanizing. The electroplating of zinc upon iron or steel.

Electroless plating. Immersion plating where a chemical reducing agent changes metal ions to metal.

Electrolysis. Chemical change resulting from the passage of an electric current through an electrolyte.

Electrolyte. (1) An ionic conductor. (2) A liquid, most often a solution, that will conduct an electric current.

Electrolytic cell. An assembly, consisting of a vessel, electrodes, and an electrollyte, in which electrolysis can be carried out.

Electrolytic cleaning. Removing soil from work by electrolysis, the work being one of the electrodes. The electrolyte is usually alkaline.

Electrolytic copper. Copper that has been refined by electrolytic deposition, including cathodes that are the direct product of the refining operation, refinery shapes cast from melted cathodes, and, by extension, fabricators' product made therefrom. Usually when this term is used alone, it refers to electrolytic tough-pitch copper without elements other than oxygen being present in significant amounts.

Electromachining. (1) Electrical discharge machining. (2) Electrolytic machining.

Electromotive series. A list of elements arranged according to their standard electrode potentials. In corrosion studies, the analogous but more practical galvanic series of metals is generally used. The relative position of a given metal is not necessarily the same in the two series.

Electroplating. Electrodepositing metal (may be an alloy) in an adherent form upon an object serving as a cathode.

Elongation. In tensile testing, the increase in the gauge length, measured after fracture of the specimen within the gage length, usually expressed as a percantage of the original gauge length.

Elutriation. Powdered metal. Classification of powder particles by means of a rising stream of gas or liquid.

Embrittlement. Reduction in the normal ductility of a metal due to a physical or chemical change.

Emery. An impure mineral of the corundum or aluminum oxide type used extensively as an abrasive before the development of electric-furance products.

Emulsifier. (1) *See* Emulsifying agent. (2) In penetrant inspection, a material that is added to some penetrants, after the penetrant is applied, to make a water-washable mixture.

Emulsifying agent. A material that increases the stability of a dispers of one liquid in another.

Emulsion. A suspension of one liquid phase in another.

Emulsion cleaner. A cleaner consisting of organic solvents dispersed in an aqueous medium with the aid of an emulsifying agent.

Erosion. Destruction of metals or other materials by the abrasive action of moving fluids, usually accelerated by the presence of solid particules or matter in suspension. When corrosion occurs simultaneously, the term *erosion-corrosion* is often used.

Eutectic. (1) An isothermal reversible reaction in which a liquid solution is converted into two or more intimately mixed solids on cooling, the number of solids formed being the same as the number of components in the system. (2) An alloy having the composition indicated by the eutectic point on an equilibrium diagram. (3) An alloy structure of intermixed solid constituents formed by an eutectic reaction.

Eutectic melting. Melting of localized microareas whose composition corresponds to that of the eutectic in the system.

Eutectoid. (1) An isothermal reversible reaction in which a solid solution is converted into two or more intimately mixed solids on cooling, the number of solids formed being the same as the number of components in the system. (2) An alloy having the composition indicated by the eutectoid point on an equilibrium diagram. (3) An alloy structure of intermixed solid constituents formed by an eutectoid reaction.

Extensometer. An instrument for measuring changes in a linear dimension of a body caused by stress.

Extrusion. Conversion of a billet into lengths of uniform cross section by forc-ing the plastic metal through a die orifice of the desired cross-sectional outline. In "direct extrusion," the die and ram are at opposite ends of the billet, and the product and ram travel in the same direction. In "indirect extrusion" (rare), the die is at the ram end of the billet, and the product travels through and in the opposite direction to the hollow ram. A "stepped extrusion" is a single proudct with one or more abrupt cross-section changes and is obtained by interrupting the extrustion by die changes. "Impact extrustion" (cold ex-trusion) is the process or resultant product of a punch striking an unheated slug in a confining die. The metal flow may be either between the punch and die or through another opening. (*See* Hooker process, which uses a pierced slug.) "Hot extrusion" is similar to cold extrusion except that a preheated slug is used and the pressure application is slower.

Extrusion billet. A cast or wrought metal slug used for extrusion.

Extrusion defect. A defect of flow in extruded products caused by the oxidized outer surface of the billet flowing into the center of the extrusion. It normally occurs in the last 10 % to 20% of the extruded bar. Also called *pipe* or *core*.

Extrusion ingot. A solid or hollow cylindrical casting used for extruding into rods, bars, shapes, or tubes.

Fatigue. The phenomenon leading to fracture under repeated or fluctuating stresses having a maximum value less than the tensile strength of the material. Fatigue fractures are progressive, beginning as minute cracks that grow under the action of the fluctuating stress.

Fatigue life. The number of cycles of stress that can be sustained prior to failure for a stated test condition.

Fatigue limit. The maximum stress below which a material can presumably en-dure an infinite number of stress cycles. If the stress is not completely reversed, the value of the mean stress, the minimum stress, or the stress ratio should be stated.

Fatigue ratio. The ratio of the fatigue limit for cycles of reversed flexural stress to the tensile strength.

Fatigue strength. The maximum stress that can be sustained for a specified number of cycles without failure, the stress completely reversed within each cycle unless otherwise stated.

Fatigue-strength reduction factor (K_f). The ratio of the fatigue strength of a member or specimen with no stress concentration to the fatigue strength with stress concentration. K_f has no meaning unless the geomertry, size, and ma-terial of the member or specimen and stress range are stated.

Faying surface. The surface of a piece of metal (or a member) in contact with another to which it is to be joined.

Ferrite. (1) A solid solution of one or more elements in body-centered cubic iron. Unless otherwise designated (for instance, as chromium ferrite), the solute is generally assumed to be carbon. On some equilibrium diagrams, there are two ferrite regions separated by an austenite area. The lower area is alpha ferrite; the upper delta ferrite. If there is no designation, alpha ferrite is assumed. (2) In the field or magnetics, substances having the general formula:

$$M{+}{+}O \cdot M_2{+}{+}{+}O_3,$$

the trivalent metal often being iron.

File hardness. Hardness as determined by the use of a file of standardized hardness on the assumption that a material that cannot by cut with the file is as hard as, or harder than, the file. Files covering a range of hardnesses may be employed.

Fillet. (1) A radius (curvature) imparted to inside meeting surfaces. (2) A concave cornerpiece used on foundry patterns.

Fillet weld. A weld, approximately triangular in cross section, joining two surfaces essentially at right angles to each other in a lap, tee, or corner joint.

Fine silver. Silver with a fineness of 999; equivalent to a minimum content of 99.9% silver with the remaining content not restricted.

Flame annealing. Annealing in which the heat is applied directly by a flame.

Flame straightening. Correcting distortion in metal structure by locatized heating with a gas flame.

Flash welding. A resistance butt-welding process in which the weld is produced over the entire abutting surface by pressure and heat, the heat being produced by electric arcs between the numbers.

Flow brazing. Brazing by pouring molten filler metal over a joint.

Fluorescence. The emission of characteristics electromagnetic radiation by a substance as a result of the absorption of electromagnetic or corpuscular radiation having a greater unit energy than that of the fluorescent radiation. It occurs only as long as the stimulus responsible for it is maintained.

Fluorescent magnetic-particle inspection. Inspection with either dry magnetic particles or those in a liquid suspension, the particles being coated with a fluorescent substance to increase the visibility of the indications.

Fluoroscopy. An inspection procedure in which the radiographic image of the subject is viewed on a fluorescent screen, normally limited to low-density materials or thin section of metals because of the low light output of the fluorescent screen at safe levels of radiation.

Flute. (1) As applied to drills, reamers, and taps, the channels or grooves formed in the body of the tool to provide cutting edges and to permit passage of cutting fluid and chips. (2) As applied to milling cutters and hobs, the chip space between the back of one tooth and face of the following tooth.

Flux. (1) In metal refining, a material used to remove undesirable substances, such as sand, ash, or dirt, as a molten mixture. It is also used as a protective covering for certain molten-metal baths. Lime or limestone is generally used to remove sand, as in iron smelting; sand, to remove iron oxide in copper refining. (2) In brazing, cutting, soldering, or welding, material used to prevent the formation of, or to dissolve and facilitate removal of, oxides and other undesirable substances.

Flux density. (1) In magnetism, the number of flux lines per unit area passing through a cross section at right angles. It is given by $B = \mu H$, where μ and H are permeability and magnetic field intensity, respectively. (2) In neutron radiation, the neutron flux, total.

Forge welding. Welding hot metal by pressure or blows only.

Forging. Plastically deforming metal, usually hot, into desired shapes with compressive force, with or without dies.

Forming. Making a change, with the exception of shearing or blanking, in the shape or contour of a metal part without intentionally altering the thickness.

Foundry. A commercial establishment or building where metal castings are produced.

Fracture stress. (1) The maximum principal true stress at fracture. Usually refers to unnotched tensile specimens. (2) The (hypothetical) true stress, which will cause fracture without further deformation at any given strain.

Free carbon. The part of the total carbon in steel or cast iron that is present in the elemental form as graphite or temper carbon.

Free cyanide. The cyanide not combined in complex ions.

Free ferrite. Ferrite that is structurally separate and distinct, as may be formed without the simultaneous formation of carbide when cooling hypoeutectoid austenite into the critical temperature range. Also proeutectoid ferrite.

Free fit. Various clearance fits for assembly by hand and free rotation of parts. *See* Running fit.

Free machining. Pertains to the machining characterisitcs of an alloy to which an ingredient has been introduced to give small broken chips, lower power consumption, better surface finish, and longer tool life; among such additions are sulfur or lead to steel, lead to brass, lead and bismuth to aluminum, sulfur or selenium to stainless steel.

Fretting (fretting corrosion). Action that results in surface damage, especially in a corrosive environment, when there is relative motion between solid surfaces in contact under pressure. *See* Chafing fatigue.

Full annealing. Annealing a ferrous alloy by austenitizing and then cooling slowly through the transformation range. The austenitizing temperature for hypoeutectoid steel is usually above Ac_1; and for hypereutectoid steel, usually between Ac_1 and Ac_{cm}.

Fusion welding. Welding, without pressure, in which a portion of the base metal is melted.

Galling. Developing a condition on the rubbing surface of one or both mating parts where excessive friction between high spots results in localized welding with subsequent spalling and a further roughening of the surface.

Galvanic cell. A cell in which chemical change is the source of electrical energy. It usually consists of two dissimilar conductors in contact with each other and with an electrolyte or of two similar conductors in contact with each other and with dissimilar electrolytes.

Galvanic corrosion. Corrosion associated with the current of a galvanic cell consisting of two dissimilar conductors in an electrolyte or two similar conductors in dissimilar electrolytes. Where the two dissimilar metals are in contact, the resulting reaction is referred to as "couple action."

Galvanic series. A series of metals and alloys arranged according to their relative electrode potentials in a specified environment.

Gamma. (1) In photography, the slope of the straight-line portion of a film's characteristic curve. (2) Also used inexactly in photography to refer to film contrast, gradient, or average gradient.

Gammagraphs. A radiograph produced by gamma rays.

Gamma iron. The face-centered cubic form of pure iron, stable from 1670 to 2550°F.

Gamma ray. Short-wavelength electromagnetic radiation of nuclear origin with a range of wavelengths from about 10^{12} to 10^{9} cm.

Gas-shielded arc welding. Arc welding in which the arc and molten metal are shielded from the atmosphere by a stream of gas, such as argon, helium, argon-hydrogen mixtures, or carbon dioxide.

Gold-filled. Covered on one or more surfaces with a layer of gold alloy to form a clad metal. By commercial agreement, a quality mark showing the quantity and fineness of gold alloy may be affixed, which shows the actual proportional weight and karat fineness of the gold-alloy cladding. For example, "$\frac{1}{10}$ 12K Gold Filled" means that the article consists of base metal covered on one or more surfaces with a gold alloy of 12-karat fineness comprising one-tenth part by weight of the entire metal in the article. No article having a gold alloy coating of less than 10-karat fineness may have any quality mark affixed. No article having a gold-alloy portion of less than one-twentieth by weight may by marked "Gold Filler," but may be marked "Rolled Gold Plate" provided the proportional fraction and fineness designation precede. These standards do not necessarily apply to watchcases.

Grain. An individual crystal in a polycrystalline metal or alloy.

Grain-fineness number. A weighted-average grain size of a granular material.

The AFS grain-fineness number is calculated with prescribed weighting factors from the standard screen analysis.

Grain growth (*coarsening*). An increase in the size of grains in polycrystalline metal, usually effected during heating at elevated temperatures. The increase may be gradual or abrupt, resulting in either uniform or nonuniform grains after growth has ceased. A mixture of nonuniform grains is sometimes termed *duplexed*. Abnormal grain growth (exaggerated grain growth) implies the formation of excessively large grains, uniform or nonuniform. The abrupt form of abnormal grain growth is also termed *germinative grain growth* when a critical amount of strain or other nuclei are present to promote the growth. Secondary recrystallization is the selective grain growth of a few grains only, as distinct from uniform coarsening, when the new set of grains resulting from primary recrystallization is subjected to further annealing.

Grain size. (1) For metals, a measure of the areas or volumes of grains in a polycrystalline material, usually expressed as an average when the individual sizes are fairly uniform. Grain sizes are reported in terms of number of grians per unit area or volume, average diameter, or as a grain-size number derived from area measurements. (2) For grinding wheels, *see* the preferred term, Grit size.

Granular fracture. A type of irregular surface produced when metal is broken, which is characterized by a rough, grainlike appearance as differentiated from a smooth silky, or fibrous, type. It can be subclassified into transgranular and intergranular forms. This type of fracture is frequently called *crystalline fracture*, but the inference that the metal has crystallized is not justified.

Granulation. The production of coarse metal particles by pouring the molten metal through a screen into water or by agitating the molten metal violently during its solidification.

Grit size. Nominal size of abrasive particles in a grinding wheel corresponding to the number of openings per linear inch in a screen through which the particles can just pass. Sometimes called *grain size*.

Hard chromium. Chromium deposited for engineering purposes, such as increasing the wear resistance of sliding metal surfaces, rather than as a decorative coating. It is usually applied directly to basis metal and is customarily thicker than a decorative deposit.

Hardener. An alloy, rich in one or more alloying elements, added to a melt to permit closer composition control than possible by addition of pure metals or to introduce refractory elements not readily alloyed with the base metal.

Hardness. (1) Resistance of metal to plastic deformation usually by indentation. However, the term may also refer to stiffness or temper, or to resistance to scratching, abrasion, or cutting. Indentation hardness may be measured by various hardness tests, such as Brinell, Rockwell, and Vickers. (2) For grinding wheels, same as grade.

Heat treatment. Heating and cooling a solid metal or alloy in such a way as to obtain desired conditions or properties. Heating for the sole purpose of hot working is excluded from the meaning of this definition.

Hooke's law. Stress is proportional to strain. The law holds only up to the proportional limit.

Hydrogen brazing. Brazing in a hydrogen atmosphere, usually in a furnace.

Hydrogen embrittlement. A condition of low ductility in metals resulting from the absorption of hydrogen.

Impedance. (1) Acoustical impedance is the complex ratio of the sound pressure to the product of the product of the sound velocity and the area at a given surface. It is frequently approximated by only the product of the density and velocity. (2) Electrical impedance is the complex property of an electrical circuit, or the components of a circuit, that opposes the flow of an alternating current. The real part represents the resistance, and the imaginary part represents the reactance of the circuit.

Impregnation. (1) The treatment of porous castings with a sealing medium to stop pressure leaks. (2) The process of filling the pores of a sintered compact, usually with a liquid such as a lubricant. (3) The process of mixing particles of a nonmetallic substance in a matrix of metal powder, as in diamond-impregnated tools.

Indirect extrusion. See Extrusion.

Induction brazing. Brazing with induction heat.

Inert-gas shielded-arc cutting. Metal cutting with the heat of an arc in an inert gas such as argon or helium.

Inert-gas shielded-arc welding. Arc welding in an inert gas such as argon or helium.

Intergranular corrosion. Corrosion occurring preferentially at grain boundaries.

Intermediate annealing. Annealing wrought metals at one or more stages during manufacture and before final treatment.

Intermittent weld. A weld in which the continuity is broken by recurring unwelded spaces.

Investment casting. (1) Casting metal into a mold produced by surrounding (investing) an expendable pattern with a refractory slurry that sets at room temperature after which the wax, plastic, or frozen mercury pattern is removed through the use of heat. Also called *precision casting* or *lost-wax* process. (2) A casting made by the process.

Investment compound. A mixture of a graded refractory filler, a binder, and a liquid vehicle, used to make molds for investment casting.

Ion. An atom, or group of atoms, that has gained or lost one or more outer electrons and thus carries an electric charge. Positive ions, or cations, are deficient in outer electrons. Negative ions, or anions, have an excess of outer electrons.

Iridium-192. A radioisotope with a half-life of 74 days and 12 dominant, characteristic, gamma-radiation energies ranging from 0.14 to 0.65 MeV. It is suitable as a gamma radiation source, mostly in radiography.

Iron. (1) Element No. 26 of the periodic system, the average atomic weight of the naturally occurring isotopes being 55.85. (2) Iron-base materials not falling into the steel classifications. *See* Gray cast iron, Ingot iron, Malleable cast iron, Nodular cast iron, White cast iron, and Wrought iron.

Isothermal annealing. Austenitizing a ferrous alloy and then cooling to and holding at a temperature at which austenite transforms to a relatively soft ferrite carbide aggregate.

Isothermal transformation. A change in phase at any constant temperature.

Isotropy. Quality of having identical properties in all directions.

Izod test. A pendulum type of of single-blow impact test in which the specimen, usually notched, is fixed at one end and broken by a falling pendulum. The energy absorbed, as measured by the subsequent rise of the pendulum, is a measure of impact strength or notch toughness.

Jig boring. Boring with a single-point tool where the work is positioned upon a table that can be located so as to bring any desired part of the work under the tool. Thus, holes can be accurately spaced. This type of boring can be done on milling machines or "jig borers."

Joggle. An offset in a flat plane consisting of two parallel bends in opposite directions by the same angle.

Karat. A 24th part, used to designate the fineness of gold. Abbreviated K, Kt. Sometimes spelled "carat."

Knoop hardness. Microhardness determined from the resistance of metal to indentation by a pyramidal diamond indenter, having edge angles of $172°$, $30'$, and $130°$, making a rhombohedral impression with one long and one short diagonal.

Knurling. Impressing a design into a metallic surface, usually by means of small, hard rollers that carry the corresponding design on their surfaces.

Latent heat. Thermal energy absorbed or released when a substance undergoes a phase change.

Lay. Direction of predominant surface pattern remaining after cutting, grinding, lapping, or other processing.

Liquid honing. Polishing metal by bombardment with an air-ejected liquid containing fine solid particles in suspension. If an impeller wheel is used to propel the suspension, the process is called *wet blasting*.

Liquidus. In a constitution or equilibrium diagram, the locus of points representing the temperatures at which the various compositions in the system begin to freeze on cooling or to finish melting on heating.

Lost-wax process. An investment casting process in which a wax pattern is used.

Lubricant. Any substance used to reduce friction between two surfaces in contact.

Machining. Removing material, in the form of chips, from work, usually through the use of a machine.

Macro-etch. Etching of a metal surface for accentuation of gross structural details and defects for observation by the unaided eye or at magnifications not exceeding 10 diameters.

Macrograph. A graphic reproduction of the surface of a prepared specimen at a magnification not exceeding 10 diameters. When photographed, the reproduction is known as a *photomacrograph.*

Macroscopic. Visible at magnifications from 1 to 10 diameters.

Macrostructure. The structure of metals as revealed by examination of the etched surface of a polished specimen at a magnification not exceeding 10 diameters.

Magnesite wheel. A grinding wheel bonded with magnesium oxychloride.

Magnetic-particle inspection. A nondestructive method of inspection for determining the existence and extent of possible defects in ferromagnetic materials. Finely divided magnetic particles, applied to the magnetized part, are attracted to and outline the pattern of any magnetic-leakage fields created by discontinuities.

Malleability. The characteristic of metals that permits plastic deformation in compression without rupture.

Martempering. Quenching an austenitized ferrous alloy in a medium at a temperature in the upper part of the martensite range, or slightly above that range, and holding it in the medium until the temperature throughout the alloy is substantially uniform. The alloy is then allowed to cool in air through the martensite range.

Martensite. (1) In an alloy, a metastable transitional structure intermediate between two allotropic modifications whose abilities to dissolve solute differ considerably, the high-temperature phase having the greater solubility. The amount of the high-temperature phase transformed to martensite depends to a large extent upon the temperature attained in cooling, there being a rather distinct beginning temperature. (2) A metastable phase of steel, formed by a transformation of austenite below the M (or Ar") temperature. It is an interstitial supersaturated solid solution of carbon in iron having a body-centered tetragonal lattice. Its microstructure is characterized by an acicular, or needlelike, pattern.

Martensitic transformation. A reaction that takes place in some metals on cooling, with the formation of an acicular structure called *martensite*.

Metal. (1) An opaque lustrous elemental chemical substance that is a good conductor of heat and electricity and, when polished, a good reflector of light. Most elemental metals are malleable and ductile and are, in general, heavier than other elemental substances. (2) As to structure, metals may be distinguished from nonmetals by their atomic binding and electron availability. Metallic atoms tend to lose electrons from the outer shells, the positive ions thus formed arc held together by the electron gas produced by the separation. The ability of these free electrons to carry an electric current, and the fact that the conducting power decreases as temperature increases, establish one of the prime distinctions of a metallic solid. (3) From the chemical viewpoint, an elemental substance whose hydroxide is alkaline. (4) An alloy.

Metal-arc welding. Arc welding with metal electrodes. Commonly refers to shielded metal-arc welding using covered electrodes.

Metallurgy. The science and technology of metals. Process (chemical) metallurgy is concerned with the extraction of metals from their ores and with the refining of metals; physical metallurgy, with the physical and mechanical properties of metals as affected by composition, mechanical working, and heat treatment.

Micrograph. A graphic reproduction of the surface of a prepared specimen, usually etched, at a magnification greater than 10 diameters. If produced by photographic means, it is called a *photomicrograph* (not a microphotograph).

Microshrinkage. A casting defect, not detectable at magnifications lower than 10 diameters, consisting of interdendritic voids. This defect results from contraction during solidification where there is not an adequate opportunity to supply filler material to compensate for shrinkage. Alloys with a wide range in solidification temperature are particularly susceptible.

Misrun. A casting not fully formed, resulting from the metal solidifying before the mold is filled.

Modulus of elasticity. A measure of the rigidity of metal. Ratio of stress, within proportional limit, to corresponding strain. Specifically, the modulus obtained in tension or compression is Young's modulus, stretch modulus, or modulus of extensibility; the modulus obtained in torsion or shear is modulus of rigidity, shear modulus, or modulus of torsion; the modulus covering the ratio of the mean normal stress to the change in volume per unit volume is the bulk modulus. The tangent modulus and secant modulus are not restricted within the proportional limit; the former is the slope of the stress-strain curve at a specified point; the latter is the slope of a line from the origin to a specified point on the stress-strain curve. Also called *elastic modulus* and *coefficient of elasticity*.

Modulus of rigidity. See Modulus of elasticity.

Modulus of rupture. Nominal stress at fracture in a bend test or torsion test. In bending, modulus of rupture is the bending moment at fracture divided by the section modulus. In torsion, modulus of rupture is the torque at fracture divided by the polar section modulus.

Monotectic. An isothermal reversible reaction in a binary system, in which a liquid on cooling decomposes into a second liquid of a different composition and a solid. It differs from a eutectic in that only one of the two products of the reaction is below its freezing range.

Natural aging. Spontaneous aging of a supersaturated solid solution at room temperature. *See* Aging. *Compare with* Artificial aging.

Nitriding. Introducing nitrogen into a solid ferrous alloy by holding at a suitable temperature (below Ac_1 for ferritic steels) in contact with a nitrogenous material, usually ammonia or molten cyanide of appropriate composition. Quenching is not required to produce a hard case.

Noble metal. (1) A metal whose potential is highly positive relative to the hydrogen electrode. (2) A metal with marked resistance to chemical reaction, particularly to oxidation and to solution by inorganic acids. The term as often used is synonymous with *precious metal.*

Nondestructive inspection. Inspection by methods that do not destroy the part to determine its suitability for use.

Nondestructive testing. *See* Nondestructive inspection.

Normalizing. Heating a ferrous alloy to a suitable temperature above the transformation range and then cooling in air to a temperature substantially below the transformation range.

Optical pyrometer. An instrument for measuring the temperature of heated material by comparing the intensity of light emitted with a known intensity of an incandescent lamp filament.

Overaging. Aging under conditions of time and temperature greater than those required to obtain maximum change in a certain property, so that the property is altered in the direction of the initial value. *See* Aging.

Particle size. Powdered metal. The controlling lineal dimension of an individual particle, as determined by analysis with screens or other suitable instruments.

Parting. (1) In the recovery of precious metals, the separation of silver from gold. (2) The zone of separation between cope and drag portions of mold or flask in sand casting. (3) A composition sometimes used in sand molding to facilitate the removal of the pattern. (4) Cutting simultaneously along two parallel lines or along two lines that balance each other in the matter of side thrust. (5) A shearing operation used to produce two or more parts from a stamping.

Parting line. A plane on a pattern or a line on a casting corresponding to the separation between the cope and drag portions of a mold.

Passivation. he changing of the chemically active surface of a metal to a much less reactive state.

Passivity. A condition in which a piece of metal, because of an impervious covering of oxide or other compound, has a potential much more positive than where the metal is in the active state.

Pearlite. A lamellar aggregate of ferrite and cementite, often occurring in steel and cast iron.

Penetrant. A liquid with low surface tension used in penetrant inspection to flow into surface openings of parts being inspected.

Penetrant inspection. A method of nondestructive testing for determining the existence and extent of discontinuities that are open to the surface in the part being inspected. The indications are made visible through the use of a dye or fluorescent chemical in the liquid employed as the inspection medium.

Penetration. (1) In founding, a defect on a casting surface caused by metal running into voids between sand grains. (2) In welding, the distance from the original surface of the base metal to that point at which fusion ceased.

Percussion welding. Resistance welding simultaneously over the entire area of abutting surfaces with arc heat, the pressure being applied by a hammerlike blow during or immediately following the electrical discharge.

Peritectoid. An isothermal reversible reaction in which a solid phase reacts with a second solid phase to produce yet a third solid phase on cooling.

Permanent mold. A metal mold (other than an ingot mold) of two or more parts that is used repeatedly for the production of many castings of the same form. Liquid metal is poured in by gravity.

Permanent set. Plastic deformation that remains upon releasing the stress that produces the deformation.

Permeability. (1) *Founding.* The characteristics of molding materials that permit gases to pass through them. *Permeability number* is determined by a standard test. (2) *Powdered metal.* A property measured as the rate of passage under specified conditions of a liquid or gas through a compact. (3) *Magnetism.* A general term used to express various relationships between magnetic induction and magnetizing force. These relationships are either "absolute permeability," which is the quotient of a change in magnetic induction divided by the corresponding change in magnetizing force, or "specific (relative) permeability," the ratio of the absolute permeability to the permeability of free space.

pH. The negative logarithm of the hydrogen ion activity; it denotes the degree of acidity or basicity of a solution. At $25°C$, seven is the neutral value. Acidity increases with decreasing values below seven; basicity increases with increasing values above seven.

Plaster molding. Molding wherein a gypsum-bonded aggregate flour in the form of a water slurry is poured over a pattern, permitted to harden, and, after removal of the pattern, thoroughly dried. The technique is used to make smooth nonferrous castings of accurate size.

Plastic deformation. Deformation that does or will remain permanent after removal of the load that caused it.

Platen. (1) The sliding member or slide of a hydraulic press. *See* Slide and ram. (2) A part of a resistance welding, mechanical testing, or other machine with a flat surface, to which dies, fixtures, backups, or electrode holders are attached, and which transmits pressure.

Plating. Forming an adherent layer of metal upon an object.

Plug. (1) A rod or mandrel over which a pierced tube is forced. (2) A rod or mandrel that fills a tube as it is drawn through a die. (3) A punch or mandrel over which a cup is drawn. (4) A protruding portion of a die impression for forming a corresponding recess in the forging. (5) A false bottom in die. Also called a *peg*.

Plug tap. A tap with chamfer extending from three to five threads.

Poisson's ratio. The absolute value of the ratio of the transverse strain to the corresponding axial strain, in a body subjected to uniaxial stress; usually applied to elastic conditions.

Polarization. In electrolysis, the formation of a film on an electrode such that the potential necessary to get a desired reaction is increased beyond the reversible electrode potential.

Precious metal. One of the relatively scarce and valuable metals: gold, silver, and the platinum-group metals.

Precipitation hardening. Hardening caused by the precipitation of a constituent from a supersaturated solid solution. *See* Age hardening, Aging.

Precipitation heat treatment. Artificial aging in which a constituent precipitates from a supersaturated solid solution.

Proof stress. (1) The stress that will cause a specified small permanent set in a material. (2) A specified stress to be applied to a member or structure to indicate its ability to withstand service loads.

Proportional limit. The maximum stress at which strain remains directly proportional to stress.

Pyrometallurgy. Metallurgy involved in winning and refining metals where heat is used, as in roasting and smelting.

Pyrometer. A device for measuring temperatures above the range of liquid thermometers.

Quarter hard. *See* Temper of copper, Copper alloys.

Quench aging. Aging induced by rapid cooling after solution heat treatment.

Quench annealing. Annealing an austenitic ferrous alloy by solution-heat treatment.

Quench hardening. Hardening a ferrous alloy by austenitizing and then cooling rapidly enough so that some or all of the austenite transforms to martensite. The austenitizing temperature for hypoeutectoid steels is usually above Ac_3; for hypereutectoid steels, it usually ranges from Ac_1 and Ac_{cm}.

Quenching. Rapid cooling. When applicable, the following more specific terms should be used: *direct quenching, fog quenching, hot quenching, interrupted quenching, selective quenching, spray quenching,* and *time quenching.*

Quench time. In resistance welding, the time from the finish of the weld to the beginning of temper. Also called *chill time.*

Radiation energy. The energy of a given photon or particle in a beam of radiation, usually expressed in electron volts.

Radioactive element. An element that has at least one isotope that undergoes spontaneous nuclear disintegration to emit positive alpha particles, negative beta particles, or gamma rays.

Radioactivity. The spontaneous nuclear disintegration with emission of corpuscular or electromagnetic radiation.

Radiograph. A photographic shadow image resulting from uneven absorption of radiation in the object being subjected to penetrating radiation.

Radiographic contrast. The difference in density between an image, or part of an image, and its immediate surroundings on a radiograph. Radiographic contrast depends upon both subject contrast and film contrast.

Radiographic equivalence factor. The reciprocal of the thickness of a given material taken as a standard. It not only depends on the standard, but also on the radiation quality.

Radiograhpic sensitivity. A measure of quality of radiographs whereby the minimum discontinuity that may be detected on the film is expressed as a percentage of the base thickness. It depends on subject and film contrast and on geometrical and film graininess factors.

Radiography. A nondestructive method of internal examination in which metal or other objects are exposed to a beam of X-ray or gamma radiation. Differences in thickness, density, or absorption, caused by internal discontinuities, are apparent in the shadow image either on a fluorescent screen or on photographic film place behind the object.

Radioisotope. An isotope that emits ionizing radiation during its spontaneous decay. *See* Cesium-137, Cobalt-60, Iridium-192, Radium, Radon, Strontium-90, and Thulium-170, which are used commercially.

Radium. A radioactive element. It is found in nature as radium-226, which has a half-life of 1620 years. In equilibrium with its daughter products, it emits 11 principal gamma rays, ranging from 0.24 to 2.20 MeV. It is used as a gamma-radiation source, especially in radiography and therapy.

Rare earth metal. One of the group of 15 similar metals with atomic numbers 57 through 71.

Recrystallization. (1) The change from one crystal structure to another, as occurs on heating or cooling through a critical temperature. (2) The formation of a new, strain-free grain structure from that existing in cold-worked metal, usually accomplished by heating.

Recrystallization annealing. Annealing cold-worked metal to produce a new grain structure without phase change.

Recrystallization temperature. The approximate minimum temperature at which complete recrystallization of a cold-worked metal occurs within a specified time.

Reduction of area. (1) Commonly, the difference, expressed as a percentage of original area, between the original cross-sectional area of a tensile test specimen and the minimum cross-sectional area measured after complete separation. (2) The difference, expressed as a percentage of original area, between original cross-sectional area and that after straining the specimen.

Refractory. (1) The material of very high melting point with properties that make it suitable for such uses as furnace linings and kiln construction. (2) The quality of resisting heat.

Refractory alloy. (1) A heat-resistant alloy. (2) An alloy having an extremely high melting point. *See* Refractory metal. (3) An alloy difficult to work at elevated temperatures.

Refractory metal. A metal having an extremely high melting point. In the broad sense, it refers to metals having melting points above the ranges of iron, cobalt, and nickel.

Resistance brazing. Brazing by resistance heating, the joint being part of the electrical circuit.

Resistance welding. Welding with resistance heating and pressure, the work being part of the electrical circuit. Example: resistance spot welding, resistance seam welding, projection welding, and flash butt welding.

Rockwell hardness test. A test for determining the hardness of a material based upon the depth of penetration of a specified penetrator into the specimen under certain arbitrarily fixed conditions of test.

Rough machining. Machining without regard to finish, usually to be followed by a subsequent operation.

Roughness. Relatively finely spaced, surface irregularities, of which the height, width, and direction establish the predominant surface pattern.

Running fit. Any clearance fit in the range used for parts that rotate relative to each other. Actual values of clearance resulting from stated shaft and hole tolerances are given for nine classes of running and sliding fits for 21 nominal shaft sizes in ASA B4.1-1955.

Runout. (1) The unintentional escape of molten metal from a mold, crucible, or furnace. (2) The defect in a casting caused by the escape of metal from the mold. (3) *See* Axial runout, Radial runout.

Rust. A corrosion product consisting of hydrated oxides of iron. Applied only ferrous alloys.

Salt fog test. An accelerated corrosion test in which specimens are exposed to a fine mist of a solution usually containing sodium chloride but sometimes modified with other chemicals. For testing details, see ASTM B117 and B287.

Sand. A granular material resulting from the disintegration of rock. Foundry sands are mainly silica. *Bank sand* is found in sedimentary deposits and contains less than 5% clay. *Dune sand* occurs in windblown deposits near large bodies of water and is very high in silica content. *Molding sand* contains more than 5% clay, usually between 10 and 20%. *Silica sand* is a granular material containing at least 95% silica and often more than 99%. *Core sand* is nearly pure silica. *Miscellaneous sands* include zircon, olivine, calcium carbonate, lava, and titanium minerals.

Scleroscope test. A hardness test when the loss in kinetic energy of a falling metal "tup," absorbed by indentation upon impact of the tup on the metal being tested, is indicated by the height of rebound.

Seam welding. (1) Arc or resistance welding in which a series of overlapping spot welds is produced with rotating electrodes, rotating work, or both. (2) Making a longitudinal weld in sheet metal or tubing.

Shear. (1) That type of force that causes, or tends to cause, two contiguous parts of the same body to slide relative to one another in a direction parallel to their plane of contact.

Shear fracture. A fracture in which a crystal (or a polycrystalline mass) has separated by sliding or tearing under the action of shear stresses.

Shearing strain (shear strain). *See* Stain.

Shear modulus. *See* Modulus of elasticity.

Shear plane. A confined zone along which shear takes place in metal cutting. It extends from the cutting edge to the work surface.

Shear strength. The stress required to produce fracture in the plane of cross section, the conditions of loading being such that the directions of force and of resistance are parallel and opposite although their paths are offset a specified minimum amount.

Shielded-arc welding. Arc welding in which the arc and the weld metal are protected by a gaseous atmosphere, the products of decompositon of the electrode covering, or a blanket of fusible flux.

Shielded metal-arc welding. Arc welding in which the arc and the weld metal are protected by the decomposition products of the covering on a consumable metal electrode.

Shore hardness test. *See* Scleroscope test.

Shrink fit. A fit that allows the outside member, when heated to a practical temperature, to assemble easily with the inside member.

Sigma phase. A hard brittle nonmagnetic intermediate phase with a tetragonal crystal structure, containing 30 atoms per unit cell, space group $P4_2/mnm$, occurring in many binary and ternary alloys of the transition elements. The composition of this phase in the various systems is not the same, and the phase usually exhibits a wide range in homogeneity. Alloying with a third transition elemeht usually enlarges the field of homogeneity and extends it deep into the ternary section.

Silver brazing. Brazing with silver-base alloys as the filler metal.

Silver brazing alloy. Filler metal used in silver brazing.

Silver solder. *See* Silver brazing alloy.

Sliding fit. A series of nine classes of running and sliding fits of 21 nominal shaft sizes defined in terms of clerance and tolerance of shaft and hole in ASA B4.1-1955.

Smelting. Thermal processing wherein chemical reactions take place to produce liquid metal from a beneficiated ore.

Snug fit. A loosely defined fit implying the closest clearances that can be assembled manually for firm connection between parts and comparable to one or more of the 11 classes of clearance locational fits given in ASA B4.1-1955.

Soldering. Similar to brazing, with the filler metal having a melting temperature range below an arbitrary value, generally $800°F$. Soft solders are usually lead-tin alloys.

Solidus. In a constitution or equilibrium diagram, the locus of points representing the temperatures at which various compositions finish freezing or cooling or begin to melt on heating.

Spalling. The cracking and flaking of particles out of a surface.

Spot welding. Welding of lapped parts in which fusion is confined to a relatively small circular area. It is generally resistance welding, but may also be gas-shielded tungsten-arc, gas-shielded metal-arc, or submerged-arc welding.

Steel. An iron-base alloy, malleable in some temperature range as initially cast, containing manganese, usually carbon, and often alloying elements. In carbon steel and low-alloy steel, the maximum carbon is about 2.0%; in high-alloy steel, about 2.5%. The dividing line between low-alloy and high-alloy steel is generally regarded as being at about 5% metallic alloying elements.

Steel is to be differentiated from two general classes of "irons": the cast irons on the high-carbon side, and relatively pure irons such as ingot iron, carbonyl iron, and electrolytic iron on the low-carbon side. In some steels containing extremely low carbon, the magnanese content is the principal differentiating factor, steel usually containing at least 0.25%; ingot iron contains considerably less.

Strain. A measure of the change in the size or shape of a body, referred to its original size or shape. *Linear strain* is the change per unit length of a linear dimension. *True strain* (or *natural strain*) is the natural logarithm of the ratio of the length at the moment of observation to the original gauge length. *Conventional strain* is the linear strain referred to the original gauge length. *Shearing strain* (or *shear strain*) is the change in angle (expressed in radians) between two lines originally at right angles. When the term *strain* is used alone, it usually refers to the linear strain in the direction of the applied stress.

Strain aging. Aging induced by cold working. *See* Aging.

Strain hardening. An increase in hardness and strength caused by plastic deformation at temperature below the recrystallization range.

Stress. Force per unit area, often thought of as force acting through a small area within a plane. It can be aivided into components, normal and parallel to the plane, called *normal stress* and *shear stress*, respectively. *True stress* denotes the stress where force and area are measured at the same time. *Conventional stress*, as applied to tension and compression tests, is force divided by the original area. *Nominal stress* is the stress computed by simple elasticity formula, ignoring stress raisers and disregarding plastic flow. In a notch-bend test, for example, it is bending moment divided by minimum section modulus.

Stress-corrosion cracking. Failure by cracking under combined action of corrosion and stress, either external (applied) or internal (residual). Cracking may be either intergranular or transgranular, depending on metal and corrosive medium.

Stress relieving. Heating to a suitable temperature, holding long enough to reduce residual stresses and then cooling slowly enough to minimize the development of new residual stresses.

Stud welding. Welding a metal stud or similar part to another piece of metal, the heat being furnished by an arc between the two pieces just before pressure is applied.

Superalloy. An alloy developed for very-high-temperature service where relatively high stresses (tensile, thermal, vibrator, and shock) are encountered and where oxidation resistance is frequently required.

Supercooling. Cooling below the temperature at which an equilibrium phase transformation can take without actually obtaining the transformation.

Superficial Rockwell hardness test. Form of Rockwell hardness test using relatively light loads, which produce minimum penetration. Used for determining surface hardness or hardness of thin sections or small parts, or where large hardness impression might be harmful.

Superfinishing. A form of honing in which the abrasive stones are spring-supported.

Superheating. (1) Heating a phase above a temperature at which an equilibrium can exist between it and another phase having more internal energy, without obtaining the high-energy phase. (2) Heating molten metal above the normal casting temperature so as to obtain more complete refining or greater fluidity.

Taper tap. A tap with a chamber of seven to nine threads in length.

Tarnish. Surface discoloration of a metal caused by formation of a thin of corrosion product.

Temper. (1) In heat treatment, reheating hardended steel or hardened cast iron to some temperature below the eutectoid temperature for the purpose of decreasing the hardeness and increasing the toughness. The process is also sometimes applied to normalized steel. (2) In tool steels, "temper" is sometimes used, but inadvisedly, to denote the carbon content. (3) In nonferrous alloys and in some ferrous alloys (steels that cannot be hardened by heat treatment), the hardness and strength produced by mechanical or thermal treatment, or both, and characterized by a certain structure, mechanical properties, or reduction in area during cold working. Refer to section G-IX.

Temper brittleness. Brittleness that results when certain steels are held within, or are cooled slowly through, a certain range of temperature below the transformation range. The brittleness is revealed by notched-bar impact tests at or below room temperature.

Tempering. Reheating a quench-hardened or normalized ferrous alloy to a temperature below the transformation range and then cooling at any rate desired.

Tensile strength. In tensile testing, the ratio of maximum load to original cross-sectional area. Also called *ultimate strength.*

Thermal fatigue. Fracture resulting from the presence of temperature gradients, which vary with time in a manner to produce cyclic stresses in a structure.

Thermal shock. The development of a steep temperature gradient and accompanying high stresses within a structure.

Thermit welding. Welding with heat produced by the reaction of aluminum with a metal oxide. Filler metal, if used, is obtained from the reduction of the appropriate oxide.

Thermocouple. A device for measuring temperatures, consisting of two dissimilar metals that produce an electromotive force roughly proportional to the temperature difference between their hot and cold junction ends.

Toughness. Ability of a metal to absorb energy and deform plastically before fracturing. It is usually measured by the energy absorbed in a notch-impact test, but the area under the stress-strain curve in tensile testing is also a measure of toughness.

Transducer. A device actuated by one transmission system and supplying related waves to another transmission system; the input and output energies may be of different forms. (Ultrasonic transducers accept electrical waves and deliver ultrasonic waves, the reverse also being true.)

Transformation ranges (transformation temperature ranges). Those ranges of temperature within which austenite forms during heating and transforms during cooling. The two ranges are distinct, sometimes overlapping, but never coinciding. The limiting temperatures of the ranges depend on the composition of the alloy and on the rate of change of temperature, particularly during cooling. *See* Transformation temperature.

Transformation temperature. The temperature at which a change in phase occurs. The term is sometimes used to denote the limiting temperature of a transformation range. The following symbols are used for iron and steel:

Ac_{cm}. In hypereutectoid steel, the temperature at which the solution of cementite in austenite is completed during heating.

Ac_1. The temperature at which austenite begins to form during heating.

Ac_3. The temperature at which transformation of ferrite to austenite is completed during heating.

Ac_4. The temperature at which austenite transforms to delta ferrite during heating.

$Ac_{cm}, Ae_1, Ae_3, Ae_4$. The temperatures of phase changes at equilibrium.

Ac_{cm}. In hypereutectoid steel, the temperature at which precipitation of cementite starts during cooling.

Ar_1. The temperature at which transformation of austenite to ferrite, or to ferrite, or to ferrite plus cementite is completed during cooling.

Ar_3. The temperature at which austenite begins to transform to ferrite during cooling.

Ar_4. The temperature at which delta ferrite transforms to austenite during cooling.

M_s (or Ar''). The temperature at which transformation of austenite to martensite starts during cooling.

M_f. The temperature at which martensite formation finishes during cooling.

Note: All these changes except the formation of martensite occur at lower temperatures during cooling than heating, and depend on the rate of change of temperature.

Transitional fit. A fit, which may have clearance of interference resulting from specified tolerances on hole and shaft as given by six classes of transition locational fits of 13 nominal shaft sizes in ASA B4.1-1955.

Tungsten-arc welding. Inert-gas, shielded-arc welding using a tungsten electrode.

Ultimate strength. The maximum conventional stress (tensile, compressive, or shear) that a material can withstand.

Ultrasonic beam. A beam of acoustical radiation with a frequency higher than the frequency range for audible sound.

Ultrasonic cleaning. Immersion cleaning aided by ultrasonic waves which cause microagitation.

Ultrasonic frequency. A frequency, associated with elastic waves, that is greater than the higest audible frequency, generally regarded as being higher than 15 kilocycles per second.

Ultrasonics. The acoustic field involving ultrasonic frequencies.

Undercut. A groove melted into the base metal adjacent to the toe of a weld and left unfilled.

Vacuum deposition. Condensation of thin metal coatings on the cool surface of work in a vacuum.

Vacuum melting. Melting in a vacuum to prevent contamination from air, as well as to remove gases already dissolved in the metal; the solidification may also be carried out in a vacuum or at low pressure.

Vacuum refining. See Vacuum melting.

Vapor blasting. See Liquid honing.

Vapor degreasing. Degreasing work in vapor over a boiling liquid solvent, the vapor being considerably heavier than air. At least one constituent of the soil must be soluble in the solvent.

Vapor plating. Deposition of a metal or compound upon a heater surface by reduction or decomposition of a volatile compound at a temperature below the melting points of the deposit and the basis material. The reduction is usually accomplished by a gaseous reducing agent such as hydrogen. The decomposition process may involve thermal dissociation or reaction with the basis material. Occasionally used to designate deposition on cold surfaces by vacuum evaporation. *See* Vacuum deposition.

Vickers hardness test. See Diamond pyramid hardness test.

Welding. (1) Joining two or more pieces of material by applying heat, pressure, or both, with or without filler material, to produce a localized union through fusion or recrystallization across the interface. The thickness of the filler material is much greater than the capillary dimensions encountered in brazing. (2) May also be extended to include brazing.

Welding current. The current flowing through a welding circuit during the making of a weld. In resistance welding, the current used during preweld or postweld intervals is excluded.

Welding cycle. The complete series of events involved in making a resistance weld. Also applies to semiautomatic mechanized fusion welds.

Welding force. See Electrode force.

Welding generator. A generator used for supplying current for welding.

Welding ground. *See* Work lead.

Welding lead (welding cable). A work lead or an electrode lead.

Welding machine. Equipment used to perform the welding operation; for example, spot-welding machine, arc-welding machine, seam-welding machine.

Welding procedure. The detailed methods and practices, including joint-welding procedures, involved in the production of a weldment.

Welding rod. Filler metal in rod or wire form used in welding.

Welding schedule. A record of all welding machine settings plus identification of the machine for a given material, size, and finish.

Welding sequence. The order of welding the various component parts of a weldment or structure.

Welding stress. Residual stress caused by localized heating and cooling during welding.

Welding technique. The details of a welding operation that, within the limitations of a welding procedure, are performed by the welder.

Welding tip. (1) A replaceable nozzle for a gas torch that is especially adapted for welding. (2) A spot-welding or projection-welding electrode.

Weld line. The junction of the weld metal and the base metal, or the junction of the base-metal parts when filler metal is not used.

Weldment. An assembly whose component parts are joined by welding.

Weld metal. That portion of a weld that has been melted during welding.

Wetting. A phenomenon involving a solid and a liquid in such intimate contact that the adhesive force between the two phases is greater than the cohesive force within the liquid. Thus, a solid that is wetted, on being removed from the liquid bath, will have a thin continuous layer of liquid adhering to it. Foreign substances such as grease may prevent wetting. Addition agents, such as detergents, may wetting by lowering the surface tension of the liquid.

Wetting agent. A surface-active agent that produces wetting by decreasing the cohesion within the liquid.

Work hardening. *See* Strain hardening.

Wringing fit. A fit of zero to negative allowance comparable to fits assigned to the first six nominal shaft sizes listed under class LN2 of interference locational fits in ASA B4.1-1955.

Wrought iron. A commercial iron consisting of slag (iron silicate) fibers entrained in a ferrite matrix.

Xeroradiography. A process utilizing a layer of photoconductive material on an aluminum sheet upon which an electrical charge is placed. After X-ray exposure, the electrical potential remaining on the plate in the form of a latent electrical pattern is developed by contact with a cloud of finely dispersed powder.

X-ray. Electromagnetic radiation, of wavelength less than about 500 angstrom units, emitted as the result of deceleration of fast-moving electrons (*bremsstrahlung*, continuous spectrum) or decay of atomic electrons from excited orbital states (characteristic radiation). Specifically, the radiation produced when an electron beam of sufficient energy impinges upon a target of suitable material.

Yield point. The first stress in a material, usually less than the maxium attainable stress, at which an increase in strain occurs without an increase in stress. Only certain metals exhibit a yield point. If there is a decrease in stress after yielding, a distinction may be made between upper and lower yield points.

Yield strength. The stress at which a material exhibits a specifed deviation from proportionlity of stress and strain. An offset of 0.2% is used for many metals.

Young's modulus. See Modulus elasticity.

Ziron sand. A very refractory mineral, composed chiefly of zironium silicate of extreme fineness, having low thermal expansion and high thermal conductivity.

TERMS RELATING TO PHYSICAL PROPERTIES OF MATERIALS

The following compilation is included to provide the engineer or designer with a brief, easily visualized conception of the meaning of various terms relating to physical properties (principally optical, magnetic, and thermal properties) of materials to assist in material selection. References to ASTM, federal, and TAPPI standards and specifications are given in many instances so that these documents may be consulted for details on conditioning, sampling procedures, statistical interpretations, testing techniques, etc.

Analysis, modulation. Instrumentation method used in electromagnetic testing which separates responses due to factors influencing the total magnetic field by separating and interpreting individually, frequencies or frequency bands in the modulation envelope of the (carrier frequency) signal. (ASTM E-268)

Acoustic impedance. Ratio of sound pressure to product of sound velocity and area at a given surface. Acoustic impedance is frequently approximated as the product of only the density and velocity.

Acoustic impedance, specific normal. Ratio of sound pressure to component of particle velocity normal to the surface (ASTM C-384).

Acoustic reactance, resistance. Components of acoustic impedance of a material.

Anisotropy. If physical properties of a material differ along different directions, material is said to be anisotropic. Some crystals are easier to magnetize along one axis than along another and are therefore anisotropic.

Arc resistance. A measure of resistance of the surface of an electrical insulting material to breakdown under electrical stress. The time in seconds during which an arc of increasing severity is applied intermittently to the surface until failure occurs. Failure may be one of four general types: (1) material becomes incandescent and hence capable of conducting current, regaining its insulting qualities upon cooling; (2) material bursts into flame, although no visible conducting forms; (3) a thin wiry line ("tracking") forms between electrodes; or (4) surface carbonizes until there is sufficient carbon to carry current (ASTM D-495).

Autoradiography. Inspection technique in which radiation spontaneously emitted by a material is recorded photographically. The radiation is emitted by radio-isotopes that are produced in or added to a material. The technique serves to locate the position of the radioactive element or compound.

Bearing strength. The maximum bearing load at failure divided by the effective bearing area. In a pinned or riveted joint, the effective area is calculated as the product of the diameter of the hole and the thickness of the bearing member. (ASTM D-953, ASTM-E-238)

Betatron. A device for accelerating electrons by means of magnetic induction.

Black light. Electromagnetic radiation not visible to the human eye. The portion of the spectrum generally used in fluorescent inspection falls in the ultraviolet region between 3300 to 4000 angstrom units, with the peak at 3650 angstrom units.

Bond strength. Stress, i.e., tensile load divided by area of bond, required to rupture a bond formed by an adhesive between two metal blocks. (ASTM D-952)

Breakdown ratio, surface. Ratio of arc resistance after tracking to arc resistance before tracking. Also called *surface breakdown voltage ratio.* Can be determined by allowing specimen to cool after an arc-resistance test, then repeating the test. (ASTM D-495)

Breakdown voltage. Voltage at which a material fails in a dielectric strength test.

Breaking load. The load that causes fracture in a tension, compression, flexure, or torsion test.

Brittleness temperature. Temperature at which plastics and elastomers exhibit brittle failure under impact conditions specified in ASTM D-746. A test for determining brittleness temperature in plastic film is presented in ASTM D-1790.

Brightness. Relative amount of light reflected by a material. Measured by the reflectance of a material.

Bulk factor. Ratio of volume of powdered material to volume of solid piece. Also, ratio of density of solid material to apparent density of loose powder.

Bursting strength. A measure of the ability of materials in various forms to withstand hydrostatic pressure. (ASTM D-1180, ASTM D-751)

Capacitance. Capacity of material for storing electrical energy. Is often used in calculating dielectric constant of a material. (ASTM D-150)

Capacity, specific. An alternate term for *dielectric constant.*

Charpy test. A pendulum-type single-blow impact test in which the specimen, usually notched, is supported at both ends as a simple beam and broken by a falling pendulum. The energy absorbed, as determined by the subsequent rise of the pendulum, is a measure of impact strength or notch toughness. (ASTM E-23, ASTM A-327, ASTM D-256, ASTM D-758)

Chroma. Color intensity or purity of tone, being a degree of freedom from gray.

Chromaticity coordinates. Two of three parameters commonly used in specifying color and describing color difference.

Chromaticity diagram. Plot of chromaticity coordinates useful in comparing color of materials.

Coefficient of expansion. Fractional change in length (or sometimes volume) of a material for a unit change in temperature. Values for plastics range from 0.01 to 0.2 mil/in./$^{\circ}$C.

Coercvie force. A magnetizing force (HC) required to bring induction of a material to zero, when the material is in a symmetrically, cyclically magnetized condition. (ASTM E-269)

Coercive force, intrinsic. A magnetizing force (H_{ci}) required to bring instrinsic induction of a magnetic material to zero, when the material is in a symmetrically, cyclically magnetized condition.

Coercive force, relaxation. Reversed magnetizing force (H_{cr}) of such value that when it is reduced to zero, induction becomes zero.

Coercivity. Maximum coercive force (H_{cs}) for a material.

Cold cracks. Straight or jagged lines, usually continuous throughout their length. Cold cracks generally appear singly and start at the surface. (ASTM E-192)

Color. Property of light by which an observer may distinguish between two structure-free patches of light of the same size and shape. Neutral color qualities such as black, white, and gray which have a zero saturation or chroma are called *achromatic colors.* Color that have a finite satuation or chroma are said to be *chromatic* or *colored.* Color has three attributes: hue, lightness, and saturation. (ASTM D-307 and D-791)

Color difference. Difference in color between two materials. Both quantitative and qualitative tests are used. (ASTM D-1365)

Color, mass. Color, when viewed by reflected light, of a pigmented coating of such thickness that it completely obscures the background. Sometimes called *overtone* or *masstone.*

Compressive strength. The maximum compressive stress that a material is capable of developing, based on original area of cross section. In the case of a material that fails in compression by a shattering fracture, the compressive strength has a very definite value. In the case of materials that do not fail is compression by a shattering fracture, the value obtained for compressive strength is an arbitrary value depending upon the degree of distortion that is regarded as indicating complete failure of the material.

Conductivity, electrical. Ability of a material to conduct an electric current. The reciprocal of resistivity electrical conductivity is often expressed in percent, based on a value of 100 for the International Annealed Copper Standard (IACS), which has a resistivity of 10.371 ohm-cir mil/foot or 1.7241 microhm-cm at 68°F.

Contact scanning. In ultrasonics, a planned systematic movement of the beam relative to the object being inspected, the search unit being in contact with and coupled to this object by a thin film of coupling material.

Contrast ratio. Measure of hiding power or opacity. Ratio of the reflectance of a material having a black backing to its reflectance with a white backing.

Continuous method. A method of magnetic particle testing in which the indicating medium is applied while the magnetizing force is present.

Core loss. Active power (watts) expended in a magnetic circuit in which there is a cyclically alternative induction. Measurements are usually made with sinusoidally alternating induction. (ASTM A-346)

Core loss, specific appparent. Product of rms-induced voltage and rms-exciting current for a ferromagnetic core, where induced voltage is approximately sinusoidal. Specific apparent core loss in apparent W/lb or kg is apparent core loss divided by core weight.

Core loss, standard. Specific core loss at an induction of 10 kilogausses and a frequency of 60 Hz, designated P/10/60.

Covering power. Alternative term for *opacity* or *hiding power of paints.*

Cracks, base metal. Discontinuity resulting from very narrow separations of metal. (ASTM E-99)

Cracks, welds, transverse, longitudinal. Discontinuity resulting from very narrow separation of metal. (ASTM E-99)

Creep. Deformation that occurs over a period of time when a material is subjected to constant stress at constant temperature. In metals, creep usually occurs only at elevated temperatures. Creep at room temperature is more common in plastic materials.

Current-flow method. A method of magnetizing by passing current through a component via prods or contact heads. The current may be alternating, rectified alternating, or direct. (ASTM E-269)

Currie point. Temperature at which ferromagnetic materials can no longer be magnetized by outside forces, and at which they lose their residual magnetism (approximately 1200° to 1600°F for many metals).

Demagnetization. Reduction in degree of residual magnetism in ferromagnetic materials to an acceptable level.

Demagnetizing coefficient. Ratio (D_D) of extent to which an applied magnetizing force, as measured in a vacuum, exceeds the magnetizing force in a material, to the extent to which the induction in a material exceeds the induction in a vacuum for the same magnetizing force (i.e., intrinsic induction).

Demagnetizing curve. Part of normal induction curve or hysteresis loop that lies between the residual induction point B_r, and the coercive force point, H_c.

Density. Fundamental property of matter that is a measure of the compactness of its particles. Density is expressed as the ratio of mass to volume and depends on the composition of the specimen, its homogeneity, temperature, and, especially in the case of gases, on pressure. It is equal to specific gravity multiplied by the difference in weight between a unit volume of air and a unit volume of water at the same temperature.

Density, apparent. Most commonly measured for powders. It is a measure of their fluffiness or bulk, and is useful in calculating bulk factor for determining proper charges to the mold. (ASTM D-212 for metal powder, D-1182 for thermoplastics, D-1457 for TFE powders, and B-329 for refractory metal powders.)

Density, bulk. Mass per unit volume of a powdered material as determined in a reasonably large volume. Recommended test for plastic molding powders is ASTM D-1182. Term is also used for refractory brick (ASTM D-134), fired whiteware ceramics (ASTM C-373), and granular refractories (ASTM C-357).

Density, green or pressed. Density of an unsintered compact.

Density, tap. Apparent density of a powder obtained by measuring volume in a receptacle that is tapped or vibrated during loading in a specified manner.

Density, true. Alternate term for density, as opposed to apparent density.

Depth of penetration. Depth at which magnetic field strength or intensity of induced eddy currents has decreased to 37 percent of its surface value. Also known as standard depth of penetration or skin depth. (ASTM E-268)

Dielectric constant. Normally the relative dielectric constant. For practical purposes, the ratio of capacitance of two electrodes separated solely by an insulating material to its capacitance when electrodes are separated by air (ASTM D-150). Also called relative are permittivity and specific capacity. Low values are desirable when material is used as an insulator, high values used in a capacitor. Dielectric constant generally increases with temperature, humidity, exposure to weather, and deterioration. For most materials, dielectric constant varies considerably with frequency, and to a lesser extent with voltage as a result of polarization.

Dielectric proof voltage test. An acceptance test, nondestructive to acceptable material (ASTM D-1389). An insulating material passes between two electrodes at a uniform rate. Proof voltage across electrodes is generally selected as percentage of dielectric strength (short-term) of material, or as a multiple of dielectric breakdown of an air gap of equivalent thickness. Results are reported in terms of frequency of breakdown occurrance.

Dielectric strength. (1) That property of an insulating material that enables it to withstand electric stresses successfully. (2) The highest electric stress that an insulating material can withstand for a specified time without occurrence of electrical breakdown by any path through its bulk. Its value is given by potential difference in volts divided by thickness of test specimen in thousandths of an inch, i.e., in volts per mil (ASTM D-149). (3) The highest potential difference that a specimen of insulating material of given thickness can withstand for a specified time without occurrence of electrical breakdown through its bulk. Four basic tests are used: short-time, step-by-step, slow rate of rise (generally an alternate to step-by-step), and long-time.

Diffuse light transmission factor. Ratio of transmitted to incident light for translucent reinforced plastic building panels (ASTM D-1494). Property is an arbitrary index of comparsion and is not related directly to luminous transmittance.

Diffusion value. Measure of reflective and transmissive diffusion characteristics of plastics (ASTM D-636). Index of scattering or diffusion of light by a material compared to a theoretically perfect light-scattering material (rated as 1.0).

Dimensional stability. Ability of a material to retain precise shape in which it was molded, fabricated, or cast.

Discontinuity. Any interruption in the normal physical structure or configuration of a part such as cracks, laps, seams, inclusions, or porosity.

Dry method. Magnetic-particle inspection in which particles employed are in dry-powder form. (ASTM E-269)

Ductility. The extent to which a material can sustain plastic deformation without rupture. Elongation and reduction of area are common indexes of ductility.

Eddy currents. Currents caused to flow in an electrical conductor by time or space variation, or both, of an applied magnetic field. (ASTM E-268)

Eddy current loss. The part of core loss that is due to current circulation in magnetic materials as a result of electromotive forces induced by varying induction.

Eddy-current testing. Nondestructive testing method in which eddy-current flow is induced in the object being tested. Changes in flow caused by variations in the specimen are reflected into coils for subsequent analysis.

Elasticity. The ability of a material to return to its original configuration when the load causing deformation is removed.

Electromagnetic testing. A nondestructive test method for engineering materials, including magnetic materials, which uses electronmagnetic energy having frequencies less than those of visible light to yield information regarding the quality of the tested material. (ASTM E-268)

Elongation. In tensile testing, the increase in gauge length, measured after fracture of the specimen within the gauge length, usually expressed as a percentage of the original gauge length.

Excitation rms. The rms alternating current required to produce a specified induction in a material. For inductions of less than 10 kilogausses, rms excitation (H_z) is usually expressed in oersteds; for high inductions, in amp-turns/inch.

False indication. In nondestructive inspection, an indication that may be interpreted erroneously as a discontinuity.

Fatigue. Permanent structural change that occurs in a material subjected to fluctuating stress and strain. In general, fatigue failure can occur with stress levels below the elastic limit.

Fatigue strength. The magnitude of fluctuating stress required to cause failure in a fatigue test specimen after a specified number of loading cycles.

Ferromagnetic. A term applied to materials that can be magnetized or strongly attracted by a magnetic field. (ASTM E-269)

Film contrast. A qualitative expression of the slope of the characteristic curve of a film; that property of a photographic material that is related to magnitude of density difference resulting from a given exposure difference. (ASTM E-94)

Flash point. Temperature at which vapor of a material will ignite when exposed to a flame in a specially designed testing apparatus.

Flash radiography. High-speed radiography in which exposure times are sufficiently short to give an ublurred photograph of moving objects such as fired projectiles or high-speed machinery.

Flexivity. Index of change in flexure with temperature of thermostat bimetals (ASTM D-106). Change of curvature of longitudinal centerline of specimen per unit temperature change per unit thickness.

Fluorescence. Property of emitting visible light as a result of, and only during absorption of, radiation from other energy sources. (ASTM E-270)

Fluorescent liquid penetrant. Highly penetrating liquid used in performance of liquid penetrant testing and characterized by its ability to fluoresce under black light. (ASTM E-270)

Fluorescent magnetic particle inspection. A magnetic particle inspection process employing a powdered ferromagnetic inspection medium coated with material that fluoresces when activated by light of suitable wavelength. (ASTM E-269)

Flux density. The strength of a magnetic field. (ASTM E-269)

Flux leakage. Magnetic lines of force that leave and enter the surface of a part due to a discontinuity that forms poles at the surface of the part. (ASTM E-268)

Freezing point. Temperature at which a material solidifies on cooling from molten state under equilibrium conditions.

Gas holes. Round or elongated, smooth-edged dark spots, which may occur individually, in clusters, or distributed throughout a casting section. Gas holes are usually caused by trapped air or mold gases. (ASTM E-310, ASTM E-272)

Gel point. Stage at which a liquid begins to exhibit pseudoelastic properties. Gel-point stage may be conveniently observed from inflection point on a viscosity-time plot.

Glassy transition. Change in an amorhpous ploymer or amorphous regions of partially crystalline polymer from (or to) viscous or rubbery condition to (or from) a hard and relatively brittle one.

Gloss. Ratio of light reflected from surface of a material to light incident on surface when angles of incidence and reflectance are equal numerically but opposite in sign.

Gravity, absolute specific. Ratio of weight in a vacuum of a given volume of material to weight in a vacuum of an equal volume of gas-free distilled water. Sometimes called simply *specific gravity*, but it is not the same as specific gravity, which is based on weight measurements in air. For practical purposes, *specific gravity* is more commonly used.

Hall effect. Deflection by a magnetic field of an electric current traveling through a thin film. Force experienced by the current is perpendicular to both magnetic field and direction of current flow.

Hardness. A measure of the resistance of a material to localized plastic deformation. Hardness is tested and expressed in numerous ways (Brinell hardness, Rockwell hardness, scleroscope, durometer, etc.). A table relating various hardness values appears in ASTM E-140.

Haze. A measure of the extent to which light is diffused in passing through a transparent material. Percentage of transmitted light that, in passing through the material, deviates from the incident beam by forward scattering. (ASTM D-1003)

Heat distortion point. Temperature at which a standard test bar (ASTM D-048) deflects 0.010 inches under a stated load of either 86 or 264 psi.

Heat of fusion. Amount of heat per unit weight absorbed by a material in melting.

Heat, specific. Component of thermal diffusivity of a material. Quantity of heat required to change temperature of unit mass of material $1°$, commonly expressed in Btu/lb/$°$F. Mean specific heat is average specific heat over specified heat over specified temperature range. (ASTM C-351)

Hiding power. Ability of paint to obscure surface. Also called *opacity*. Usually measured by contrast ratio.

Hole size. Diameter of the hole in a reference block, which determines the area of the hole bottom. (ASTM E-127)

Hot tears. Ragged dark lines of variable width and numerous branches. Hot tears have no denitie lines of continuity and may exist in groups. They may originate internally or at the surface. (ASTM E-192)

Hue. Attribute of color that determines its position in wavelength spectrum; i.e., whether it is red, yellow, green, etc.

Hysteresis. Retardation of magnetic effect when magnetizing forces acting on a ferromagnetic body are changed. (ASTM E-269)

Hysteresis loop, magnetic. Curve showing relationship between magnetizing force and magnetic induction in a material in a cyclically magnetized condition. For each value and direction of magnetizing force, there are two values of induction; when the magnetizing force is increasing, and when the magnetizing force is decreasing. Result is actually two smooth curves joined at ends to form a loop. Area within loop represents energy expended in material as heat. Where alternating current is used, the loop indicates amount of energy transformed into heat during each cycle. Magnetic hysteresis loop is commonly determined by normal induction measurements giving values of residual induction (B_r) and coercive force (H_c).

Hysteresis loss, magnetic. Power (P_a) expended in magnetic material as result of magnetic hysteresis when induction is cyclic. Enclosed area of magnetic hysteresis loop. It is one component of specific core loss and is a function of coercivity and retentivity.

Impact strength. The energy required to fracture a specimen subjected to shock loading as in an impact test.

Impedance analysis. A type of signal processing sometimes used in eddy-current analysis. A signal that represents a change, in both amplitude and phase, of an impedance vector is resolved into any pair of its components separated $90°$ in phase.

Inclusions. Isolated, irregular, or elongated variations of magnetic particles occurring singly, in a linear distribution, or scattered randomly in feathery streaks. (ASTM E-125)

Incomplete fusion. Fusion that is less than complete. Failure of weld metal to fuse completely with base metal or preceding leads. (ASTM E-99)

Indication. That which marks or denotes presence of a discontinuity. In liquid-penetrant inspection, the presence of detectable bleedout of liquid penetrant from material discontinuities (ASTM E-270). In magnetic-particle inspection, the detectable magnetic particle buildup resulting from interruption of the magnetic field. (ASTM E-269)

Indications. Eddy-current signals caused by any change in the uniformity of a tube. (ASTM E-215)

Induction, intrinsic. Extent to which induction in magnetic material exceeds induction in vacuum for a particular magnetizing force (B_1). The measured induction minus the product of magnetizing force and permeability of a vacuum. In the cgs electromagnetic system (gausses, oersteds), permeability of a vacuum is arbitrarily taken as unity, so that intrinsic induction is numerically, equal to normal induction minus magnetizing force. Intrinsic induction is sometimes plotted against magnetizing force.

Induction, normal. Limiting induction (in either directional) or material in a symmetrically, cyclically magnetized condition. Induction is measured for both increasing and decreasing magnetizing force, and for magnetizing force in both directions, to develop magnetizing and demagnetizing curves. Complete magnetizing and demagnetizing curves from a magnetic hysteresis loop. Other important properties that can be determined from the normal induction curve include residual induction, coercive force, and normal permeability. (ASTM A-341)

Induction, residual. Measure of permanence of magnetization. Magnetic induction (B_r) corresponding to zero magnetizing force in a magnetic material in a symmetrically, cyclically magnetized condition. Induction at either of two points where magnetic hysteresis loop intersects the B axis. Determined from normal induction data.

Induction, saturation. Maximum intrinsic induction possible in a material (B_s).

Iron loss. Alternate term for *specific core loss.*

Lamination factor. Measure of effective volume of laminated structure, which is composed of strips of magnetic material. Ratio of volume of structure as caluclated from weight and density of strips, to measured solid volume of structure under pressure. (ASTM A-344)

Latent heat. Heat that must be applied to material to effect change in state without change in temperature. For example, latent heat of fusion of ice water is 80 cal/gm.

Leak testing. Technique of liquid-penetrant testing wherein penetrant is applied to one side of a material and observation is made on the opposite side to ascertain the presence of voids extending throughout the material. (ASTM E-270)

Lightness. Attribute of color that permits any color to be classified as equivalent to one of a series of grays ranging between black and white. The term *shade* is often used to describe differences in lightness. Lightness difference is measured as part of the color-difference test.

Linear discontinuities. Ragged lines of variable width that may appear as a single jagged line or exist in groups. They may or may not have a definite line of continuity, often originate at the casting surface, and usually become smaller as a function of depth. (ASTM E-215)

Liquid temperature. Temperature at which an alloy finishes melting during heating or starts freezing during cooling, under equilibrium conditions. For pure metals, same as solidus temperature and known simply as *melting point.* Effective liquids temperature is raised by fast heating and lowered by fast cooling.

Loss angle. Measure of electrical power losses in insulating material subjected to alternating current. The arc-tangent of dissipationg factor and, thus, the angle between a material's parallel resistance and its total parallel impedance in a vector diagram when the material is used as dielectric in a capacitor. Complement of phase angle. Sometimes called *phase defect angle.* Conventionally designated as δ.

Loss factor. Measure of electrical-power loss in insulating material subjected to alternating current. Product of dissipation factor and dielectric constant. Loss factor is expressed in the same units as dissipation factor. Low loss factor is generally desirable. Loss factor generally increases with humidity, weathering, deterioration, and exponentially with temperature.

Magnetic energy product, maximum. Maximum external energy that contributes to magnetization of material. Corresponds to maximum value of the abscissa for magnetic-energy-product curve.

Magnetic-energy-product curve. Curve obtained by plotting product of induction and demagnetizing force against corresponding values of induction; i.e., product of coordinates of demagnetizing curve (B_d, H_d) as abscissa versus induction (B_d) as ordinate.

Magnetic field strength. The measured intensity of a magnetic field at any given point, usually expressed in oersteds. (ASTM E-269)

Magnetic hysteresis. In a magnetic material, such as iron, a lagging in the values of the resulting magnetization due to a changing magnetic force. (ASTM E-269)

Magnetic-inspection flaw indications. Accumulation of ferromagnetic particles along areas of flaws or discontinuities due to distortion of magnetic lines of force in those areas. (ASTM E-269)

Magnetic-particle inspection. A nondestructive method for detecting cracks or other discontinuities at or near the surface of ferromagnetic material. (ASTM Method E-138, ASTM E-269)

Melt index. Amount, in grams, of a material (usually a thermoplastic) that can be forced through a 0.0825-inch orifice when subjected to a 2160-gram force for 10 minutes at 190°C.

Melting point. For a pure metal, the temperature at which liquefaction occurs on heating or solidification occurs on cooling under equilibrium conditions. Alloys and other materials have a melting range, and melting point is often desired as a temperature near the bottom of this range at which observable change caused by melting occurs. In general, melting point is determined by heating a small specimen to a temperature not too far below the melting point, then raising temperature slowly (a few degrees a minute) and watching closely for first indication of liquefaction.

Melting range. Range of temperature between solidus temperature and allow liquidus temperature.

Moisture-vapor transmission. Rate at which water vapor penetrates film over given time at specified temperature and relative humidity (e.g., gm-mil/24 hr/100 in.^{2}).

Molecular weight, average. Modecular weight of polymeric materials determined by viscosity of the polymer in solution at a specific temperature. Gives average molecular weight of molecular chains in a polymer independent of specific chain length.

Neutron absorption cross section. Measure of probability that a single nucleus of material subjected to nuclear bombardment will intercept and interact with an incoming neutron. When expressed in square centimeters, it is the fraction of neutrons contained in a beam of 1-cm^2 cross section that can be expected to be intercepted by a single nucleus. Common units are barns/atom, where 1 barn = 10^{-24} cm^2. Low cross section is desirable for nuclear-reactor core materials. High value for reactor shielding. Effective neutron-absorption cross section depends on level of incident energy; for test neutron scattering, it approximates theoretical area of nucleus.

Opacity. Degree to which material or coating obstructs transmittance of visible light. Term is used primarily with nearly opaque materials, for which opacity is generally reported by contrast ratio. Opacity of light-transmitting materials is usually expressed by its opposite or complementary property; i.e., total transmittance of diffuse light transmission factor.

Optical density. Term indicating absorption of light. Numerically equal to the logarithm to base 10 of reciprocal of transmittance of material.

Optical distance. Product of actual length of light path and refractive index of material.

Optical distortion. Any apparent alternation of geometric pattern of an object when seen through a plastic or as reflection from plastic surface.

Peel strength. A measure of the strength of an adhesive bond. The average load per unit width of bond line required to part bonding materials where the angle of separation is 180° and the separation rate is 6 inch/minute. (ASTM D-903)

Penetrant. A liquid that has unique properties that render it highly capable of entering small openings. This characteristic makes the liquid especially suitable for detecting discontinuities in material. (ASTM E-270)

Penetrant inspection. A method of nondestructive testing for determining the existence and extent of discontinuities that are open to the surface of the material being inspected. The indications are made visible through the use of of a dye or fluorescent chemical in the liquid employed as the inspection medium.

Permeability. (1) Passage or diffusion of gas, vapor, liquid, or solid through a barrier without affecting it. (2) Rate of such passage.

Permeability, ac. Ratio of maximum value of induction to maximum value of magnetizing force for material in a symmetrically, cyclically magnetized condition.

Permeability, apparent impedance. Index of ease of magnetization. Ratio of induction to rms excitation in material that is symmetrically, cyclically, magnetized.

Permeability, effective ac. Denotes capacitance needed to balance effective induction of a material when magnetized as a core.

Permeability, incremental ac. When compared with normal ac permeability, an index of increase in mean induction that can be expected when ac magnetization is superimposed upon dc magnetization. (ASTM A-343)

Permeability, initial. Slope of normal induction curve at zero magnetizing force.

Permeability, intrinsic. Ratio of intrinsic induction to corresponding magnetizing force.

Permeability, magnetic. A factor, characteristic of material, that is proportional to magnetic flux density (magnetic induction) B, produced in material by the magnetic field divided by the intensity of field H. Different names are given to permeabilities measured under different circumstances: *initial, incremental, reversible.*

Permeability, maximum. Maximum value of normal permeability for a magentic material.

Permeability, normal. Ratio of normal induction to corresponding magnetizing force. Slope of magnetizing or normal induction curve at specified value of magnetizing force.

pH value. Numerical expression to describe hydrogen ion concentration. Therefore, pH is simply a number denoting degree of acidity or alkalinity.

Phase angle. Angle by which voltage leads current in material subjected to ac current. Angle between the parallel reactance of the material and its total parallel impedance in a vector diagram when the material is used as dielectric in capacitor. Complement of loss angle.

Phase transition. Abrupt change in physical properties as temperature is changed continuously. Freezing of water at $0°C$ is an example.

porosity. Relative extent of volume of open pores in a material. Ratio of pore volume to overall volume of the material in percent. (ASTM D-328, D-116, E-125)

Power factor. Ratio of power expended in circuit to the product of emf acting in circuit and current in it; i.e., ratio of watts to volt-amperes. Because energy loss is directly proportional to frequency, low power factor is essential in materials used at high frequencies.

Quality factor. Reciprocal of dissipation factor or loss tangent. Also called *storage factor.* When materials have approximately the same dielectric constant, a higher quality factor indicates less power loss.

Radiograph. A permanent visible image on a recording medium produced by penetrating radiation passing through the material being tested. (ASTM E-142)

Radiographic inspection. The use of X-rays or nuclear radiation, or both, to detect discontinuities in a material, and to present their images on a recording medium. (ASTM E-142)

Reactive power, specific. For a specified normal induction in a material, the component of applied ac power, which is "reactive"; i.e., returned to the source, when polarity is reversed. Product of induced voltage and reactive current per unit weight of material; i.e., product of voltage, current, and sine of phase angle. (ASTM A-346)

Reactive power, specific incremental. Specific reactive power in a magnetic material when subjected simultaneously to a unidirectional ("biasing") and an alternating magnetizing force. This can be determined by the test used for incremental ac permeability. (ASTM A-343)

Reflectance, infrared. Measure of infrared brightness. Two direct optical methods are used. In each, a standard beam of light is thrown on a specimen normal to the surface and reflectance is measured from the same position, using filters to select specified wavelengths for viewing in the infrared range. (Fed 141-6341, 141-6242)

Reflectance, luminous. Measure of brightness. Ratio of light reflected to light incident. Luminous reflectance of a material of such thickness that any increase in thickness would fail to change the value, is called *luminous reflectivity*. Calculated from spectral reflectance (including spectral component) and spectral luminosity, it is a function of spectral and angular distribution of incident light energy. Also called *total luminous reflectance* to distinguish it more readily from luminous directional reflectance.

Reflectance, luminous directional. Ratio of brightness to brightness that an ideally diffusing, completely reflecting light surface would have when illuminated and viewed in the same manner. Luminous directional reflectance is commonly determined for an opaque material illuminated as 45° and viewed normal to the surface.

Reflectance, spectral. Measure of brightness. Ratio of light reflected by a specimen to homogeneous light energy incident on it. It depends on the angular distribution of the incident energy. The term *spectral reflectivity* is used to indicate the inherent property of a material of such thickness that any increase in thickness would fail to change the spectral reflectance. It may be expressed in the form of a curve of luminous directional reflectance versus wavelength.

Reflectance, spectral directional. Measure of brightness. The spectral reflectance which an ideal diffusing surface would need to appear the same as the specimen under test when illuminated and viewed the same way. I depends on the angular distribution of incident light and on the direction of viewing. The term *spectral directional reflectivity* is used to denote an inherent property of the material as opposed to that of a particular object.

Reflectivity, apparent. Term used for *luminous directional reflectance* (reflectivity) of paper. (TAPPI T442m)

Refraction, index of. Also called *refractive index*. Ratio of velocity of light in vacuum to its velocity in the material. Ratio of the sine of the angle of incidence of light to the sine of the angle of refraction. (ASTM D-542)

Reluctivity. Reciprocal of magnetic normal permeability of a material.

Remanence. Magnetization remaining when ferromagnetic material is removed from a magnetic field. Conversely, a material that exhibits remanence is ferromagnetic.

Residual method. A method of magnetic particle testing in which the indicating medium is applied after the magnetizing force has been discontinued. (ASTM E-138)

Resistance, insulation. A measure of the ability of a nonconducting material to resist flow of electric current. Ratio, in ohms or megohms, of direct voltage applied between two electrodes in contact with material, to total current between electrodes. It a property of form, not material, and this useful only in comparing materials having the same form. (ASTM D-257)

Resistance, interlamination. Measure of resistance of a laminated magnetic structure to stray, intersheet power losses. Generally, a measure of the effectiveness of sheet surface oxides or coatings in reducing intersheet losses. (ASTM A-344)

Resistivity. Measure of electrical resistance properties of a conducting material, given by the resistance of a unit length of unit cross-section area. Reciprocal of conductivity. ASTM tests include D-193 for electrical conductor materials; D-63 for materials used in resistors, heating elements, and electrical contacts; and A-344 for magnetic materials. The term *resistivity* and *volume resistivity* are used interchangeably for conductors.

Resistivity, electrical. The electrical resistance of a material: (1) per unit length and unit cross-section area or (2) per unit length and unit weight.

Resistivity, insulation. A measure of the ability of a material to resist flow of electric current. Insulation resistance per unit volume. This property makes no distinction between volume resistivity and surface resistivity, even where no such distinction is made in the test. However, it is more common to use the term *volume resistivity*.

Resistivity, surface. A measure of the ability of surface of a dielectric material to resist flow of electric current. Ratio of potential gradient parallel to current along surface of material, to current per unit width of surface. Together with volume resistivity, surface resistivity is often used to check purity of an insulating material during development and its uniformity during processing. (ASTM D-257)

Resistivity, volume. A measure of the ability of material to resist flow of an electric current. The ratio, of electrical potential gradient parellel to current in a material, to current density. Although volume resistivity is of interest, knowledge of its changes with temperature and humidity are even more important, and curves plotted against temperature are sometimes particularly desirable. (ASTM D-257)

Resistivity, weight. A measure of the ability of a material to resist flow of electric current. Also called *mass resistivity*. Electrical resistance of a material per unit length and unit weight. A term generally reserved for metallic materials, i.e., conductors. Weight resistivity in ohm-lb/mile^2 or ohm-gm/m^2 is equal to $WR/L_1 L_2$ where W is the weight of the specimen in pounds or grams, R is the measured resistance in ohms, L_1 is the gauge length in miles or meters used to determine R, and L_2 is the length of the specimen.

Retentivity. The maximum value of residual induction in magnetic material (B_{rs}). A measure of the permanence of magnetization. Low values are desirable for electromagnetic devices; high values, for permanent magnets, relays, and magnetos.

Rupture strength. The nominal stress developed in a material at rupture. This is not necessarily equal to ultimate strength. Since necking is not taken into account in determining rupture strength, it seldom indicates true stress at rupture.

Saturation. Attribute of a color that determines the extent to which it differs from a gray of the same lightness.

Saturation, magnetic. If a ferromagnetic material is magnetized by an external field, magnetization cannot increase indefinitely with increasing external field. The maximum value of magnetization is saturation magnetization.

Shade. Term used to characterize difference in lightness between two surface colors, the other attributes of color being essentially constant. A lighter shade of a color has higher lightness but about the same hue and saturation.

Shear strength. The maximum shear stress that can be sustained by a material before rupture. The ultimate strength of the material when subjected to shear loading. For shear strength in plastics, see ASTM-D-732; for structural adhesives, ASTM E-229; for timber, ASTM D-143 and ASTM D-198.

Sheen. The 85° specular gloss or nometallic materials.

Softening point. Indication of maximum temperature to which a nonmetallic material can be heated without loss of its normal "body." Minimum temperature at which a specified deformation occurs under a specified load. Softening point is often determined for materials that have no definite melting point, i.e., materials that gradually change from brittle or very thick and slow-flowing materials, to softer materials or loss viscous liquids. Vicat softening point is commonly used for some plastics. (ASTM D-1525)

Softening range. Range of temperature in which a plastic changes from rigid to soft state. Values depend on the method of test. Sometimes referred to as *softening point.*

Solidus temperature. Temperature at which an alloy begins to melt during heating or finishes freezing during cooling. For pure metals, same as liquids temperature and known simply as *melting point.*

Sound-absorption coefficient, normal incidence. A measure of the effectiveness of material in absorbing sound energy. Fraction of normally incident sound energy asborbed by a material assumed to have an infinite surface. (ASTM C-384)

Sound-absorption coefficient, statistical. A measure of the effectiveness of a material in absorbing sound energy. Fraction of incident sound energy absorbed by a material under conditions where it is subject to equal sound energy from all directions over a hemisphere, i.e., under reverberant sound conditions. Measurement is expensive and time-consuming, requiring a specially constructed reverberation chamber. Thus, property is sometimes estimated from specific normal sound-absorption coefficient or specific normal acoustic impedance. The normal coefficient is about half statistical coefficient for very low values and approaches equality with it at very high values. Maximum numerical difference occurs at intermediate values and is about 0.25 to 0.35.

Specific gravity. Ratio of the mass of a unit volume of a material at a stated temperature to the mass of same volume of gass-free distilled water at a stated temperature. If material is a solid, volume shall be that of the impermeable portion. Two basic methods are used for solids: liquid displacement (ASTM D-792, D-176, C-135, and C-329), and change in weight (D-792, D-311, and D-328). Three methods are commonly used for liquids: plummet displacement (D-176), volumetric (D-115), and hydrometer (D-901).

Sponge shrinkage. Found in heavier sections (generally more than 2 inches thick). Sponge shrinkages appear on radiographs as dark areas, lacy in texture, usually with a diffuse outline. (ASTM E-310, ASTM E-272)

Storage factor. Reciprocal of dissipation factor or loss tangent. Also called *quality factor.*

Storage life. Period of time during which resin or adhesive can be stored under specified temperature conditions and remain suitable for use. Storage life is sometimes called *shelf life.*

Stress corrosion cracking. Failure of a material due to the combined effects of corrosion and stress. Generally, stress corrosion cracking refers to the phenomenon by which stress increases the corrosion rate.

Subsurface discontinuity. Any defect that does not open onto the surface of the part in which it exists. (ASTM E-269)

Surge method. Inspection by first using a high surge of magnetizing force, followed by a reduced magnetic field during application of a powdered ferromagnetic inspection medium. (ASTM E-269)

Tear resistance. A measure of the ability of sheet or film materials to resist tearing. For paper, it is the force required to tear a single ply or paper after the tear has been started (ASTM D-698). Three standard methods are available for determining tear resistance of plastic films. ASTMD-1004 details a method for determining tear resistance at low rates of loading. A test in ASTM D-1922 measures the force required to propagate a precut slit across the sheet specimen. ASTM D-1038 gives a method for determining tear-propagation resistance that is recommended for acceptance testing only. Tear resistance of rubber is force required to tear a specimen 1 inch in thickness under conditions outlined in ASTM D-624. For tear resistance of textiles, see ASTM D-1424.

Temperature-resistance. Relationship between the temperature of a material and its electrical resistance, expressed by a multivalue table, by a graph, by a calculated temperature coefficient, or by calculated values for constants in a standard mathematical equation. Provides information needed for design and use of resistance heating elements and precision resistors in electrical and electronic circuits.

Tensile strength. The ultimate strength of a material subjected to tensile loading. The maximum stress developed in a material in a tension test.

Thermal absorptivity. Fraction of the heat impinging on a body that is absorbed.

Thermal conductivity. Rate of heat flow in a homogeneous material, under steady conditions, through unit area, per unit temperature gradient in direction perpendicular to the area. Thermal conductivity is usually expressed in English units as $Btu/ft^2/hr/°F$ for a thickness of 1 inch. (ASTM C-177)

Thermal diffusivity. Rate at which temperature diffuses through a material. Ratio of thermal conductivity to product of density and specific heat, commonly expressed in ft^2/hr.

Thermal emissivity. Ratio of heat emitted by a body to heat emitted by a black body as same temperature.

Thermal expansion. When heated or cooled, materials undergo a reversible change in dimensions that depends on the original size of the body and the temperature range studied. In addition to the change in dimensions, a change in shape may occur; i.e., the expansion or contraction may be anisotropic.

Thermal expansion, cubical coefficient of. Rate at which material increases in volume when heated. Unit increase in volume of material per unit rise in temperature over specified temperature range. Mean coefficient in mean slope of this curve over specified range of temperature.

Thermal expansion, linear coefficient of. Extent to which a material elongates when heated. Unit increase in length per unit rise in temperature over specified temperature range. Slope of the temperature to dilation curve at a specified temperature. Mean coefficient is mean slope between two specified temperatures. (ASTM D-1037)

Thermal insulating efficiency. Ratio of heat saved by an insulation to the heat that would be lost without insulation.

Thermal reflectivity. Fraction of heat impinging on a body that is reflected.

Transmittance, diffuse. Ratio of light scattered by material to light incident on it. Ratio depends on spectral distribution of incident energy. The analogous spectral ratio is diffuse luminous transmittance.

Transmittance, diffuse luminous. Ratio of light scattered by a material to light incident on it.

Transmittance, luminous. Ratio of light transmitted by material to light incident on it. Calculated from spectral transmittance and spectral luminosity, it is a function of spectral distribution of incident light energy. In test for haze, it is called *total luminous transmittance* to distinguish it from its diffuse component, *diffuse luminuous transmittance.*

Transmittance, thermal. Rate at which heat it transmitted through a material by combined conduction, convention, and radiation. Overall coefficient of heat transfer. Term used particularly for textile fabrics and batting where heat transfer between opposite surfaces is not confined to conduction. For solid materials, thermal conductivity is a measure of thermal transmittance. (ASTM D-1518)

Transmittance, total. Ratio of light transmitted by material to light incident on it. The ratio depends on spectral distribution of incident light energy.

Ultrasonic. In ultrasonic nondestructive testing, pulses of sound energy are created by electrically exciting a crystal, called a *transducer*, which has piezoelectric properties, causing the crystal to vibrate. Sound energy is transmitted into the material being tested through a couplant, which can be oil, water, or glycerin. Sound that returns to the crystal as reflected energy is converted to electrical signals by the transducer and can be monitored on an oscilloscope. The time and amplitude relation of the reflected signals can be interpreted to show location of internal reflecting surfaces in three dimensions.

Viscosity. Resistance to flow exhibited within a material. Can be expressed in terms of relationship between applied shearing stress and resulting rage of strain in shear. Viscosity usually means "Newtonian viscosity," in which case the ratio of shearing stress to rate of shearing strain in constant. In non-Newtonian behavior (which occurs with plastics), the ratio varies with the shearing stress. Such ratios are often called the *apparent viscosities* at the corresponding shearing stresses.

Viscosity, absolute. In a fluid, the tangential force on a unit area of either of two parallel planes at unit distance apart when the space between planes is filled with the fluid in question and one of the planes moves with unit differential velocity in its own plane.

Viscosity coefficient. Shearing stress necessary to induce a unit velocity flow gradient in a material. The viscosity coefficient of a material is obtained from ratio of shearing stress to shearing rate.

Voltage ratio. Alternate term for surface-breakdown rate.

Volume, specific. Reciprocal of density.

Water-vapor permeability. Water vapor transmission through a homogeneous body under unit vapor pressure difference between two surfaces, per unit thickness.

Water-vapor transmission. Steady-state time rate of water-vapor flow through unit area of humidity at each surface.

Xeroradiography. A process utilizing a layer of photoconductive material on an aluminum sheet upon which is placed an electrical charge. After X-ray exposure, the electrical potential remaining on the plate in the form of a latent electrical pattern is developed by contact with a cloud of finely dispersed powder.

X-ray. Electromagnetic radiation, of wavelength less than about 500 Angstrom units, emitted as the result of deceleration of fast-moving electrons (*brems-*

strahlung, continuous spectrum) or decay of atomic electrons from excited orbital states (characteristic radiation). Specifically, the radiation produced when an electron beam of sufficient energy impinges upon a target of suitable material.

Yield point. The first stress in a material, usually less than the maximum attainable stress, at which an increase in strain occurs without an increase in stress. Only certain metals exhibit a yield point. If there is a decrease in stress after yielding, a distinction may be made between upper and lower yield points.

FLAT-METAL MANUFACTURING PROCESSES

The terms defined below shall be used in process sheets, reports, or whenever specification or notation of a particular process is required to describe the fabrication of flat metals. While these terms are more generally associated with metals, their use does not preclude flat materials of a nonmetallic nature such as sheet plastics and fibers. It should be noted, however, that not every term or its explanation is applicable to all flat nonmetallic materials.

The flat-metal manufacturing process terms are listed in the following by similarity of the operation involved.

Material-Removal Methods

1. By cleavage due to single or repetitive impact:
   Shaving
   Shearing
   Stamping
     Blanking
     Dinking
     Lancing*
     Nibbling
     Perforating
     Piercing
     Punching
     Trimming
2. By successive cutting:
   Boring
   Broaching
   Counterboring
   Countersinking
   Drilling
     Drilling, bottom
   Hobbing

*Material not necessarily removed.

Milling
Planing
Reaming
Routing
Sawing
Serrating
Shaping
Slitting
Slotting
Spotfacing
Tapping
    Tapping, bottom
3. By abrasion:
Blasting
    Peening, shot
    Sand
Buffing
Grinding
Lapping
Polishing
Superfinishing
Tumbling, barrel
4. Other:
Cutting, flame
Etching, acid

Material-Forming Methods

1. By compressing:
Bulging
Burnishing
Coining
Embossing
Flattening
Forging, cold
Forging, hot
Heading, cold
Ironing
Squeezing
Swaging, cold
Swaging, hot
Swedging
Upsetting, hot

2. By stretching:
Beading (or curling)
Bending
Dimpling
Drawing, shallow
Drawing, deep
Extruding, impact
Flanging
Forming
  Contour
  Roll
  Rubber die**
  Stretch
Hydroforming
Necking
Sizing
Spinning

**Guering process, Marform process.

ALPHABETICAL LIST OF DEFINITIONS

Following is an alphabetical list of terms used in manufacturing processes for flat metals and a brief definition of each term.

Beading. The operation of rolling over the edge of circular-shaped material, either by stamping or spinning.

Bending. The operation of forming flat materials, usually sheet metal or flat wire, into irregular shapes by the action of a punch forcing the material into a cavity or depression. Such forming is done in a punch press. When the operation involves long straight bends, the material is usually bent in a press brake.

Blanking. The operation of separating a flat piece of a certain shape a flat product preparatory to other processes or operations necessary to finalize the piece into a finished part.

Blasting. The operation of removing material by the action of abrasive material directed under air pressure on the surface of the piece part through controlled orifice openings at the end of a flexible hosing.

Blasting, sand. Same as Blasting except that the abrasive material is fine sand of some definite grit.

Boring. The operation of enlarging a round hole.

Broaching. The operation of cutting away the material by a succession of cutting teeth on a tool that is either pushed or pulled along the surface of the workpiece. Cutting may be applied to either holes or the outside edges of the piece.

Buffing. Essentially, the same as polishing except wheels and belts are made of softer material and finer abrasive is combined with a lubricant binder.

Bulging. The operation of expanding the metal below the opening of a drawn cup or shell.

Burnishing. The operation of producing smooth-finished surfaces by compressing the outer layer of the metal, either by the application of highly polished tools or by the use of steel balls, in rolling contact with the surface of the piece part.

Coining. The operation of forming metallic material in a shaped-cavity die by impact. All the material is restricted to the cavity resulting in fine-line detail of the piece part.

Counterboring. The operation of enlarging some part of a cylindrical bore or hole.

Countersinking. The operation of producing a conical entrance to a bore or hole.

Curling. See Beading.

Cutting, flame. The operation of severing ferrous metals by rapid oxidation from a jet of pure oxygen directed at a point heated to the fusion point.

Dimpling. The operation of forming semispherical-shaped impressions in flat material.

Dinking. The operation of blanking a piece out of a soft flat material by the use of sharp beveled cutting edges arranged to produce a part of a specific configuration. Dinking is similar to steel rule die.

Drawing. The operation of forming cylindrical and other shaped parts from flat stock by action of a punch pushing the metal into a cavity having the same shape as the punch.

Drawing, deep. Essentially the same as Drawing except that the final shape is the result of a series of redraws.

Drawing, shallow. Essentially the same as Drawing except that the operation is restricted to shallow pieces and operation is usually accomplished by a stamping single-action process.

Drilling. The operation of producing cylindrically shaped holes, usually a cone shape at the bottom.

Drilling, bottom. Same as Counterboring except performed with square-ended drill instead of the conical tip.

Embossing. The operation of production raised patterns on the surface of materials by use of dies or plates brought forcibly upon the material to be embossed.

Etching, acid. *Same as* Chemical etching except material used for metal removing is confined to action of acids.

Etching, chemical. The operation of removing material from the surface of a metal by the action of chemicals upon the surface.

Extruding, impact. The operation of producing cup-shaped parts by a single blow against a confined slug of cold metal until the desired length and wall thickness is reached.

Flanging. Essentially the same as Beading or Curling except the portion of metal rolled over is left in a flat form. The operation is only performed on parts with trimmed edges since the normal drawing operation will always begin with a flat piece forming the flange, which is later trimmed to size.

Forging, cold or hot. The operation of forming relatively thick sections of metallic material into a desired shape. Material may be worked cold or hot between shaped forms or shaped rolls. In the case of roll dies, the material is squeezed into shape. In the case of shaped dies, the material is formed by a succession of blows forcing the material into the shaped cavity, or the material may be squeezed into the shaped cavity by a steady pressure of the moving form on the metal.

Forming, contour. The operation of reforming material sections of all types, rolled formed sections, cold-drawn or rolled shapes, extrusions, etc., normally received in straight lengths, into various contours. (Section shape remains unchanged for full length of piece.)

Forming, roll. The operation of shaping flat-strip material into a desired form by passing it through a series of shaped rolls. Each roll brings the shape closer to its final form.

Forming, rubber die, Guerin process, Marform process. Essentially the same as Hydroforming except the rubber pad (or other suitable material) is not backed up with any pressure.

Forming, stretch. The operation of forming sheet or strip by gripping two sides or ends (depending on the shape of unformed blank or required contour) until the material takes a set.

Grinding. The operation of separating material from the surface of a material by means of abrasion. When restricted to flat materials, the term is more generally known as *surface grinding.* Abrasion may be by means of an abrasive wheel or belt. When performed by a belt, it is more commonly known as *abrasive belt grinding.*

Heading, cold. The operation of gathering sufficient stock at one end of a bar of metallic material to form a desired shape. Gathering the material is done by impact of the material against a shaped die.

Hobbing. The operation of producing teeth on the outside edge of the workpiece. Tool and workpiece move with respect to each other.

Hydroforming. The operation of drawing sheet metal using a shaped punch that is forced against the material and a rubber pad backing the material, while hydraulic pressure on the opposite side is applied.

Ironing. The operation of reducing the walls of drawn shells to assure uniform thickness.

Lancing. The operation of piercing the material in a shape that will leave the portion cut affixed to the part.

Lapping. The operation of refining a surface from undulations, roughness, and toolmarks left by previous operations.

Milling. Essentially the same as Sawing except the cutting operation leaves a trough of some specific shape depending on the form of the cutter used.

Necking. The operation of reducing the opening of a drawn cup or shell.

Nibbling. The operation of removing flat material by means of a rapidly reciprocating punch to form a contour of a desired shape.

Peening, shot. The operation of compressing a selective surface of a piece part by a rain of metal shot impelled from rotating blades of a wheel or air blast.

Perforating. The operation of piercing, in rows, large numbers of small (usually round) holes in a part that may have been previously formed.

Piercing. Essentially the same as Blanking except that the operation may be done on a previously formed piece part and that portion that is separated is usually scrap.

Planing. Essentially the same as Shaping except the operation is primarily intended to produce flat surfaces.

Polishing. The operation of smoothing the surface of a piece part by action of abrasive particles, which are adhered to the surface of moving belts or wheels, coming in contact with the piece part being polished.

Punching. The operation of separating or forming a flat piece of material in some particular shape.

Reaming. The operation of enlarging and finishing a hole to an accurate dimension.

Routing. The operation of cutting away material overhanging a templete by means of a rotating cylindrical cutter engaging the edge of the material to be profiled or cut. (The term *routing* is also applied to the operation of making shaped cavities and is more commonly known as *end milling.*)

Sawing. The operation of separating material by teeth performing successive cuts in material. Operation may follow a straight line or contour.

Serrating. The operation of producing teeth on the inside or outside edge of the workpiece by broaching. Workpiece is fixed, and tool moves.

Shaping. The operation of removing material from the surface of a material mass by a series of horizontally adjacent cuts. The tool is moved in forward strokes while the material moves in preset increments in a plane at right angles to the tool.

Shaving. The operation of finalizing a dimensional requirement on the shape of a part by cutting away a very small portion around its configuration.

Shearing. The operation of cutting the material by means of striaght blades, one of which is fixed and the other moving progressively toward it until the material between is finally cleaved. Shearing may be in a straight line, circular, or other shape.

Sizing. The operation of finalizing a dimension on a part by direct pressure, impact, or a combination of both.

Slitting. Essentially the same as Milling except the trough produced is usually very narrow, as in the case of screw heads.

Slotting. Essentially the same as Shaping except the tool is moved vertically. Sometimes referred to as *vertical shaping*.

Spinning. The operation of forming sheet metal into circular shapes by means of a lathe, forms, and hand tools, which press and shape the metal about the revolving form.

Spotfacing. The same as Counterboring except that the depth of the hole is restricted to just breaking through the surface of the material.

Squeezing. The operation of forming material by compressing it into a desired shape. The material may be confined in a shaped cavity or may be free-flowing.

Stamping. The process of stamping generally refers to that portion of the press working and forming fields, which is considered as including the following operations:

1. Punching
   a. Blanking
   b. Piercing
   c. Lancing
2. Bending
3. Forming
   a. Beading or curling
   b. Embossing
4. Shallow drawing
5. Extruding
6. Swedging
7. Coining
8. Shaving
9. Trimming

Superfinishing. Essentially the same as Lapping except that the operation results in a higher degree of surface finish.

Swaging, cold. The operation of forming or reducing a metallic material by successive blows of a pair of dies or hammers. Material flows at right angles to the pressure.

Swaging, hot. Essentially the same as Cold swaging except that the metal is heated and the arrangement of the dies causes them to be carried in a slow rotary motion. This operation is also known as *rotary swaging.*

Swedging. The operation of forming or shaping by a single blow or squeeze of a die.

Tapping. The operation of cutting a screw thread into a round hole.

Tapping, bottom. The same as Tapping except extending the threads to the bottom of a hole.

Trimming. The operation of removing excess material on the ends or edges of articles resulting from some kind of forming operation.

Tumbling, barrel. The operation of removing burrs, fins, scale, and roughness by an abrasive material falling against finished piece parts in a revolving barrel.

Upsetting, hot. Essentially the same as Cold beading except that the material is heated in order to aid ductility in forming.

INDEX